INTRODUCTION TO BIOSTATISTICS

[A TEXTBOOK OF BIOMETRY]

INTRODUCTION TO BIOSTATISTICS

A TEXTBOOK OF BIOMETRY

For the Undergraduate & Postgraduate Students of Biological Sciences

Dr. PRANAB KUMAR BANERJEE
M.Sc.(C.U.), Ph.D.(C.U.), FZS, FZSEI
Associate Professor
Chairperson, Department of P.G. Studies in Zoology
Serampore College, Serampore, Hooghly
Department of Microbiology, R.K. Mission Vidyamandira
Belur Math, Howrah

S Chand And Company Limited
(ISO 9001 Certified Company)

S Chand And Company Limited

(ISO 9001 Certified Company)

Head Office: D-92, Sector–2, Noida – 201301, U.P. (India), Ph. 91-120-4682700

Registered Office: A-27, 2nd Floor, Mohan Co-operative Industrial Estate, New Delhi – 110 044, Phone: 011-49731800

www.**schandpublishing.com;** e-mail: **info@schandpublishing.com**

Marketing Offices:

Chennai : Ph: 23632120; chennai@schandpublishing.com
Guwahati : Ph: 2738811, 2735640; guwahati@schandpublishing.com
Hyderabad : Ph: 40186018; hyderabad@schandpublishing.com
Jalandhar : Ph: 4645630; jalandhar@schandpublishing.com
Kolkata : Ph: 23357458, 23353914; kolkata@schandpublishing.com
Lucknow : Ph: 4003633; lucknow@schandpublishing.com
Mumbai : Ph: 25000297; mumbai@schandpublishing.com
Patna : Ph: 2260011; patna@schandpublishing.com

First Edition 2004
Subsequent Editions and Reprints 2005, 2006, 2007, 2008, 2009 (Twice)
Fourth Revised and Enlarged Edition 2011
Reprints 2013, 2014 (Twice), 2015, 2016 (Twice), 2017 (Twice), 2018 (Twice), 2019, 2020, 2021 (Twice), 2022 (Twice), 2023, 2024

Reprint 2025

ISBN: 978-81-219-2329-3 **Product Code:** H6BSS68BSTT10ENAD11O

PRINTED IN INDIA

By Vikas Publishing House Private Limited, Plot 20/4, Site-IV, Industrial Area Sahibabad, Ghaziabad – 201 010 and Published by S Chand And Company Limited, A-27, 2nd Floor, Mohan Co-operative Industrial Estate, New Delhi – 110 044.

PREFACE TO THE FOURTH REVISED AND ENLARGED EDITION

To begin with I acknowledge whole heartedly to my beloved students as well as to my colleagues, friends of different colleges and universities, throughout India for the admiration and reception they have shown to my book "**Introduction to Biostatistics**" (1st to 3rd edition). I do hope, they will also extend their good wishes and appreciation for this enlarged, revised and elegant edition (4th) of this book.

To enhance the utility of this book, a thorough revision and recast along with the addition of numerous examination oriented solved problems as well as a number of topics viz Set theory, Binomial expansion, Permutation, Combination and non parametric statistics have been incorporated. Theoretical discussion as well as solution of problems have been represented in a simple, lucid and unambiguous language so as to cater to the needs of students of all streams of Biosciences. (Zoology, Botany, Physiology, Microbiology & Biotechnology etc.).

I claim no originality for the matter presented in this book but the method of presentation and illustration is my own. I have tried my level best to present the entire text in such as manner that kindles the interest of a student in Biostatistics.

In preparing this book I have taken profuse help from several books. I have expressed my acknowledgment in the bibliography. I am grateful to all those teachers who have helped a lot by giving their valuable suggestions. I convey my respect and pronams to Swami Atmopriyanandji Maharaj (Vice Chancellor **R.K. Mission Vivekananda University** Belurmath, Howrah). I extend my respect and gratitude to my teacher Prof. Rabindra Nath Chatterjee (**Genetics Research Unit Dept.** of Zoology, University of Calcutta), Prof. Dhrubojyoti Chatterjee (**Pro-Vice Chancellor**, Academic, University of Calcutta), Prof. Chandra Sekhar Chakraborty, (**Vice Chancellor**, West Bengal University of Animal and Fishery Sciences) and Dr. Tarit Kumar Banerjee (Associate Professor Dept. of Zoology R.P.M. College Uttarpara).

I also extend my thanks to my Departmental Colleagues (both teaching and non teaching) for their encouragement and constant inspiration.

My sincerest thanks are forwarded to management, DTP and Editorial team of S. Chand & Company Ltd. for their encouragement and neat execution to the revised & enlarged edition of this book in a suitable form.

Last but not the least I extend my sincere thanks to Mrs. Mandira Banerjee (**Head Mistress** Baidyabati Charusila Bose Balika Vidyalaya) and Miss Debdatta Banerjee (Daughter) for their endurance and active assistance during the preparation of this book.

Dr. Pranab Kumar Banerjee
Dept. of Zoology
Serampore College

PREFACE TO THE FIRST EDITION

This book entitled "Introduction to Biostatistics" is based on my experience of teaching Biostatistics to the students of B.Sc. (Zoology Botany) Courses of Indian Universities. Biostatistics is the study of the application of statistical methodology to analyse biological variations, correlation and regression in biological measurement which is also known as "Biometry". I claim no originality for the matter presented in the text but the method of presentation is my own. This book has been written in a clear, lucid manner to cover theoretical, practical and applied aspect of statistics. It will help the graduate and post graduate students of biological sciences (Zoology, Botany & Cytogenetics), Psychology, Education of Indian universities. Research workers and teachers of Biological and Medical Sciences may also get help from this book. I believe large number of illustrative and various types of examples would make this book easy to understand without having any external help.

In preparing this book, I have been greatly helped by several books. I express my acknowledgment in the bibliography. I take the opportunity to express my indebtedness to Mr. R. M. Nath, Manager, S. Chand & Co. Ltd. and my students C. Chowdhary whose encouragement has driven me to prepare this book.

The enthusiastic inspiration from Swami Athmaprianandaji Maharaj (Principal R.K. Mission Vidyamandira, Belur Math) Dr. Lalchun Nunga (Principal Serampore College) Prof. R.N. Chatterjee (Head of the Department Zoology, C.U.) Dr. T.K. Banerjee (Reader, Dept. of Zoology, R.P.M. College, Uttarpara), and Mr. R. Das (Dept. of Mathematics, Serampore College) was always a booster for the preparation of this book.

I shall be grateful for bringing out the mistakes and misprints top my notice which shall be removed in the next generation. Comments criticisms and suggestions from students as well as from many teachers friends for corrections and improvement of this book will be greatly acknowledged.

My heartful, thanks to Mr. Ravindra. Kr. Gupta, Managing Director & Mr. Navin Joshi, General Manager of S. Chand & Co. Ltd. for kind co-operation in the preparation and brought out the book in time & in nice form.

Last but not least I must sincerely thanks to Mrs. Mandira Banerjee for her continuous and untiring active assistance during this dedicated work.

Dr. Pranab Kumar Banerjee
Dept. of Zoology
Serampore College

CONTENTS

CONTENTS

CHAPTER

PRELIMINARY CONCEPT

INTRODUCTION TO BIOSTATISTICS

Statistics:

It refers to the subject of scientific activity which deals with the theories and methods of collection, analysis and interpretation of such data.

Biostatistics:

This term is used when tools of statistics are applied to the data that is derived from biological organisms.

Characteristics of Statistics:

1. Statistics are the aggregate of facts.
2. Statistics are numerically expressed.
3. Statistics are affected by multiplicity of causes and not by single cause.
4. Statistics must be related to some field of inquiry.
5. Statistics should be capable of being related to each other, so that some cause & effect relationship can be established.
6. The reasonable standard of accuracy should be maintained in statistics.

Importance and Usefulness of Statistics:

1. Statistics help in presenting large quantity of data in a simple and classified form.
2. It gives the methods of comparison of data.
3. It enlarges individual mind.
4. It helps in finding the conditions of relationship between the variables.
5. It tries to give material for the business man as well as the administrators so as to serve as a guide in planning and shaping future policies and programmes.
6. It proves useful in number of fields viz. railways, Banks, Army, etc.

Limitation of Statistics:

1. Statistics laws are held to be true on the average and in the long run.
2. Statistics can be used to analyse only collective matters not individual events
3. It is applicable only to quantitative data.
4. Statistical results are ascertained by samples. If the selection of samples is biased, errors will accumulate and results will not be reliable.

5. The greatest limitation of statistics is that only one who has an expert knowledge of statistical methods can efficiently handle statistical data.

Application and Uses of Biostatistics:

1. In Physiology and Anatomy

(*i*) To define what is normal or healthy in a population and to find limits of normality in variables.

(*ii*) To find the difference between the means and proportions of normal at two places or in different periods.

(*iii*) To find out correlation between two variables *X* and *Y* such as height and weight.

2. In Pharmacology

(*i*) To find out the action of drug—a drug is given to animals & humans to observe the changes produced are due to the drug or by chance.

(*ii*) To compare the action of different drugs or two successive dosages of the same drug.

(*iii*) To find out the relative potency of a new drug with respect to a standard drug.

3. In Medicine

(*i*) To compare the efficacy of a particular drug. For this, the percentage of cured & died in the experiment & control groups.

(*ii*) To find out an association between two attributes such as cancer and smoking.

(*iii*) To identify signs and symptoms of a disease or syndrome. Cough & typhoid is found by chance and fever is found in almost every case.

4. In Community Medicine and Public Health

(*i*) To test usefulness of sera and Vaccines in the field-the percentage of attacks or deaths among the vaccinated subject is compared with that among the unvaccinated ones to find whether the difference observed is statistically significant.

(*ii*) In epidemiological studies—the role of causative factors is statistically tested.

(*iii*) In public health, the measures adopted are evaluated.

DATA:

Data is a collection of observations expressed in numerical figures. The collection may be done in *two* ways.

(*a*) by complete enumeration and

(*b*) simple survey method.

Data is always in collective sense and never be used singular.

Types of Data:

The statistical data can be divided into *two* broad categories:

(*a*) Qualitative (*b*) Quantitative.

Qualitative Data :

In this type of data, there is no numerical relation with one another.

Example: Skin colour—brown, black, white
Eye colour—blue, brown
Sex—Male, Female.

Quantitative Data:

1. In this type of data, there is numerical relation with one another.
2. It may be continuous or discrete.

Example: Discrete = Number of books, number of students.

Continuous = Height or Weight of person.

Qualitative data	*Quantitative data*
1. Always Discrete.	1. Discrete or continuous.
2. No magnitude.	2. Have magnitude.
3. Persons with same character are counted to form groups.	3. Arranged by both character and frequency.
4. Results are expressed as ratio or proportion.	4. Such data are analysed through statistical method *e.g.*, mean, median, mode, S.D. etc.

A. According to source of data collection:

(*a*) **Primary data:** Directly from field or experiment.

(*b*) **Secondary data:** Obtained from primary data or review.

B. According to variable:

(*a*) Univariable.

(*b*) Bivariable.

(*c*) Multivariable.

C. According to compilation:

(*a*) **Raw data:** Data before compilation.

(*b*) **Derived data:** Calculated from primary value of data.

PRIMARY DATA:

These data are collected directly from the field of enquiry for a specific purpose. These are raw data or data in original nature, and directly collected from population. The collection of primary data may be made through either by complete enumeration or sampling survey methods.

SECONDARY DATA:

These are numerical information which have been already collected by some agency for a specific purpose and are subsequently compiled from that source for application in different connection.

In other words, data used by any other agency than the collecting authority will be termed as secondary data.

COLLECTION OF PRIMARY DATA:

The following methods are generally used for collection of primary data:

(*a*) Direct personal observation.

(*b*) Indirect oral investigation.

(*c*) Questionnaires sent by mail.

(*d*) Schedules sent through investigators.

QUESTIONNAIRE:

It is a proforma containing a sequence of questions relevant to a statistical enquiry. It is used for collection of primary data from individual persons through their response to the set of questions.

RELATIVE ADVANTAGES OF PRIMARY DATA:

1. Primary data provides with detailed information but in secondary data some information may be suppressed.

2. Primary data is free from transcribing errors and estimation errors where as a secondary may contain such errors.
3. Secondary data normally do not contain information regarding methods of procuring data where as primary data often include them.
4. Cost effectiveness is a vital plus point for using secondary data. Thus time, cost suitability and accuracy are the essential factors whether we would use primary or secondary data.

POPULATION:

It is an entire group of people or study elements-persons, things or measurements having some common fundamental characteristics.

(*a*) **Finite:** If a population consist of fixed number of value, it is said to be finite *e.g.*, number of days in a week.

(*b*) **Infinite:** If a population consist of an endless succession of values, it is said to be infinite *e.g.*, number of animals in ocean.

SAMPLING:

The technique of obtaining information about the whole group by examining only the part of the whole group is called *sampling.*

Types of Sampling:

(*a*) Random sample (Probability sample).

(*b*) Non-random sample (Non-probabilities sample).

Objectives of Sampling:

1. Estimation of population parameter (mean, SD etc.) from the sample statistics.
2. To test hypothesis about the population from which the sample or samples are drawn.

SAMPLE:

It is a relatively small group of selected number of individuals or objects or cases drawn from a particular population and is used to throw light on the population characteristics.

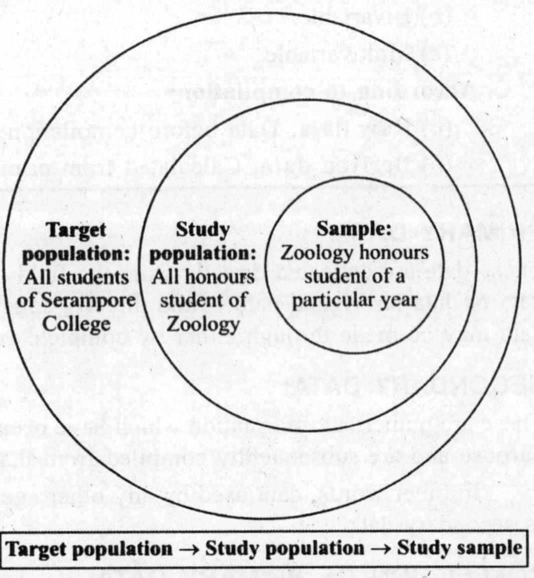

Fig. 1.1

RANDOM SAMPLE:

It is a sample chosen in a very specific way and has been selected in such a way that every element in the population has an equal opportunity (unbiased) of being included in the sample.

CHARACTERISTIC:

The term 'characteristic' means a quality possessed by an individual (*i.e.*, object, item of population). Height, weight, age etc. are characteristics.

In statistics, characteristics are of *two* kinds:

(*a*) Non-measurable 'characteristics' (attributes)

(*b*) Measurable 'characteristics' (variables).

(*a*) **Attributes:** Attributes are the non-measurable characteristics which can not be numerically expressed in terms of unit. These are qualitative object.

For example—Religion, Nationality, Illiteracy etc.

(b) Variables: Variables are the measurable characteristics which can be numerically expressed in terms of some unit. These are quantities which are capable of being measured by quantitative methods directly.

For example—Height in inches, cm, weight in kg, pound, marks in examination etc.

(i) Discrete Variables (Discontinuous/meristic): There are the quantities which can be measured in whole integral values. It does not take fractional value.

Example—Number of books marks in examination.

(ii) Continuous Variables: These are quantities which can take any value in specified range. Thus it can take integral and fractional values both.

Example Heights, weights etc.

Discrete Variables:

Specimen No.	*Number of bristles in Drosophila*
1.	3
2.	7
3.	10
4.	8
5.	6

Continuous Variable:

Specimen No.	*Body Weight of some crabs*
1.	5.026 gram
2.	3.732 gram
3.	4.875 gram
4.	3.781 gram
5.	6.023 gram

STATISTICAL ERROR:

In statistical terminology the word 'error' is used in special sense. Error shows the extent to which the observed value of a quantity exceeds the true value.

Error = Observed value – True value.

TYPES:

Statistical error may be classified as:

(a) *Biased errors:* (which arise due to personal prejudices or bias of investigator & informants)

(b) *Unbiased errors:* (which enter into statistical enquiry due to chance causes).

ARRAY:

The presentation of data in ascending order of magnitude is called *array.*

TALLY:

(i) A tally mark is an upword slanted stroke (*I*) which is put against each occurrence of value.

(ii) When value occurs more than four times the fifth occurrence is denoted by a cross (\) tally mark, running diagonally, across the four tally marks. This facilitates the counting of tally marks at the end.

(*iii*) The total count of tally against each value is called its *frequency*.

(*iv*) A frequency distribution with individual values is called *simple frequency distribution*.

Example: Form a frequency table for the following variables:

51, 59, 52, 51, 60, 68, 63, 64, 65, 66, 68, 52, 59 60, 58, 51, 54, 55, 56, 61, 62, 69, 70, 58, 69, 65 67, 63, 63, 62, 61, 51, 59, 63, 68, 67, 69, 53, 53 51, 59, 56, 55, 70, 65, 62, 65, 66, 69, 70, 52, 55 64, 65, 69, 61, 63, 54, 64, 61, 61, 62, 51, 52, 52, 54, 55, 52, 52, 66.

Solution:

Values of variables	*Tally*	*Frequency*
51	𝍸 I	6
52	𝍸 II	7
53	II	2
54	III	3
55	IIII	4
56	II	2
58	II	2
59	IIII	4
60	II	2
61	𝍸	5
62	IIII	4
63	𝍸	5
64	III	3
65	𝍸	5
66	III	3
67	II	2
68	III	3
69	𝍸	5
70	III	3
Total		70

CLASSIFICATION: It is the process of arranging the collected statistical information under different categories or classes according to some common characteristics possessed by an individual member.

Types of Classification:

There are four types of classification.

(*a*) **On qualitative basis:** Here non measurable characteristics are classified.

(*b*) **On quantitative basis:** Here measurable characteristics are classified.

(*c*) **On time basis:** Here the statistical data are arranged in order of their time of occurrence.

(*d*) **On the geographical basis:** The total population of a country may be classified by states, or districts. The basis of classification in such cases is by geographical regions.

METHOD OF PRESENTATION OF STATISTICAL DATA:

Statistical data are presented in *three* processes:

(*a*) **Textual Presentation:**

(*i*) Numerical data presented in a descriptive form are called *textual presentation.*

(*ii*) It is lengthy. Some words may repeat several times in the text.

(*iii*) It becomes difficult to grasp salient points in a textual presentation.

(*b*) **Tabular Presentation:**

(*i*) The logical and systematic presentation of numerical data in rows and columns designed to simplify the presentation and facilitate comparison is termed as tabulation.

(*ii*) Tabulation is thus a form of presenting quantitative data in condensed and coincise form so that numerical figures are capable of easy & quick reception by the eyes.

(*iii*) It is more convenient than textual presentation.

(*c*) **Graphical Presentation:** The presentation of quantitative data by graphs and charts are termed as graphical presentation.

Tabulation:

It may be defined as the logical and systematic presentation of numerical data in rows and columns designed to simplify the presentation and facilitate comparisons.

The Advantages of tabulation are:

(*i*) It enables the significance of data readily understood and leaves a lasting impression than textual impression.

(*ii*) It facilitates quick comparison of statistical data shown between rows and columns.

(*iii*) Errors and omissions can be readily detected when data are tabulated.

(*iv*) Repetition of explanatory terms and phrases can be avoided, and the concise tabular form clearly reveals the characteristics of data.

Types of Tabulation:

There are *two* types of tabulation:

(*a*) **Simple tabulation:** It contains data in respect of one characteristic only.

(*b*) **Complex tabulation:** It contains data of more than one characteristics.

Example: (Simple tabulation): Number of students in three colleges.

Name of the Colleges	*No. of students*
1. Raja Peary Mohan College	2750
2. Serampore College	3400
3. R.K. Mission Vidyamandira	1400

Example: (Complex tabulation)

Name of the colleges	*Number of Students*				
	B.A. (Hons)	*B.SC. (Hons)*	*B.A.*	*B.Sc.*	*Total*
1. R.P.M College	450	400	850	1050	2750
2. Serampore College	600	500	800	1500	3400
3. R.K. Mission Vidyamandira	200	200	300	700	1400

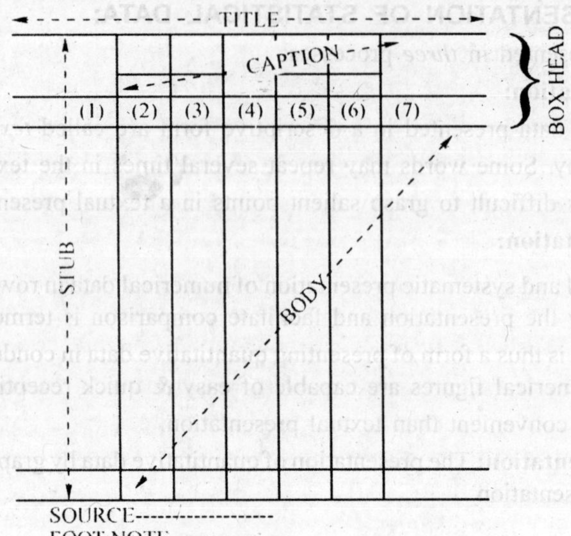

Fig. 1.2. Different parts of table.

Statistical Tables:

Statistical table is a systematic arrangement of quantitative data under appropriate heads in rows and columns. After the data have been collected, they should be tabulated that is put in the form of a table, so that whole information can be had at a glance.

Parts of a Table:

(I) Title:

(*a*) This is a brief description of the contents of the table along with time, place and category of item if required.

(*b*) The title should be clear and precise.

(*c*) It should be at the top of the table.

(II) Stub:

(*a*) The extreme left part of the table where descriptions of the rows are shown in called *stub*.

(*b*) It must be precise and clear.

(III) Caption and Box head:

(*a*) The upper part of the table which shows the description of columns and sub columns is called *Caption*.

(*b*) The whole of the upper part including caption units of measurement and column number if any is called *Boxhead*.

(IV) Body:

(*a*) It is the main part of the table except the title stub and captions.

(*b*) It contains numerical information which are arranged in the table according to the descriptions of the rows and columns given the stub and caption.

(V) Source and foot note:

(*a*) It is customary that source of data from which information has been arrived should be given at the end of the table.

(*b*) Foot note is the part below the body where the source of data and any explanation are shown.

Essential features of a good table:

1. A table must have a title giving clear and precise idea about the contents of the table.
2. Units of measurements adopted in a table must be shown clearly in the top of the column.
3. It is a necessity that an investigator prepares a table well proportioned in length and breadth.
4. For a compatible comparison, column of relevant figures must be kept as close as possible.
5. Distinction is preferred in columns and sub columns. It can be made by distinct ruling (viz double ruling, single ruling etc.).
6. Totals of columns may be shown in the bottom of the table. In cases where row totals are useful, they should also be shown.
7. Table must contain necessary details.
8. Source of information must be disclosed at the end of the table.
9. Any ambiguous or confusing entry in the table should bear a special note at the end of the table for experiment.
10. The arrangement of items in the table should have a logical sequence.

Satatistic (s)	***Parameter***
(*i*) Any statistical measure calculated on the basis of sample observations is called a statistic. Example : Sample mean, Sample S.d (*ii*) It characterise samples. (*iii*) It can be directly worked out. (*iv*) Statistics is a variate i.e. the values of statistics varies from sample to sample.	(*i*) Any statistical measure based on all units in the population in called a parameter. Example - Population mean, population sd. (*ii*) It characterise population. (*iii*) It is not directly worked out. (*iv*) A parameter is fixed quantity, *i.e.* the values of parameter is constant.

Measure	***Statistic (s)***	***Parameter***
• Mean	$\overline{x}$	μ
• Standard deviation	S	σ
• Variance	S^2	σ^2
• Correlation coefficiant	r	P

Find whether the variable is continuous or discontinuous in the following cases:

(*i*) Number of individuals in a family.

(*ii*) Time of flight of a missile.

(*iii*) Number of gallons of water in a washing machine.

(*iv*) Life time of television tubes produced in a company. **(B.A. Punjab University, 1967)**

Ans: (*i*) Discrete, (*ii*) Continuous, (*iii*) Discrete, (*iv*) Continuous.

CHAPTER

FREQUENCY DISTRIBUTION

Values of Variable of Frequency:

The distinct observation is known as *values of variable*. The values of variable obtained by observations are termed as *observed values* or observation.

If a value repeats more than once, the number of times the value is repeated will be termed as frequency.

Frequency Distribution:

Frequency Distribution is a statistical table which shows the values of variable arranged in order of magnitude either individually or in groups and also the corresponding frequencies side by side.

Types of Frequency Distribution:

There are *two* types of frequency distribution:

(a) Simple frequency distribution.

(b) Grouped frequency distribution.

Simple frequency distribution shows the values of variable individually where as groups frequency distribution shows the values of the variable in groups or intervals.

Table 2.1 Simple Frequency Distribution

Number of class test	*Marks obtained (20)*
1	14
2	13
3	15
4	12
5	14
6	12
7	11
8	17
Total	108

Table 2.2 Grouped Frequency Distribution

Age in years	*Frequency (No. of persons)*
15–19	37
20–24	81
25–29	43
30–34	24
35–44	9
45–59	6
Total	200

Notes to Remember for Forming Frequency Distribution:

(*i*) Each class must be clearly defined.

(*ii*) Each class must be exhaustive *i.e.*, each raw data must be included in the classes.

(*iii*) Classes must be exclusive *i.e.*, non-overlapping.

(*iv*) It is expected that normally classes should be made of equal width.

(*v*) The number of classes should neither be too large nor too small.

Terms Associated with Grouped Frequency Distribution:

(*i*) Class or class interval.

(*ii*) Class limit.

(*iii*) Class boundaries.

(*iv*) Class mark.

(*v*) Class width.

(*vi*) Class frequency, total frequency, percentage frequency, frequency density.

(*vii*) Cumulative Frequency.

Class Interval or Class:

When a large number of observations varying in a wide range are available, these are classified in several groups according to the size of values. Each of these groups defined by an interval is called *class interval* or *class*.

Class Intervals are of two types:

(*a*) **Continuous Class Interval:** A Class interval which does not contain the upper boundary of the class will be called *continuous class interval*.

A class interval of the form 10–20 in continuous class will contain values from 10 to less 20. An example is of the form:

Class	*Range*
0–10	From zero less than 10
10–20	From 10 less than 20
20–30	From 20 less than 30
30–40	From 30 less than 40

(*b*) **Discontinuous class interval:** A class interval where each class includes the end values will be called *discontinuous class interval*. A class interval of the form 0–9 in discontinuous class will contain values from 0 to 9 both inclusive. An example is of the form.

Class	*Range*
0–9	From 0 to 9
10–19	From 10 to 19
20–29	From 20 to 29

Continuous class intervals are formed generally with continuous type values or non-integral values *e.g.*, Rupees, Kg. etc. Discontinuous class intervals are formed generally with discrete or integral values *e.g.*, marks.

Open end Class:

When one end of a class is not specified, the class is called *open end class*. A frequency distribution may have either one or two open end classes.

Income (Rs.)	*Frequency*
0–50	90
50–100	150
100–150	100
150–200	80
200–250	10

Class Limits:

In the construction of groups' frequency distribution, the class interval must be defined by pairs of numbers such that the upper end of one class does not coincide with the lower end of the immediate following class.

The two numbers used to specify the limits of a class interval for the purpose of tallying the original observation into the various classes, are called *class limits.*

(*i*) The smaller of the pair is known as *lower class limit.*

(*ii*) The larger of the pair is called as *upper class limit.*

Class Boundaries:

In most of the measurement of continuous variables, all data are recorded nearest to a certain unit or integer value. The most extreme values which would ever be included in a class interval are called *class boundaries*. Infact it is the actual or real limits of a class interval.

(*i*) The lower extreme point is called *lower class boundary.*

(*ii*) The upper extreme point is called *upper class boundary.*

Calculation:

If α is the gap between the upper class limit of any class or class interval and the lower class limit of the next class or class interval.

$$\text{Lower class boundary} = \text{lower class limit} - \frac{1}{2}\alpha.$$

$$\text{Upper class boundary} = \text{upper class limit} + \frac{1}{2}\alpha.$$

Class limits are used only for the construction of the grouped frequency distribution but in all statistical calculations and diagrams involving end points of classes (*e.g.* median, mode, histogram and ogive etc.) Class boundaries are used.

Class Mark (or midvalue or midpoint):

(*i*) It is the midvalue of a class or class interval exactly at the middle of the class or class interval.

(*ii*) It lies half way between the class limits or between the class boundaries

$$\textbf{Class mark} = \frac{\text{Lower class limit} + \text{Upper class limit}}{2}.$$

(*iii*) It is used as representative value of the class interval for the calculation of means, & standard deviation, mean deviation etc.

Class Width:

It is range or length of a class interval or difference between the upper and lower class boundaries.

Width of class = Upper class boundary − lower class boundary.

Class Frequency and Total Frequency:

- The number of observations falling within a class is called its *class frequency* or *simple frequency*.
- The sum of all the class frequencies is called *total frequency*.

Relative Frequency:

It is the ratio of the frequency of the class to the total frequency.

(*i*) It is not expressed in percentage. *Relative frequency of a class* $= \dfrac{\text{Frequency of the class}}{\text{Total frequency}}$.

(*ii*) Relative frequencies are used to compare two or more frequency distributions or two or more items in the same frequency distribution.

Percentage Frequency:

Percentage of class interval or class is the frequency of the class interval (class) expressed as percentage of the total frequency distribution.

$$\textit{Percentage frequency of a class} = \frac{\text{Frequency of the class}}{\text{Total frequency}} \times 100.$$

Table 2.3 Class limit, class boundaries, class width, frequency density, Relative frequency

Class interval	Class frequency	Class limits		Class boundaries		Class marks	Class width	Frequency density	Relative frequency
		Lower	Upper	Lower	Upper				
1	2	3	4	5	6	7	8	9	10
15–19	18	15	19	14.5	19.5	17	5	3.6	.18
20–24	34	20	24	19.5	24.5	22	5	6.8	.34
25–29	21	25	29	24.5	29.5	27	5	4.2	.21
30–34	12	30	34	29.5	34.5	32	5	2.4	.12
35–39	9	35	44	34.5	39.5	39.5	5	1.8	.09
40–44	6	45	59	39.5	44.5	52	5	1.2	.06
Total	100	—	—	—	—	—	—	—	1.00

Frequency Density:

Frequency density of a class interval is its frequency per unit width. It shows the concentration of frequency in a class.

(*i*) *Frequency density:* $\dfrac{\text{Class frequency}}{\text{Width of the class}}$.

(*ii*) It is used in drawing histogram when the classes are of unequal widtn.

Cumulative Frequency Distribution:

Cumulative frequency corresponding to a class is the sum of all the frequency up to and including that class.

I. It is obtained by adding to the frequency of that class and all the frequencies of the previous classes

II. Cumulative frequencies are of two types:

(*a*) Less than Cumulative Frequency: The number of observations 'upto' a given value is called *less than* cumulative frequency.

(*b*) **More than Cumulative Frequency:** The number of observations 'greater than' a value is called the *more than* cumulative frequency.

Table 2.4

Class interval	Frequency	Cumulative frequency	
		Less than	More than
30–40	8	8	100
40–50	12	20	92
50–60	20	40	80
60–70	25	65	60
70–80	18	83	35
80–90	17	100	17
Total	100		

Uses:

1. To find out the number of observations less than or more than any given value.
2. To find out the number of observations falling between any two specified values of the variable.
3. To find out median, quartiles & pentiles.

Problem: Form a frequency table (class limit, class boundaries, class width, class density, relative frequency) from the following data:

Marks in Biostatistics	Number of students
Under 30	0
Under 35	10
Under 40	15
Under 45	8
Under 50	5

Solution:

Class interval	Class frequency	Class limits		Class boundaries		Class marks	Class width	Frequency density	Relative frequency
		Lower	Upper	Lower	Upper				
1	2	3	4	5	6	7	8	9	10
30–35	10	30	35	29.5	35.5	32.5	6	1.66	0.25
35–40	15	35	40	35.5	40.5	37.5	6	2.5	0.375
40–45	8	40	45	40.5	45.5	42.5	6	1.33	0.2
45–50	7	45	50	45.5	50.5	47.5	6	0.833	0.175
Total	40								

CHAPTER

GRAPHICAL REPRESENTATION OF DATA

Graphical Representation of Data

The representation of quantitative data suitably through charts and diagrams is known as *Graphical Representation of Data* (*Statistical Information*). Graph includes both charts and diagrams. The main object of diagrammatic representation is to emphasis the relative position of different subdivisions and not simply to record details.

Advantages of Graphical Representation:

(*i*) It is easily understood by all.

(*ii*) The data can be presented in a more attractive form.

(*iii*) It shows the trend and tendency of values of the variable.

(*iv*) Diagrammatic representations are useful to detect mistakes at the time of data computations.

(*v*) It shows relationship between two or more sets of figures.

(*vi*) It has the universal applicability.

(*vii*) It is helpful in assimilating the data readily and quickly.

Disadvantages of Graphical Representation:

(*i*) It does not show (details) or all the facts.

(*ii*) Graphical representation can reveal only the approximate position.

(*iii*) It takes a lot of time to prepare of graph.

The Different Types of Diagrammatic Representation:

There are various types of graphs in the form of charts and diagrams. Some of them are:

1. Line Diagram or Graph.
2. Bar Diagram.
3. Pie Chart.
4. Histogram.
5. Frequency polygon.
6. Ogives (cumulative frequency polygon).

Modes of Graphical Representation of Data:

The data in the form of raw scores is known as *ungrouped data* and when it is organized into frequency distribution then it is referred to as *grouped data*.

Separate modes and methods are used to represent these *two* types of data ungrouped and grouped.

A. Graphical Representation of Ungrouped Data: For the ungrouped data, the following graphical presentations are used.

1. Line diagrams or graphs.
2. Bar diagram.
3. Pie diagram or charts.
4. Pictograms.

1. Line Diagrams or Graphs or (Historigram): It is the most common method of representing statistical information mainly used in business and commerce.

(*i*) These are drawn on the plane paper by plotting the data concerning one variable on the horizontal *x*-axis (*abscissa*) and other variable of data on *y*-axis (*ordinate*), which intersect at a point called *origin*.

(*ii*) With the help of such graphs the effect of one variable upon another variable during an experimental study may be clearly demonstrated.

(*iii*) According to data for corresponding *X*, *Y* values (in pairs), we will find a point on the graph paper. The points thus generated are then jointed by pieces of straight lines successfully. The figure thus formed is called *Line diagram* or graph.

(*iv*) Two types of line diagram are used (a) natural scale (b) ratio scale.

Example: The data of effect of practice on learning are given in the table.

Trial No.	1	2	3	4	5	6	7	8	9	10	11	12
Score	4	5	8	8	10	13	12	12	14	16	16	16

Draw a line graph for the representation and interpretation of the above data.

Solution: Plot the points (1, 4), (2, 5), (3, 8), (4, 8), (5, 10), (6, 13), (7, 12), (8, 10), (9, 14) and (10, 16).

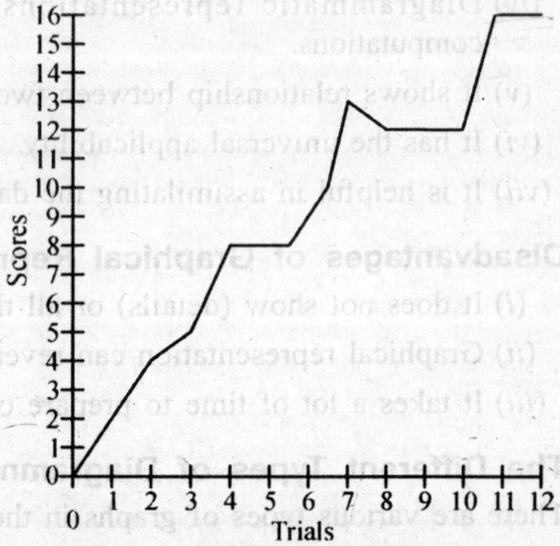

Fig. 3.1 Line Graph—The effect of practice on learning

2. Bar Diagram: A bar diagram is a graph on which the data are represented in the form of bars and is useful in comparing qualitative or quantitative data of discrete type.

(*i*) It consists of a number of equally spaced rectangular areas with equal width and originates from a horizontal base line (*X*-axis).

(*ii*) The length of the bar is proportional to the value it represents. It should be seen that the bars are neither too short nor too long.

(*iii*) They are shaded or coloured suitably.

(*iv*) The bars may be vertical or horizontal in a bar diagram. If the bars are placed horizontally, it is called *horizontal bar diagram*, when bars are placed vertically it is called a *vertical bar diagram*.

There are *three* types of bar diagram:

(*i*) Simple bar diagram

(*ii*) Multiple or grouped bar diagram

(*iii*) Component or subdivided bar diagram.

(i) Simple Bar Diagram:

(*i*) It consists of a number of equally spaced vertical bars of uniform width originating from a horizontal axis and is shaded.

(*ii*) These bars are usually arranged according to relative magnitude of bars.

(*iii*) The length of the bar is determined by the value or the amount of variables.

(*iv*) The limitation of simple bar diagram is that only one variable can be represented on it.

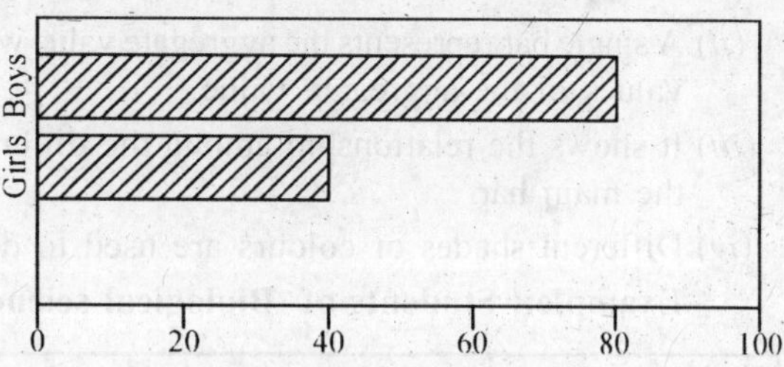

Fig. 3.2 Simple Bar Diagram showing weight in kg among Boys & Girls

Example:

The heart beat rates of four mammals are given below. Represent it with the help of bar diagram.

Mammals	*Heart beat (per minute)*
Whale	75
Horse	45
Pig	70
Cow	50

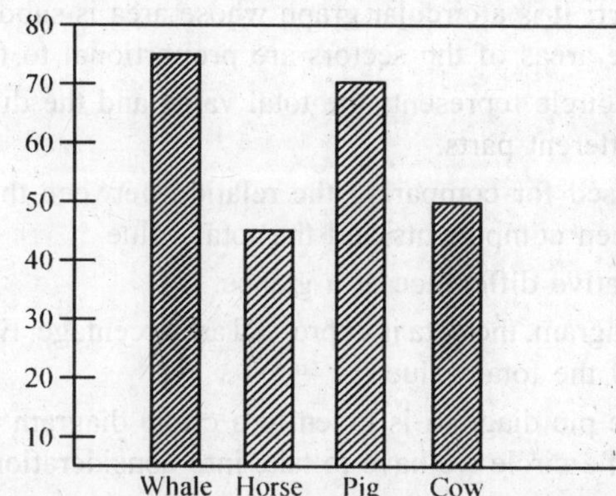

Fig. 3.3 Bar diagram showing heart beat (minute) of mammals

(ii) Multiple or Grouped Bar Diagram:

(*i*) Multiple bar diagram represents more than one type of data at a time.

(*ii*) In this case numerical values of major categories are arranged in ascending or descending order so that categories can be readily distinguished.

(*iii*) Different shades or colour are used for each category.

(*iv*) It is to be remembered that the gap between the variable must be same.

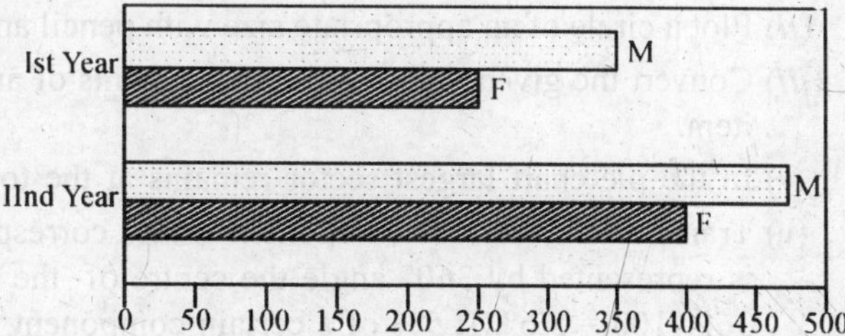

Fig. 3.4 Multiple or Grouped Bar Diagram showing the number of Male & Female students in 1st and 2nd year B.Sc. students.

(iii) Component or Subdivided Bar Diagram:

(*i*) Each bar in component bar diagram is subdivided into several component parts.

(*ii*) A single bar represents the aggregate value where as the component parts represent the component values of the aggregate value.

(*iii*) It shows the relationship among the different parts and also between the different parts and the main bar.

(*iv*) Different shades or colours are used to distinguish the various components.

Example: Students of Biological sciences of Serampore College during 2007 & 2008.

Year	Zoology (Hons)	Botany (Hons)	Physiology (Hons)	General
2007	16	22	20	52
2008	22	25	18	55

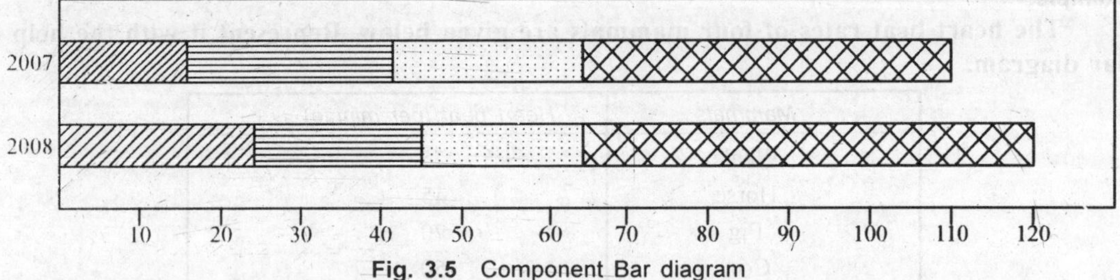

Fig. 3.5 Component Bar diagram

3. **Pie Diagram or Chart:** It is a circular graph whose area is subdivided into sectors by radii in such a way that the areas of the sectors are proportional to the angles at the centre.

(*i*) The area of the circle represents the total value and the different sectors of the circle represent the different parts.

(*ii*) It is generally used for comparing the relation between the various components of a value and between components and the total value.

(*iii*) It gives comparative difference at a glance.

(*iv*) In pie chart or diagram, the data is expressed as percentage. Each component is expressed as percentage of the total value.

(*v*) The name of the pie diagram is given to a circle diagram because in determining the circumference of a circle we have to take into consideration a quantity known as 'pie' (π).

Working Procedure:

(*i*) The surface area of circle is known to cover 2π radius or 360° (degrees). The data to be represented through a circle diagram may therefore be presented through 360°.

(*ii*) Plot a circle of an appropriate size with pencil and compass. The angle of a circle totals 360°.

(*iii*) Convert the given value of the components of an item in percentage of the total value of the item.

(*iv*) In the pie chart largest sector remains at the top and other in sequence running clockwise.

(*v*) Transpose the various component values correspond to the degree on the circle. Since 100% is represented by 360° angle the centre of the circle, therefore 1% value is represented by 360°/100 = 3.6°. If 5% of a certain component, the angle which represent the percentage of such component is (3.6 × 5) degrees.

(*vi*) Measure with protector, the points on a circle representing the size of each sector. Label each sector for identification.

$$\text{Angle} = \frac{\text{Value of one component}}{\text{Total of all the components}} \times 360$$

Example: Marks obtained in test examination by a Bioscience students of Serampore College.

Zoology	*Botany*	*Physiology*
65	60	55

Solution: Total number of components

$$= 65 + 60 + 55 = 180.$$

Subject	*Marks obtained*	*Angle*
Zoology	65	$\frac{65}{180} \times 360 = 130°$
Botany	60	$\frac{60}{180} \times 360 = 120°$
Physiology	55	$\frac{55}{180} \times 360 = 110°$

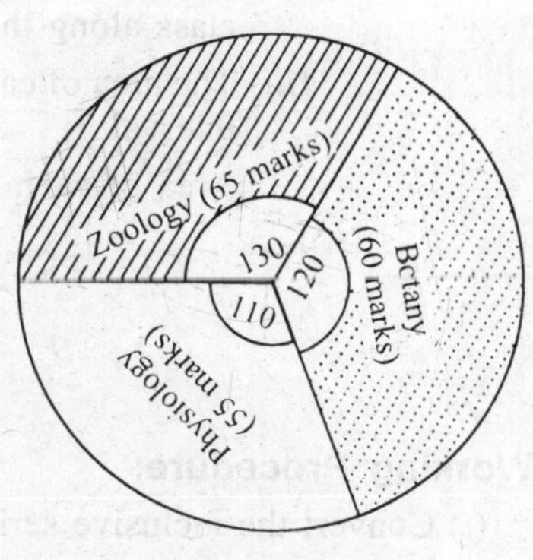

Fig. 3.6 Pie chart showing marks of three subjects

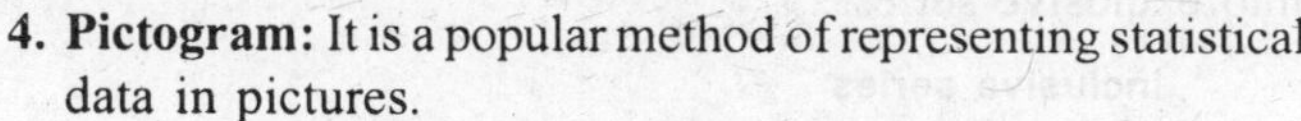

4. **Pictogram:** It is a popular method of representing statistical data in pictures.

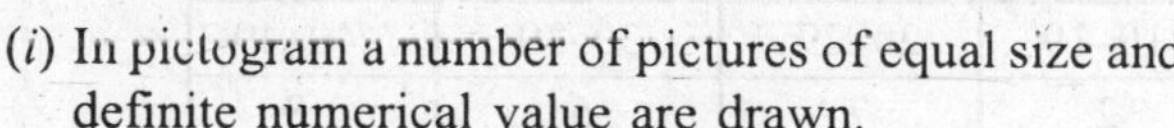

   (*i*) In pictogram a number of pictures of equal size and definite numerical value are drawn.

   (*ii*) Each picture represents a number of units.

   (*iii*) Pictures are draw side horizontally or vertically.

   (*iv*) It is widely used public and private sector.

Problem : Represent the following data of the production of books (Biostatistics) from S. Chand & Company.

Year	2006	2007	2008
Production of books	1000	2000	3000

Solution: The given data is represented by pictogram as shown in Fig. 3.7.

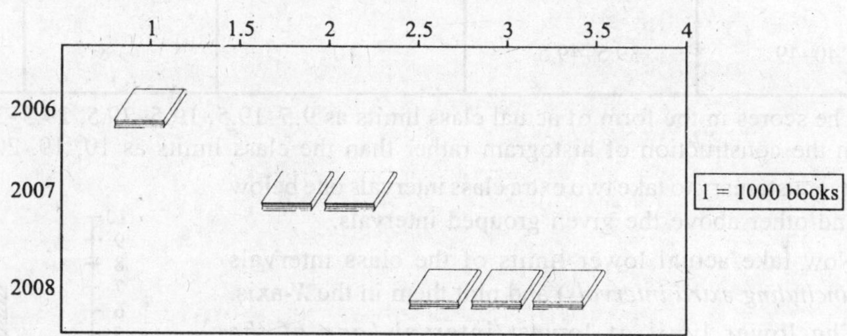

Fig. 3.7 Pictogram.

B. Graphical Representation of Grouped Data: For the grouped data, the following graphical presentation are used:

1. Histogram
2. Frequency polygon
3. Cumulative frequency curve (*or ogive*).

(*a*) **Histogram:** It is the most common form of diagrammatic representation of grouped frequency distribution of both continuous and discontinuous type.

(*i*) It consists of a set of rectangle drawn on a horizontal base line *i.e.*, *x*-axis (abscissa) and frequency (*i.e.*, number of observations) is marked on the vertical line *i.e.*, *y*-axis (*Ordinate*).

(*ii*) The width of each rectangle extends over the class boundaries of the corresponding class along the horizontal axis.

(*iii*) The area of each rectangle is proportional to the frequency in the respective class interval.

$$\begin{aligned}\text{Area of each rectangle} &= \text{width} \times \text{height}\\ &= \text{width of class} \times \text{frequency density}\\ &= \text{width of class} \times \frac{\text{class frequency}}{\text{width of class}}\\ &= \text{class frequency}\end{aligned}$$

Working Procedure:

(*i*) Convert the inclusive series into exclusive series.

Inclusive series

Age	10–19	20–29	30–39	40–49
Frequency	2	7	5	8

Exclusive Series:

Age (class interval)		*Size of interval*	*Frequency*	*Frequency density*
Score limit	*True limit*			
10–19	9.5–19.5	10	2	$\frac{2}{10} = 0.2$
20–29	19.5–29.5	10	7	$\frac{7}{10} = 0.7$
30–39	29.5–39.5	10	5	$\frac{5}{10} = 0.5$
40–49	39.5–49.5	10	8	$\frac{8}{10} = 0.8$

(*ii*) The scores in the form of actual class limits as 9.5–19.5, 19.5–29.5, 29.5–39.5 etc. are taken in the construction of histogram rather than the class limits as 10, 19, 20, 29 etc.

(*iii*) It is customary to take two extra class intervals one below and other above the given grouped intervals.

(*iv*) Now take actual lower limits of the class intervals (*including extra intervals*) and plot them in the *X*-axis. The lower limit of lowest interval (*one of the extra interval*) is taken at the intersecting point of *x*- and *y*-axes.

(*v*) Each class or interval with its specific frequency is represented by a separate rectangle. The base of each rectangle is the width of the class interval and the height is the respective frequency of that class or interval.

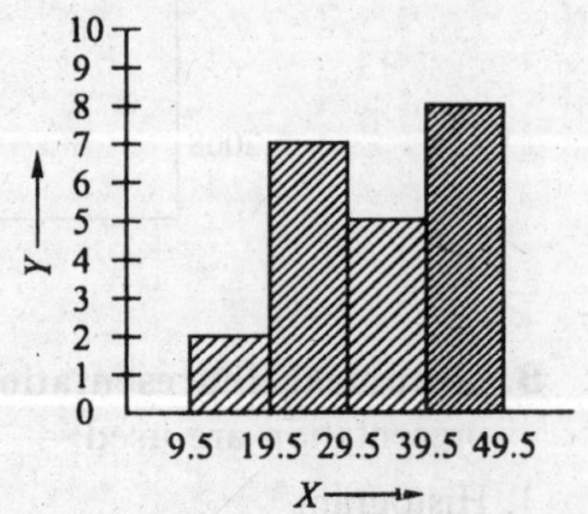

Fig. 3.8 Histogram with frequency distribution.

(*vi*) Frequencies are plotted on the *y*-axis.

(*vii*) Selection of appropriate units of representation along the *x*- and *y*-axes are essential. Both *x*- and *y*-axes should not be too short.

Types: There are two types of histograms:

(*a*) Histogram with equal class intervals. (*b*) Histogram with unequal class interval.

(*a*) Histogram with Equal Class intervals:

(*i*) Here the size of class intervals are drawn on *x*-axis with equal width and their respective frequencies on *y*-axis.

(*ii*) Class and its frequency taken together form a rectangle. The graph of rectangles is known as **histogram**.

Example: Population of carp fishes in 100 ponds are as follows:

Number of carps per ponds	0–100	100–200	200–300	300–400	400–500	500–600
No. of ponds	12	18	27	20	17	6

Solution: This is the case of Histogram with equal class interval.

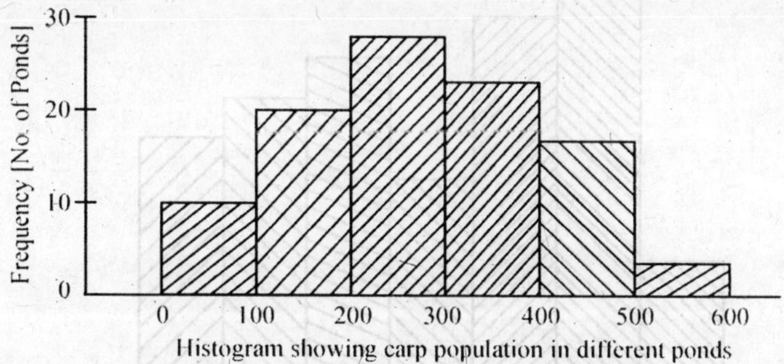

Fig. 3.9 Histogram with equal class intervals.

(*b*) Histogram with Unequal Class Intervals:

(*i*) Here the sizes of class intervals are drawn on axis with unequal width & their respective frequencies on y-axis.

(*ii*) Some have less width and some have more width. So the histogram is drawn on the basis of frequency density not on the basis of frequency.

Example: Draw the histogram of the following frequency distribution with unequal class interval:

Age group	14–15	16–17	18–20	21–24	25–29	30–34	35–39
No. of people	6	14	12	8	11	10	9

Calculation of Histogram

Age group				
Class-interval	*Class-boundary*	*Class-width*	*Frequency (No. of people)*	*Frequency density*
14–15	13.5–15.5	2	6	$\frac{6}{2} = 3$
16–17	15.5–17.5	2	14	$\frac{14}{2} = 7$
18–20	17.5–20.5	3	12	$\frac{12}{3} = 4$

Age group		Class-width	Frequency (No. of people)	Frequency density
Class-interval	Class-boundary			
21–24	20.5–24.5	4	8	$\frac{8}{4} = 2$
25–29	24.5–29.5	5	11	$\frac{11}{5} = 2.2$
30–34	29.5–34.5	5	10	$\frac{10}{5} = 2$
35–39	34.5–39.5	5	9	$\frac{9}{5} = 1.8$

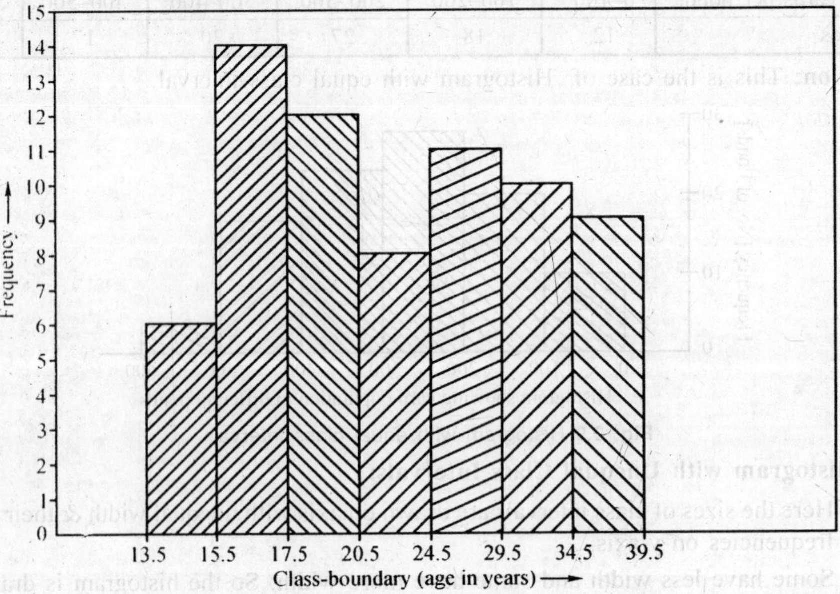

Fig. 3.10 Histogram with unequal class boundary.

Uses:

(*i*) It gives visual representation of the relative size of the various groups.

(*ii*) The surface of the tops of rectangles also give an idea of the nature of the frequency curve for the population.

(*iii*) It (histogram) may be used to find the mode graphically.

2. Frequency Polygon: It is an area diagram represented in the form of curve obtained by joining the middle points of the tops of the rectangles in a histogram or joining the mid-points of class intervals at the height of frequencies by straight lines.

It gives a polygon *i.e.*, figures with many angles.

(*i*) The frequency polygon is obtained by joining the successive points whose abscissa represent the mid values and ordinate represent the corresponding class frequencies.

(*ii*) In order to complete the drawing of a polygon, the two end points are joined to the base line at the mid points of the empty classes at each end of the frequency distribution.

(*iii*) Thus the frequency polygon has the same area as the histogiam, provided the width of all classes is the same.

Example: Construct a histogram and frequency polygon for the following data:

100–150	150–200	200–250	250–300	300–350
4	6	13	5	2

Solution: We have the case of equal class interval.

Class Interval	*Frequency*	*C.F.*
100–150	4	4
150–200	6	10
200–250	13	23
250–300	5	28
300–350	2	30

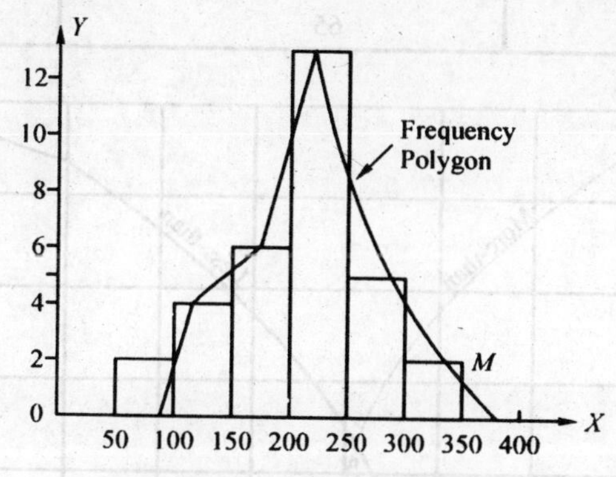

Fig. 3.11 Frequency Polygon.

Use:

(*i*) It is particularly useful in representing a simple and ungrouped frequency distribution of discrete variables.

(*ii*) It gives an approximate idea of the shape of the frequency curve.

3. Cumulative Frequency Polygon (ogive): The graphical representation of a cumulative frequency distribution where the cumulative frequencies are plotted against the corresponding class boundaries and the successive points are joined by straight lines, the diagram or curve obtained is known as ogive or cumulative frequency polygon.

Working Procedure:

(*i*) The upper limits of the classes are represented along *x*-axis.

(*ii*) The cumulative frequency of a particular class is taken along the *y*-axis.

(*iii*) The points corresponding to cumulative frequency at each upper limit of the classes are joined by a free hand curve. This curve is called a cumulative frequency curve or an ogive.

(*iv*) In case of frequency polygon, frequencies must be plotted at the upper limit of the class but in the case of an ogive, cumulative frequency is plotted at the upper limit of the class.

Example: Draw cumulative frequency polygon or ogives (both 'less than' & 'more than' types) for the following frequency distribution {diastolic blood pressure (mm Hg)}.

Class Intervals	50–59	60–69	70–79	80–89	90–99	100–109	110–119
No. of patients	8	10	16	14	10	5	2

Solution: Calculation for drawing ogives:

Class boundary	Cumulative frequency	
	Less than	More than
49.5	0	65
59.5	8	57
69.5	18	47
79.5	34	31
89.5	48	17
99.5	58	7
109.5	63	2
119.5	65	0

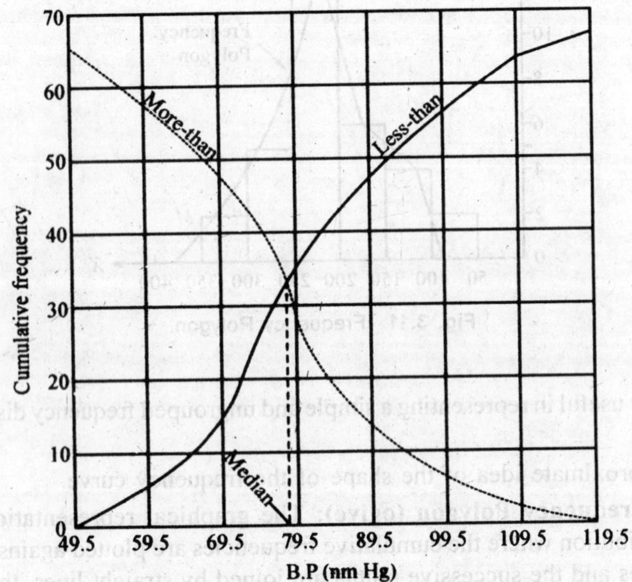

Fig. 3.12 Ogives for diastotic blood pressure.

Type:

(*a*) More than Ogive:

(*i*) It starts from the highest class boundary on the horizontal axis and gradually rising upwards end.

(*ii*) It looks an elongated letter '*S*' turned upside down.

(*b*) Less than Ogive:

(*i*) It starts from the lowest class boundary on the horizontal axis and gradually rising upwards at the highest class boundary corresponding to the cumulative frequency *i.e.*, total frequency.

(*ii*) It looks elongated letter '*S*'.

Uses:

(*i*) It is used to find the median, quartiles, decile and percentiles or value of the variables.

(*ii*) It is also useful in finding the cumulative frequency corresponding to a given value of the variable.

(*iii*) To find out the number of observation which are expected to lie between two given values.

Comparison between histogram and frequency polygon.

Histogram	*Frequency Polygon*
1. It is essentially the bar graph of the given frequency distribution.	1. The frequency polygon is a line graph of the frequency distribution.
2. It does not provide better conception of the contour of the distribution.	2. It gives much better conception of the contour of the distribution.
3. It is poorly useful.	3. It is more useful and practicable.
4. The histogram gives a very clear as well as accurate picture of the relative proportions of frequency from interval to interval.	4. In the frequency polygon, it is assumed that the frequencies are concentrated at the mid-points of the class intervals. It merely points out the graphical relationship between the mid-points and frequencies.

Difference between Diagram and Graph

Diagram	*Graph*
1. Ordinary paper can be used.	1. Graph paper must be used.
2. It is not helpful for interpolation and extra-polation technique.	2 It is helpful for interpolation and extrapolation technique.
3. The value of median and mode cannot be estimated.	3. The value of median and mode can be estimated.
4. Data are represented by bars, rectangles etc.	4. Data are represented by points or lines of different kinds—dots, dashes etc.
5. It is used for comparison only.	5. It is used for establishing mathematical relationship between two variables.
6. They are attractive & used for publicity.	6. It is useful to researchers & statisticians in data analysis.

Meta - Analysis
1. The combining of data from different research studies to gain a better overview of a topic than what was available in any single investigation.
2. The data obtained from combined studies must be comparable in order to be evaluated by this method.

BOX PLOT

- **Box Plot:** Box plot (Box-and whisker diagram or plot) is a convenient way of graphically depicting groups of numerical data through their five number summeries: smallest observation (minimum), lower quartile (Q_1), media (Q_2), upper quartile (Q_3) and largest observation (maximum) with outliers plotted individually.

- **Characteristics:**

I. A central box spans the quartiles.

II. A line in the box marks median.

III. Observations more than 1.5 × *IQR* (inter quartile range) outside the central box plotted individually as possible outliers.

IV. **Lines** called **whiskers,** extend from the box and to the smallest and the largest observations that are not the outliers.

V. Box plots can be drawn either horizontally or vertically.

VI. For symmetric distribution, box plots are symmetric but the symmetric box plot does not imply of symmetric distribution.

VII. The upper part of the box *i.e.* difference between Q_3 and M is bigger than the lowest part (difference between (M and Q_1) in case of positively skewed distribution. and smaller in case of negatively skewed.

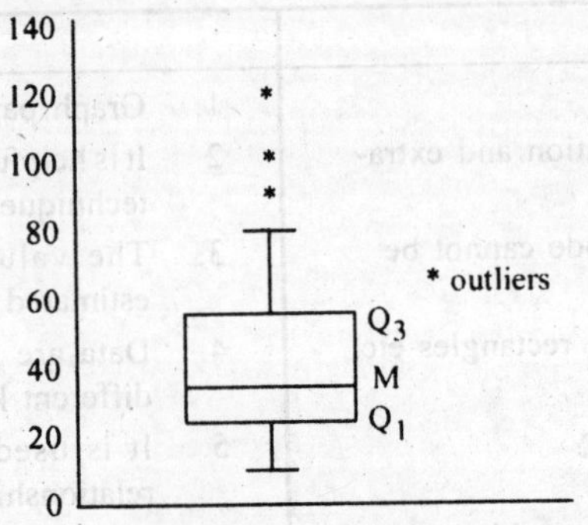

Fig. 3.13 Box plot

Uses:

I. The box plots are most useful for comparing distribution.

II. It is a quickway of examining one or more sets of data graphically.

III. The spacings between the different parts of the box indicate the degree of **dispersions** (spread) and **skewness** in the data and identify **outliers.**

- **Inter quartile range (IQR):**

  It is a measure of spread, based on quartiles. The range or extend from third quartile (Q_3) to first quartile (Q_1) is called interquartile range.

  $$IQR = Q_3 - Q_1$$

- **Semi interquartile range or quartile deviation:**

  The half of the difference between the third (Q_3) and first (Q_1) quartile is called semiquartile or quartile deviation.

- **Outlier:**

  An outlier is defined as an observation which falls more than $1.5 \times IQR$ (called step) abovc Q_3 or below Q_1.

  I. If an observation falls more than $3 \times IQR$ above Q_3 or below Q_1, then it is is known as **extreme outlier**.

  II. The observation falling between $1.5 \times IQR$ and $3 \times IQR$ above Q_3 or below Q_1 is known as a **suspect outlier**.

- **Innerfences and outerfences:**
  - The values ($Q_1 - 1.5\ IQR$, $Q_3 + 1.5\ IQR$) are known innerfences.
  - The values ($Q_1 - 3\ IQR$, $Q_3 + 3\ IQR$) are known as outerfences.

- **Five number summery**

Minimum	Q_1	M	Q_3	Maximum

CHAPTER 4

CENTRAL TENDENCY

The word 'average' denotes a representative of a whole set of observations. It is a single figure which describes the entire series of observations with their varying sizes. It is a typical value occupying a central position where some observations are larger and some others are smaller than it. Average is a general term which describes the centre of a series. The values of variable tend to concentrate around the central value. It is the central part of the distribution and therefore they are also called the measures of central tendency.

Characteristics of Central Tendency:

1. *It should be rigidly defined*:
   (*a*) An average should be properly defined so that it has one and only one interpretation.
   (*b*) The average should not depend on the personal prejudice and bias of the investigator.
2. *It should be based on all items*: The average should depend on each and every item of the series. So that if any of the item is dropped, the average itself be altered.
3. *It should be easily understood*: The desirable property of on average is that it can be readily understood and then only it can be made popular.
4. *It should not be unduly affected by the extreme value*:
   (*a*) The average should depend on each and every times, so we must be aware that no extreme observations could influence unduly on the central value.
   (*b*) Due to an extreme observation the central value changes or distorts and it can not be typical for the group values.
5. *It should be least affected by the fluctuations of the sampling*: If we select different groups of sample we should expect some central value approximately in each sample.
6. *It should be easy to interpret*: The average can become popular only because of its access for easy computation.
7. *It should be easily subjected to further mathematical calculations*: An average value could be preferred to others if it is capable to be used for further statistical computation.

Measures of Central Tendency:

The most common measures of central tendency are:

1. Mean or Arithmetic mean 2. Median 3. Mode.

Arithmetic Mean (A.M.):

It is obtained by summing up all the observations and dividing the total by the number of observations.

(*a*) **Arithmetic mean for ungrouped data or individual observations:** It $x_1, x_2, x_3, \ldots x_n$ be 'n' observations for a variable x, the arithmetic mean $\bar{x}$ is given by

$$\bar{x} = \frac{\sum_{i=1}^{n} x_i}{n} = \frac{x_1 + x_2 + x_3 \ldots + x_n}{n}$$

Steps of Calculation:

1. Add together all the values of X and get $\sum x$.
2. Divide this total by the number of observations.

Example: The following table gives the marks obtained in Statistics of 10 students of B. Com (Hons) in Serampore College.

Roll No	1	2	3	4	5	6	7	8	9	10
Marks in Statistics	67	69	66	68	72	63	76	65	70	74

Calculate the arithmetic mean of marks in statistics among these students.

Solution:

Roll No.	*Marks*
1	67
2	69
3	66
4	68
5	72
6	63
7	76
8	65
9	70
10	74
$N = 10$	$\sum x = 690$

$$\bar{x} = \frac{\sum x}{N}$$

Here $\sum x = 690 \; N = 10$

$$\bar{x} = \frac{690}{10} = 69$$

Thus arithmetic mean of marks obtained in Statistics by students is 69.

(*b*) Arithmetic mean for grouped data (discrete series): Let a variable take n values x_1, x_2, x_3 ..., x_n having corresponding frequencies $f_1, f_2, f_3 \cdots f_n$

then $\bar{x} = \dfrac{\sum_{i=1}^{n} f_i x_i}{\sum_{i=1}^{n} f_i}$ $\therefore \; \bar{x} = \dfrac{1}{N} \sum fx$, where $N = \sum f$ (sum of all frequencies)

Steps:

1. Multiply each variable value with its corresponding frequency and sum them up to get $\sum f_i x_i$.
2. Divide this total value $\sum f_i x_i$ by number of observations *i.e.*, total frequency $\sum f_i$.

Example: Find the Arithmetic mean from the frequency table.

Marks	30	40	50	60	70	80	90
No. of students	15	20	10	15	20	15	5

Solution:

Marks (x)	Number of students (f)	fx
30	15	450
40	20	800
50	10	500
60	15	900
70	20	1400
80	15	1200
90	5	450
	$\sum f = 100$	$\sum fx = 5700$

Here we apply $\quad \bar{x} = \dfrac{\sum fx}{\sum f} = \dfrac{\sum fx}{N}$

where $\quad N = \sum f = 100$ and $\sum fx = 5700$

$\therefore \quad \bar{x} = \dfrac{5700}{100} = 57$

Thus arithmetic mean of x is 57.

Calculation of Arithmetic Mean (Continuous Series) in Grouped Data

Let there be a continuous class distribution of the form $x_0 - x_1, x_1 - x_2, x_2 - x_3 - ... x_n$ etc. With the corresponding frequencies $f_1, f_2, f_3, ..., f_n$. Mid-point of each class are found & name a variable x with values of mid-points in each class.

$$\bar{X} = \frac{\sum_{i=1}^{n} f_i x_i}{\sum_{i=1}^{n} f_i} \quad \therefore \ \bar{x} = \frac{\sum fx}{N} \quad \text{where } N = \sum f, \quad f_i = \text{frequency}, \quad x_i = \text{variable.}$$

Steps:

(*i*) Evaluate the mid value of each class and denote by '*m*'.

(*ii*) Multiply the mid-points by respective frequency of each class and sum them up to $\sum f_i m_i$.

(*iii*) Divide this total of $\sum f_i m_i$ by sum of the frequencies *i.e.*, total number of observations.

Example 1. Calculate the arithmetic mean for the following data:

Class interval	10–20	20–30	30–40	40–50	50–60	60–70
Frequency	3	5	10	15	5	12

Solution:

Class interval	Mid value (m)	Frequency (f)	fm
10–20	15	3	45
20–30	25	5	125
30–40	35	10	350
40–50	45	15	675
50–60	55	5	275
60–70	65	12	780
		$\sum f = 50$	$\sum fm = 2250$

Here we apply $\bar{x} = \dfrac{\sum fm}{\sum f}$

$$\sum fm = 2250$$

$$\sum f = 50 \qquad \bar{x} = \frac{2250}{50} = 45.$$

Example 2. Calculate the arithmetic mean for the daily wages from the following data:

Wages in Rs.	10–20	20–30	30–40	40–50	50–60	60–70
Number of workers	5	10	30	20	15	10

Solution:

Class interval in Rs.	*Mid-point (m)*	*Frequency (number of workers) (f)*	*fm*
10–20	15	5	75
20–30	25	10	250
30–40	35	30	1050
40–50	45	20	900
50–60	55	15	825
60–70	65	10	650
		$\sum f = 90$	$\sum fm = 3750$

$$\bar{x} = \frac{\sum fm}{\sum f} \quad \sum f = 90 \; \sum fm = 3750$$

$$\text{So} \quad \bar{x} = \frac{3750}{90} = 41.666 = 41.67 \text{ (approx).}$$

Example 3. Calculate the arithmetic mean for the following data:

Class interval	10–20	20–30	30–40	40–50	50–60	60–70	70–80	80–90	90–100
Frequency	2	7	17	29	29	10	3	2	1

Solution:

Class interval	*Mid-point (m)*	*Frequency (f)*	*fm*
10–20	15	2	30
20–30	25	7	175
30–40	35	17	595
40–50	45	29	1305
50–60	55	29	1595
60–70	65	10	650
70–80	75	3	225
80–90	85	2	170
90–100	95	1	95
		$\sum f = 100$	$\sum fm = 4840$

$$\bar{x} = \frac{\sum fm}{\sum f} \qquad \sum fm = 4840 \qquad \sum f = 100 \qquad \bar{x} = \frac{4840}{100} \qquad \bar{x} = 48.4.$$

Short-cut method:

(*i*) Short cut method is applied when the frequencies and values of the variables are quite large. Therefore it becomes very difficult to compute the arithmetic mean.

(*ii*) Here the provisional mean or Assumed mean is taken as that values of x (mid value of the class interval) which corresponds to the middle value of the frequency distribution. This number is called *provisional mean* or *assumed mean.*

(*a*) In the case of ungrouped data:

$$\bar{x} = a + \frac{\sum d}{n} \quad \text{where } a = \text{assumed mean} \quad n = \text{number of items}$$

$$d = (x - a) = \text{deviation of any variate from } a.$$

Working procedure:

(*i*) Denote the variable of the series (x or X).

(*ii*) Take any item of series, preferably the middle one and denote it by 'a'. This number 'a' is called *assumed mean* or *provisional mean.*

(*iii*) Take the difference $X - a$ and denote 'd' or dx *i.e.*, $d = x - a$ where 'd' is the deviation of any variate from 'a'.

(*iv*) Find the sum of $\sum d$.

(*v*) Use the following formula $\bar{x} = a + \frac{\sum d}{N} \qquad N = \sum f.$

Example 1. Find the mean weight of the following students by short cut method whose weights are in kg.

67, 69, 66, 68, 63, 76, 72, 74, 70, 65

Solution: Let us take 68 as assumed mean

X	X − a = d
67	67 − 68 = −1
69	69 − 68 = +1
66	66 − 68 = −2
68	68 − 68 = +0
63	63 − 68 = −5
76	76 − 68 = +8
72	72 − 68 = +4
74	74 − 68 = +6
70	70 − 68 = +2
65	65 − 68 = −3

Total deviation *i.e.*, $\sum fd = +21 - 11 = 10 \qquad N = 10$

$$\bar{x} = a + \frac{\sum fd}{N} = 68 + \frac{10}{10}$$

$$= 68 + 1 = 69$$

Mean weight $= 69$ kg.

Example 2. Find the mean height of the 8 students by short cut method, whose heights are in centimeter.

59, 65, 69, 63, 61, 71, 73, 67

Solution: Let us take 65 as assumed mean.

X	X – a = d
59	59 – 65 = –6
65	65 – 65 = 0
69	69 – 65 = +4
63	63 – 65 = –2
61	61 – 65 = –4
71	71 – 65 = +6
73	73 – 65 = +8
67	67 – 65 = +2

Total deviation *i.e.*, $\sum fd = +20 - 12 = +8 \quad N = 8$

$$\bar{X} = a + \frac{\sum fd}{N}$$

$$= 65 + \frac{8}{8} = 65 + 1 = 66$$

Mean height $= 66$ cm.

(*b*) In the case of grouped data:

$$\bar{X} = a + \frac{\sum fd}{N}$$

fd = Product of frequency & corresponding deviation

$N = \sum f =$ the sum of all the frequencies.

Working Procedure:

(*i*) In the case of discrete series denote the variable by x or X and the corresponding frequencies by 'f' [but in the case of continuous series x is the mid value of the interval and f the frequency corresponding to that interval].

(*ii*) Take any item x series, preferably the middle one and denote it by 'a'. This number 'a' is called *assumed mean* or *provisional mean.*

(*iii*) Take the difference $x–a$ and denote it by 'd' *i.e.*, $x - a$ = deviation of any variate from 'a' the assumed mean.

(*iv*) Multiply the respective 'f' and denote the product under the column 'fd'.

(*v*) Find $\sum fd$.

(*vi*) Use the following formula to calculate the arithmetic mean:

$$\bar{X} = a + \frac{\sum fd}{\sum f} = a + \frac{\sum fd}{N}.$$

Example 3. Compute the arithmetic mean by short-cut method for the following data:

Wages in Rs.	10–20	20–40	40–50	50–70	70–80	80–100
Number of persons	5	15	25	35	12	8

Solution: Let us take assumed mean 55.

Class interval	Mid-point	No. of persons	$x - a = d$	fd
10–20	15	5	15 – 55 = –40	–200
20–40	30	15	30 – 55 = –25	–375
40–50	45	25	45 – 55 = –10	–250
50–70	60	35	60 – 55 = 5	+175
70–80	75	12	75 – 55 = 20	+240
80–100	90	8	90 – 55 = 35	+280
		$\sum f = 100,\ N = 100$		$\sum fd = -130$

$$\bar{X} = a + \frac{\sum fd}{N}$$

$$= 55 - \frac{130}{100} = 55 - 1.3 = 53.7.$$

(*c*) Step deviation method:

(*i*) When the class intervals in a grouped data are equal then the calculations can be simplified further by taking out the common factor from the deviations.

(*ii*) This common factor is equal to the width of the class interval.

(*iii*) In such cases, the deviation of variates x from the assumed mean 'a' (*i.e.*, $d = x - a$) are divided by the common factor.

$$\bar{X} = a + \frac{\sum fd}{N} \times i \quad \text{where } a = \text{assumed mean,}$$

$d = \dfrac{x - a}{i}$ deviation of any variate from 'a'

i = width of class interval

N = Number of observation.

Example 1. Find the arithmetic mean of the following:

Height	60–62	63–65	66–68	69–71	72–74
Number of student	15	54	126	81	24

Solution: Let us take 67 as assumed mean.

Height (in inch)	Mid value (m)	Frequency (f)	Deviation (d)	$d' = \frac{d}{i}$	fd'
60–62	61	15	–6	–2	–30
63–65	64	54	–3	–1	–54
66–68	67	126	0	0	0
69–71	70	81	+3	+1	+81
72–74	73	24	+6	+2	+48
		$\sum f = 300$			$\sum fd' = 45$

$$\sum fd' = 45 \quad a = 67$$

$$\sum f = N = 300 \quad i = 3$$

$$\bar{X} = a + \frac{45}{300} \times 3$$

$$= 67 + \frac{45}{300} \times 3 = 67 + .45 = 67.45 \text{ inch.}$$

Example 2. Calculate the average marks by the step deviation method.

Marks	0–10	10–20	20–30	30–40	40–50	50–60
Number of students	40	25	50	35	30	20

Solution: Let us take 35 as assumed mean.

Marks	*Midvalue m*	*No. of students (f)*	*d = X – 35*	$d' = \frac{X-35}{10}$	*fd'*
0–10	5	40	–30	–3	–120
10–20	15	25	–20	–2	–50
20–30	25	50	–10	–1	–50
30–40	35	35	0	0	0
40–50	45	30	+10	+1	+30
50–60	55	20	+20	+2	+40
		$\sum f = N = 200$			$\sum fd' = -150$

$-220 + 70 = -150$

$$a = 35 \quad N = 200 \quad i = 10$$

$$\bar{X} = a + \frac{\sum fd'}{N} \times i \qquad \sum fd' = -150$$

$$= 35 - \frac{150}{200} \times 10$$

$$= 35 - 7.5 = 27.5.$$

Example 3. In a study on patients, the following data are obtained. Find the arithmetic mean.

Age	10–19	20–29	30–39	40–49	50–59	60–69	70–79	80–89
Number of cases	1	0	1	10	17	38	9	3

Solution: Let us take 44.5 as assumed mean.

Age	*Midvalue (x)*	*No. of cases (f)*	*d = x – 44.5*	$d' = \frac{X'-d}{i}$	*fd'*
10–19	14.5	1	–30	–3	–3
20–29	24.5	0	–20	–2	0
30–39	34.5	1	–10	–1	–1
40–49	44.5	10	0	0	0
50–59	54.5	17	+10	+1	+17
60–69	64.5	38	+20	+2	+76
70–79	74.5	9	+30	+3	+27
80–89	84.5	3	+40	+4	+12
		$\sum f = N = 79$			$\sum fd' = 128$

$a = 44.5$ $\qquad \overline{X} = a + \frac{\sum fd'}{N} \times i$

$N = 79$ $\qquad = 44.5 + \frac{128}{79} \times 10$

$\sum fd' = 128$ $\qquad = 44.5 + 16.2$

$i = 10$ $\qquad = 60.7$

Mean of the Combined Distribution

When two sets of score combined into a single distribution, then the mean of the combined distribution is the weighted mean of the means of the components, the weights being total frequencies in those components.

$$\overline{X} = \frac{n_1 \overline{x}_1 + n_2 \overline{x}_2}{n_1 + n_2}$$

[$\overline{x}_1$ and x_2 are the means of the combined distribution n_1 and n_2 are the total frequencies of the component distribution]

Proof: $\frac{\sum x}{n_1} = \overline{x}_1 \quad \frac{\sum y}{n_2} = \overline{x}_2$

$$\sum x = n_1 \overline{x}, \quad \sum y = n_2 \overline{x}_2$$

Now by definition $\overline{x} = \frac{\sum x + \sum y}{n_1 + n_2} = \frac{n_1 \overline{x}_1 + n_2 \overline{x}_2}{n_1 + n_2}$.

Example 1.There are 60 students in III year B. Sc (Bio) of Serampore college of which 25 are girls. The average weight of 25 girls is 40 kg and that of 35 boys is 53 kg. Find the mean weight in kg of the entire III year students.

Solution: Let $\overline{x}_1$ and $\overline{x}_2$ denote the average weight of girls and boys respectively. Then $\overline{x}_1 = 40$, $\overline{x}_2 = 53$, $n_1 = 25$ and $n_2 = 35$.

Mean weight of the III year student

$$(\overline{x}_1) = \frac{n_1 \overline{x}_1 + n_2 \overline{x}_2}{n_1 + n_2}$$

$$= \frac{25 \times 40 + 35 \times 53}{25 + 35} = \frac{1000 + 1855}{45} = \frac{2855}{45} = 47.583 \text{ kg.}$$

Example 2.The mean age of 40 students is 16 years, and the mean age of 60 students is 20 years. Find out the mean age of all 100 students combined together.

Solution: $n_1 = 40 \quad \overline{x}_1 = 16$

$n_2 = 60 \quad \overline{x}_2 = 20$

$$\overline{X} = \frac{n_1 \overline{x}_1 + n_2 \overline{x}_2}{n_1 + n_2} = \frac{40 \times 16 + 60 \times 20}{40 + 60}$$

$$\overline{X} = \frac{640 + 1200}{100} = \frac{1840}{100} = 18.4.$$

Example 3. The mean weight of 100 carp fishes of a pond is 49.46 kg. The mean weight of 200 carp fishes is 52.32 kg. Find the combined mean weight of all the carp fishes.

Solution: $n_1 = 100 \quad \bar{x}_1 = 49.46$

$n_2 = 200 \quad \bar{x}_2 = 52.32$

$$\bar{X} = \frac{100 \times 49.46 + 200 \times 52.32}{100 + 200} = \frac{4946 + 1064}{300} = \frac{15410}{300} = 51.3666$$

$= 51.37$ kg.

Example 4. Howrah Vivekananda Institution has 4 sections in class *X*, having 40, 35, 45 and 42 students. The mean marks obtained in a chemistry test are 50, 60, 55 and 45 respectively for the 4 sections. Determine the over all average of the marks per student.

Solution: $\bar{x}_1 = 50 \quad n_1 = 40$

$\bar{x}_2 = 60 \quad n_2 = 35$

$\bar{x}_3 = 55 \quad n_3 = 45$

$\bar{x}_4 = 45 \quad n_4 = 42$

Average marks $\bar{X} = \dfrac{40 \times 50 + 35 \times 60 + 45 \times 55 + 42 \times 45}{40 + 35 + 45 + 42}$

$$= \frac{2000 + 2100 + 2475 + 1890}{162} = \frac{8465}{162} = 52.25.$$

Example 5. The mean of 25 observations was found to be 78.4. But later it was found that 96 was misread as 69. Find the correct mean.

Solution: Here $N = 25 \quad \bar{X} = 78.4 \quad \therefore \bar{X} = \dfrac{\sum x}{N}$

$$\sum X = \bar{X}N = 78.4 \times 25 = 1960$$

But this value of $\sum X$ is incorrect as 96 was misread as 69.

$\therefore$ Correct $\sum X$ = (Incorrect $\sum X$) – (sum of incorrect value) + sum of correct values

$= 1960 - 69 + 96 = 2056 - 69 = 1987$

$\therefore$ Correct mean $= \dfrac{1987}{25} = 79.48.$

Example 6. The mean of 40 observations was 160. It was detected on rechecking that the value 125 was wrongly copied as 165 for the computation of the mean. Find the correct mean.

Solution: $N = 40 \quad \bar{X} = 160 \quad \therefore \bar{X} = \dfrac{\sum X}{N}$

or $\sum X = N \times \bar{X}$

$\therefore \sum X = 40 \times 160 = 6400$

But the value of $\sum X$ is incorrect as 125 was wrongly copied as 165.

$\therefore$ Correct $\sum X$ = (Incorrect $\sum X$) – (Sum of incorrect value) + (sum of correct values)

$= 6400 - 165 + 125$

$= 6525 - 165$

$= 6360$

$\therefore$ Correct mean = $\dfrac{6360}{40} = 159.$

Example 7. The mean of 30 values was 150. It was detected on rechecking that the value 165 was wrongly copied as 135 for the computation of the mean. Find the correct mean.

Solution: $N = 30 \quad \overline{X} = 150 \quad \therefore \overline{X} = \dfrac{\sum X}{N}$ or $\sum X = N\overline{X}$

$$\sum X = 30 \times 150 = 4500$$

But the value of $\sum X$ is incorrect as 165 was wrongly copied as 135.

$\therefore$ Correct $\sum X$ = (Incorrect $\sum X$) – (Sum of incorrect value) + (sum of correct value)

$= 4500 - 135 + 165$

$= 4665 - 135$

$= 4530$

$\therefore$ Correct mean $= \dfrac{4530}{30} = 151.$

Merits, Demerits (Limitation) & Uses of Arithmetic Mean

Merits:

1. It has the simplest average formula which is easily understandable and easy to compute.
2. It is so rigidly defined by mathematical formula that every one gets same result for a single problem.
3. Its calculation is based on all the observations.
4. It is least affected by sampling fluctuations.
5. The mean is a typical *i.e.*, it balances the value on either side.
6. It is the best measure to compare two or more series (datas).
7. Mean is calculated on value and does not depend upon any position.

Demerits:

1. It can not be calculated if all the values are not known.
2. The extreme values have greater affect on mean.
3. It can not be determined for the qualitative data such as love, honesty, beauty etc.

Uses:

1. It is used in practical statistics.
2. Estimates are always obtained by mean.
3. Common people uses mean for calculating average marks obtained by students.

Median:

It is the middle most point or the central value of the variable in a set of observations when observations are arranged either in ascending or in descending order of their magnitudes.

"Median is the value of that item in a series which decides the series into two equal parts, one part consisting of all values less and the other all values greater than it".

—Prof. Ghosh & Chowdhury

Calculation of Median:

(a) Simple Series (ungrouped data)

Procedure:

(i) Arrange the data in [either ascending or descending] order of magnitude.

(ii) If the number of observations be odd, the value of the middle – most items is the median. However, if the number be even, the arithmetic mean of the two middle most items is taken as median.

When 'n' is odd. In this case $\frac{n+1}{2}$ th value is the median. $M = \frac{n+1}{2}$ th term.

When 'n' is even. In this case there are two middle terms. $\frac{n}{2}$ th and $\left(\frac{n}{2}+1\right)$ th. The median is the average of these two terms.

$$M = \frac{\frac{n}{2} + \left(\frac{n}{2}+1\right)}{2}$$

Example 1. Find the median of the following numbers:

(a) 21, 12, 49, 37, 88, 46, 55, 74, 63

(b) 88, 72, 33, 29, 70, 86, 54, 91, 61, 57. *[C.U. B.Com, 1973]*

Solution: (a) Let us arrange the data in order: 12, 21, 37, 46, 49, 55, 63, 74, 88

In this data the number of item is $n = 9$ (odd)

$\therefore$ Median $= M = \frac{(n+1)}{2} = \frac{(9+1)}{2}$ th item = 5th item

Now the 5th value in the data is 49.

$\therefore$ Median is 49.

(b) Let us arrange the data in order: 29, 33, 54, 57, 61, 70, 72, 86, 88, 91

In this data the number of item is $n = 10$ (even)

$\therefore$ Median $=$ average of $\left(\frac{n}{2}\right)$ th + $\left(\frac{n}{2}+1\right)$ terms.

Average of $\left(\frac{10}{2}\right)$ th and $\left(\frac{10}{2}+1\right)$ th terms.

= Average of 5th and 6th terms

$$M = \frac{61+70}{2} = \frac{131}{2} = 65.2$$

Median is 65.2.

Example 2. The number of runs scored by 11 players of cricket team of R.K. Mission Vidyamandira: 5, 19, 42, 11, 50, 30, 21, 0, 52, 36, 27.

Find the median.

Solution: Let us arrange the values in ascending order:

0, 5, 11, 19, 21, 27, 30, 36, 42, 50, 52.

Here item is 11 (odd)

Median $= M = \left(\frac{n+1}{2}\right)$ th item

$$= \frac{11+1}{2} = 6\text{th item } i.e., 27$$

$\therefore$ Median $= 27$ runs.

Example 3. Compute the median wage from the following 10 persons weekly wages, Rs. 75, Rs. 205, Rs. 315, Rs. 34, Rs. 340, Rs. 1025, Rs. 521, Rs. 791, Rs. 695, Rs. 2344.

Solution: Let us arrange the data in ascending order: Rs. 34, Rs. 75, Rs. 206, Rs. 315, Rs. 340, Rs. 521, Rs. 695, Rs. 791, Rs. 1,025, Rs. 2,344.

In this data, the number of items $n = 10$ which is even.

So median (M) $= \text{average of} \left(\frac{n}{2}\right)\text{th} + \left(\frac{n}{2}+1\right)\text{th}$

$$= \frac{10}{2} + \frac{10+2}{2} = \frac{1}{2}\text{ (5th + 6th) item}$$

Here 5th $= 340$. So $M = \frac{340+521}{2} = \frac{861}{2} =$ Rs. 430.50

Here 6th $= 521$

Hence median income (weekly wage) is estimated Rs 430.50.

***(b)* Grouped Data:**

1. Discrete series:

Procedure:

(*i*) Arrange the data in either ascending or descending order of magnitude.

(*ii*) A table is prepared showing the corresponding frequencies and cumulative frequencies.

(*iii*) Now median is calculated by the following formula

$$M = \left(\frac{n+1}{2}\right)\text{th } N = \sum f.$$

Example 1. Find the median for the following data:

Income Rs.	100	150	80	200	250	180	Total
No. of persons	24	26	16	20	6	30	122

Solution: Let us arrange the income in ascending order and then form cumulative frequencies.

Income in ascending order	*Number of persons (f)*	*Cumulative frequency (cf.)*
80	16	16
100	24	40
150	26	66
180	30	96
200	20	116
250	6	122

As $n = 122$ (even), So the median (M) = average of $\left(\frac{n}{2}\right)th + \left(\frac{n}{2}+1\right)th$

$$= \frac{122}{2} + \left(\frac{122+2}{2}\right)$$

$$= \frac{61 + 62}{2} = 61.5th.$$

61.5*th* lies is the interval 41 to 66. Therefore its value is 150.

Example 2. Calculate the median for the following data:

Number of students	6	16	7	4	2	8
Marks	20	25	50	9	80	40

Solution: Let us arrange the data (marks) in ascending order and then form cumulative frequencies.

Marks	*No. of students (f)*	*Cumulative frequency (cf.)*
9	4	4
20	6	10
25	16	26
40	8	34
50	7	41
80	2	43

Here $\sum f = n = 43$

Median (*M*) is $= \frac{n+1}{2} = \frac{43+1}{2} = 22\text{nd value}$

The table shows that all items from 11 to 26 have their values 25. Since 22 and item lies in this interval, therefore it value is 25.

2. Continuous series:

Procedure:

(*i*) Here data is given in the form of a frequency table with class interval.

(*ii*) Cumulative frequencies are found out for each value.

(*iii*) Median class is then calculated, where cumulative frequency $\frac{N}{2}$ lies is called *median class.*

(*iv*) Now median is calculated by applying the following formula.

$$M = L + \frac{\frac{N}{2} - C}{fm} \times i$$

L = Lower limit of the class in which median lies.

N = Total number of frequencies.

fm = Frequency of the class in which median lies.

C = Cumulative frequency of the class preceding the median class.

i = Width of the class interval in which the median lies.

Example 1. Find the median & median class of the data given below:

Class boundaries	15–25	25–35	35–45	45–55	55–65	65–75
Frequency	4	11	19	14	0	2

Solution:

Class boundary	Midvalue (m)	Frequency (f)	Cumulative frequency (cf)
15–25	20	4	4
→ 25–35	30	11	15 ← $\frac{N}{2} = 25$
35–45	40	19	34
45–55	50	14	48
55–65	60	0	48
65–75	70	2	50

$$\frac{N}{2} = \frac{50}{2} = 25$$

$$L_1 = 35 \quad fm = 19 \quad C = 15 \quad i = 10$$

It is more than cumulative frequency 15, but is less than the cf. 34. Hence the median class is 35–45.

$$M = 35 + \frac{25 - 15}{19} \times 10$$

$$= 35 + \frac{10}{19} \times 10 = 35 + 5.263 = 40.263$$

Median class 35–45.

Example 2. The following table gives the marks obtained by students in Economics. Find the median:

Marks	11–20	21–30	31–40	41–50	51–60	61–70	71–80
No. of students	42	38	125	84	45	36	30

Solution: Let us prepare the table showing mid value; frequencies & cumulative frequencies.

Marks	Class boundary	Midvalue (m)	No. of students (f)	Cumulative frequency (cf)
11–20	10.5–20.5	15.5	42	42
21–30	20.5–30.5	25.5	38	80
→ 31–40	30.5–40.5	35.5	← 125	205
41–50	40.5–50.5	45.5	84	289
51–60	50.5–60.5	55.5	45	334
61–70	60.5–70.5	65.5	36	370
71–80	70.5–80.5	75.5	30	400

It is more than cumulative frequency 80, but it is less than c.f. 205, hence the median class is 31–40.

$$\frac{N}{2} = 200 \quad L_1 = 30.5 \quad fm = 125 \quad c = 80 \quad i = 10$$

$$M = 30.5 + \frac{200-80}{125} \times 10$$

$$= 30.5 + \frac{120}{125} \times 10 = 30.5 + 9.6 = 40.1.$$

Example 3. The weekly expenditure of 100 families are given below. Find the median of weekly expenditure.

Expenditure	0–10	10–20	20–30	30–40	40–50
Frequency	14	23	27	21	15

Solution: Let us prepare the table showing frequency & cumulative frequency.

Weekly expenditure	*No. of families*	*cf*
0–10	14	14
10–20	23	37
→ 20–30	27	← 64
30–40	21	85
40–50	15	100

$$\frac{N}{2} = \frac{100}{2} = 50$$

It is more than c.f 37 but less then 64 hence median class is 20–30

$L_1 = 20 \qquad i = 10 \qquad fm = 27 \qquad C = 37$

$$20 + \frac{50-37}{27} \times 10 = 20 + \frac{13}{27} \times 10 = 24.81.$$

Example 4. An incomplete frequency distribution is given below:

Variable	10–20	20–30	30–40	40–50	50–60	60–70	70–80
Frequency	12	30	?	65	?	25	18

Given that median value is 46. Determine the missing frequencies, using median formula.

Solution: Let the frequency of the class 30–40 be f_1 and that of 50–60 be f_2. Total $\sum f = N = 229$.

Then $\qquad f_1 + f_2 = N - (f_1 + f_2) = 229 - (12 + 30 + 65 + 25 + 18)$

$\qquad = 229 - 150 = 79$

Formula: Median $\qquad = L_1 + \frac{h}{f_m}\left(\frac{N}{2} - C\right)$

L_1 = Lower limit of the median class

f_m = Frequency of the median class

h = Magnitude of the median class

C = Cumulative frequency of the class just preeceding the median class.

Here $\qquad N = 229 \quad f_m = 65 \qquad C = (12 + 30 + f_1)$

Median value 46 *i.e.*, median class 40–50

i.e., $h = 10 \quad L_1 = 40$

$$46 = 40 + \frac{10}{65}\left\{\frac{229}{2} - (12 + 30 + f_1)\right\}$$

$$46 = 40 + \left\{\frac{114.5 - (12 + 30 + f_1)}{65}\right\} \times 10$$

or $$46 - 40 = \left(\frac{114.5 - 42 - f_1}{65}\right) \times 10$$

$$6 = \frac{72.5 - f_1}{65} \times 10$$

or $$72.5 - f_1 = \frac{65 \times 6}{10} = \frac{390}{10} = 39$$

or $$f_1 = 72.5 - 39 = 33.5 \text{ or } 34 \text{ or } 33 \; [f_1 + f_2 = 79]$$

$\therefore$ $$f_2 = 79 - 33.5 = 45.5 \quad i.e., \; 45 \text{ or } 46.$$

Advantages (Merits) of Median:

1. It is easily understood although it is not so popular as mean.
2. It is not influenced or affected by the variation in the magnitude of the extreme items.
3. The value of the median can be graphically ascertained to ogives.
4. It is the best measure for qualitative data such as beauty, intelligence etc.
5. The median indicates the value of middle item in the distribution i.e. middle most item is the median.
6. It can be determined even by inspection in many cases.

Disadvantages (Demerits) of Median:

1. For the calculation of median, data must be arranged.
2. Median being a positional average cannot be dependent on each and every observations.
3. It is not subject to algebraic treatment.
4. Median is more affected or influenced by sampling fluctuations than the arithmetic mean.

Mode: Mode is considered as the value in a series which occurs most frequently and has the maximum frequency.

According Crafton and Cowden, "The mode of a distribution is the value at the point around which the items tend to be most heavily concentrated. It may be regarded as the most typical value".

Calculation of mode:

A. Ungrouped Data (simple series)

Procedure:

(*i*) In case of simple series, mode can be determined by locating that value which occurs the maximum number of times.

(*ii*) It can be determined by inspection only.

(*iii*) It is that value of the variable which corresponds to the largest frequency.

Example 1. From the data given below, find modal value:

1, 3, 1, 3, 3, 5, 3, 3, 1, 5, 3, 3, 4, 5, 4, 2, 3, 2, 3, 7, 6, 3, 2, 5, 2, 3, 3, 2, 6, 2, 3, 2, 4, 2, 3.

Solution: Let us prepare the table showing the frequency.

Value	*Number of items (f)*
1	3
2	8
3	14
4	3
5	4
6	2
7	1

Here 3 repeats 14 times and is most frequent hence 3 is the mode.

Example 2. A Khadim shop in Bombay sold 100 pairs of shoes in Khadim exclusive on certain day with the following distribution. Find the mode of the distribution.

Size of the shoe	4	5	6	7	8	9	10
Number of pairs	10	15	20	35	16	3	1

Solution: Let us prepare the table showing the frequency.

Size of the shoe	4	5	6	7	8	9	10
Number of pairs (frequency)	10	15	20	35	16	3	1

The above table shows that the size '7' has the maximum frequency –'35'. Therefore '7' is the mode of distribution.

Example 3. Find the Mean & Mode of the numbers

4, 3, 2, 5, 3, 4, 5,1, 7, 3, 2, 1.

Solution: Let us prepare the table with frequency

Value (x)	*No. of times (f)*	*fx*
1	2	2
2	2	4
3	3	9
4	2	8
5	2	10
7	1	7
	$\sum f = N = 12$	40

$$\sum fx = 40$$

$$\text{Mean} = \frac{\sum fx}{N}$$

$$\bar{x} = \frac{40}{12} = 3.33$$

The table indicates that the number '3' has the maximum frequency '3', therefore '3' is the mode of the mode of the numbers.

B. Grouped data

(a) Discrete series:

Procedure:

(*i*) In discrete series mode is determined by inspection.

(*ii*) The error of judgment is possible in these cases where the difference between the maximum frequency & the frequency preceding or succeeding it is very small and the items are heavily concentrated on ether side.

(*iii*) In these cases the value of mode is determined by preparing a *grouping table* and analysis table.

Features of Grouping Table: It has six columns

Column I: Original frequencies and the maximum frequencies is marked

Column II:

(*i*) Here frequencies of column I are combined two by two.

(*ii*) Maximum frequency is marked by bold type.

Column III:

(*i*) Leaving the first frequency of column '*I*' and combine the others two by two.

(*ii*) Again the maximum frequency is marked by bold type.

Column IV:

(*i*) Here the frequencies of the column I are combined in three by three .

(*ii*) Maximum frequency is marked by bold type.

Column V:

(*i*) Here we leave the first frequency of column 'I' & combine the others three by three.

(*ii*) Again the maximum frequency is marked as bold type.

Column VI:

(*i*) Now leave the first two frequencies of the column 'T' & combine the others three by three.

(*ii*) Mark the maximum frequency by bold type.

After preparing the grouping table, we prepare the analysis table. (I) In this table, column numbers are put on the left hand side and the probable values of mode on the right side. (II) The value which occurs maximum number of times is the mode.

Example 1. Calculate the mode of the following frequency distribution:

Height in inch	58	59	60	61	62	63	64	65	66	67	Total
No. of person	4	6	5	10	20	22	24	6	2	1	100

Solution: It is an irregular distribution in the sense that the difference between the maximum frequency 24 & frequency preceding it is very small. Let us prepare the grouping table and analysis table.

Height x	Frequency f (I)	Grouping: of two (II)	of two leaving the first (III)	of three (IV)	of three leaving the first (V)	of three leaving the first two (VI)
58	4	10		15		
59	6		11		21	
60	5	15				35
61	10		30	52		
62	20	42			66	
63	22		46			52
64	24	30		32		
65	6		8		9	
66	2	3				
67	1					

Analysis Table:

Columns	58	59	60	61	62	63	64	65	66	67
	Size of the items having maximum frequency									
I							I			
II					I	I				
III						I	I			
IV				I	I	I				
V					I	I	I			
VI						I	I	I		
Total				1	3	5	4	1		

Since the number 63 occurs maximum number of times *i.e.*, 5 times, hence mode is 63.

Example 2. Calculate the mode of following frequency distribution:

Size (x)	4	5	6	7	8	9	10	11	12	13
Frequency (f)	2	5	8	9	12	14	14	15	11	13

Solution: It is an irregular distribution in the sense that the difference between the maximum frequency 15 and frequency preceding it is very small. Let us prepare the grouping table & analysis table.

Size x	Frequency f (I)	Grouping of two (II)	of two leaving the first (III)	of three (IV)	of three leaving the first (V)	of three leaving the first two (VI)
4	2	7		15		
5	5		13		22	
6	8	17				29
7	9		21	35		
8	12	26			40	
9	14		28			43
10	14	29		40		
11	15		26		39	
12	11	24				
13	13					

Analysis Tables:

Columns	Size of the items having maximum frequency 4	5	6	7	8	9	10	11	12	13
I								I		
II							I	I		
III						I	I			
IV							I	I	I	
V					I	I	I			
VI						I	I	I		
Total					1	3	5	4	1	

Since the number 10 occurs maximum number of times *i.e.*, 5 times. Hence mode is 10.

(*b*) Continuous Series:

(*i*) Here modal class in a grouped frequency distribution is determined by inspection or with the help of grouping table.

(*ii*) If class intervals are not continuous we have to transform class limits into class boundaries.

(*iii*) If class widths are not equal, we have to make the interval equal and frequencies of such classes be adjusted considering equally distribution through out the class.

(*iv*) In some cases of class interval of unequal width, we find mode is ill defined and then we apply the empirical relationship

$$Mode = 3\ Median - 2\ Mean$$

Here all classes are of equal width.

$$\text{Mode} = L_1 + \frac{d_1}{d_1 + d_2} \times i$$

where L_1 = Lower boundary of the modal class (*i.e.*, the class containing the largest frequency).

d_1 = Difference of the largest frequency and the frequency of class just preceding the modal class.

d_2 = Difference of the largest frequency and the frequency of class just following the modal class.

i = Common width of classes.

f_m = Maximum frequency or frequency of the modal class.

f_1 = Frequency of class just proceeding the model class.

f_2 = Frequency of class just following modal class.

$d_1 = fm - f_1 \qquad d_2 = fm - f_2$

$$\text{Mode} = L_1 + \frac{fm - f_1}{(fm - f_1) + (fm - f_2)} \times i$$

$$= L_1 + \frac{fm - f_1}{2f_m - f_1 - f_2} \times i$$

Example 1. Find the mode of the following data:

Marks	1–5	6–10	11–15	16–20	21–25
No. of students	7	10	16	32	24

Solution:

Marks	*Class boundary*	*Mid value*	*No. of students (f)*
1–5	0.5–5.5	3	7
6–10	5.5–10.5	8	10
11–15	10.5–15.5	13	16
16–20	15.5–20.5	18	32
21–25	20.5–25.5	23	24

Maximum frequency is 32 & it lies in the class 15.5–20.5. Thus modal class is 15.5–20.5.

$L_1 = 15.5 \quad f_m = 32 \quad f_1 = 16 \quad f_2 = 24 \quad i = 5$

$$\text{Mode} = L_1 + \frac{fm - f_1}{2fm - f_2\ f_1} \times i = 15.5 + \frac{32 - 16}{2 \times 32\ \ 24\ \ 16} \times 5$$

$$= 15.5 + \frac{16}{64 - 40} \times 5$$

$$= 15.5 + \frac{\cancel{16}^{4}}{\cancel{24}_{6}} \times 5$$

$$= 15.5 + 3.33 = 18.83.$$

Example 2. Find the value of mode from following:

Weight (kg)	30–34	35–39	40–44	45–49	50–54	55–59	60–64
No. of students	3	5	12	18	14	6	2

Solution:

Weight	*Class boundary*	*Mid value*	*No. of students*
30–34	29.5–34.5	32	3
35–39	34.5–39.5	37	5
40–44	39.5–44.5	42	12
45–49	44.5–49.5	47	18
50–54	49.5–54.5	52	14
55–59	54.5–59.5	57	6
60–64	59.5–64.5	62	2

Maximum frequency is '18' and it lies in the classes 44.5–49.5. Thus modal classes 44.5–49.5.

$$L_1 = 44.5 \quad fm = 18 \quad f_1 = 12 \quad f_2 = 14 \quad i = 5$$

$$M = L_1 + \frac{fm - f_1}{2f_m - f_2 - f_1} \times i$$

$$= 44.5 + \frac{18 - 12}{2 \times 18 - 14 - 12} \times 5$$

$$= 44.5 + \frac{6}{36 - 26} \times 5$$

$$= 44.5 + \frac{\not{6}^{3}}{\not{10}_{2}} \times \not{5}$$

$$= 47.5.$$

Advantage & Disadvantages of Mode:

Advantages (Merits):

1. It can be obtained by inspection.
2. It is not affected by extreme values.
3. This average can be calculated from open end classes.
4. It can be easily understood.
5. It can be used to describe qualitative phenomenon.
6. The value of mode can also be found graphically.

Disadvantages (demerits):

1. Mode has no significance unless a large number of observation is available.
2. It can not be treated algebraically.
3. It is a peculiar measure of central tendency.
4. For the calculation of mode, the data must be arranged in the form of frequency distribution.
5. It is not rigidly defined measure.

Partition Values :

When we are required to divide a series into more than two equal parts, the dividing places are known as partition values.

Quartiles, Deciles and Percentiles

- **Quartiles:** Quartiles are such values which divide the total number of observation into four equal parts. Therefore there are three quartiles:

(*i*) First quartile (or Lower quartile): Q_1

(*ii*) Second quartile (or Middle quartile): Q_2

(*iii*) Third quartile (or upper quartile): Q_3.

- **Deciles:** Deciles are such values which divides the total number of observation into ten (10) equal parts. These are nine deciles viz D_1, D_2, ... D_9, called the *first decile, second decile* ... etc.
- **Percentiles:** Percentiles are such values which divide the total number of observations into hundred (100) equal parts. There are 99 percentiles P_1, P_2, ... P_{99}. Called the *first percentile second percentile* ... etc.

The term quartile is a generic term of a division point relative to any partition.

- **Relation between averages and partition values:**

1. An average (mean median, mode etc) is the representative of whole series but partition values (quartiles, deciles, percentiles) are averages of parts of the distribution (series).
2. Partition values are not averages like mean median mode.
3. Partition values are used to study the scatterednes of the values of the variable around the median.

OTHER MEASURES OF CENTRAL TENDENCY

- **Geometric mean:** The geometric mean is defined as the *n*th root of the product of *n* observations.

Geometric mean (G.M.) $= \sqrt[n]{X_1 \times X_2 \times X_3 ... X_n}$

n = Number of observations

$X_1\ X_2\ X_3\ ...$ = Variable values.

- The computation of geometric mean requires that all observation be positive *i.e.*, greater than zero.
- If zeros or negative values in the series the geometric mean can not be used.

Example: Find the G.M. of the three numbers 8, 36, 48.

Solution:

$$\begin{aligned} \text{G.M.} &= \sqrt[3]{8 \times 36 \times 48} \\ &= \sqrt[3]{2 \times 2 \times 2 \times 2 \times 2 \times 3 \times 3 \times 2 \times 2 \times 2 \times 2 \times 3} \\ &= \sqrt[3]{2^3 \times 2^3 \times 2^3 \times 3^3} \\ &= 2 \times 2 \times 2 \times 3 \\ &= 24 \end{aligned}$$

Note: The above method can be applied if there are two or three items. But if *n* is a very large number, the problem of finding the *n*th root is a tedious work. Therefore, logarithms are used to find the G.M.

$$\begin{aligned} \text{G.M.} &= \text{Antilog}\ \frac{\log X_1 + \log X_2 + \log X_3 ... \log X_n}{N} \\ &= \text{Antilog}\ \frac{\log X}{N}. \end{aligned}$$

Uses:

(*i*) It is highly useful in averaging ratios, percentage & determining ratio of change.

(*ii*) It is used in microbiology.

(*iii*) It is important in construction of index number.

Merits:

1. It is based on all observations.
2. It is rigidly defined.
3. It is capable of further algebraic treatment.
4. It is less affected by the extreme values.

Demerits:

1. It is difficult to understand.
2. It is difficult to compute & to interpret.
3. If a distribution contains both positive and negative values G.M is impossible to compute.

- **Harmonic mean:** It is defined as the reciprocal of the arithmetic mean of the reciprocals of individual observations.

Thus for observations $X_1, X_2, X_3, \ldots X_n$

$$\text{H.M.} = \frac{N}{\frac{1}{X_1} + \frac{1}{X_2} + \frac{1}{X_3} \ldots \frac{1}{X_n}}$$

or

$$\text{H.M.} = \frac{N}{\sum \frac{1}{X}}$$

Example: Find the average rate of motion in the case of a person who rides the first km at 10 km an hour, the next km at 8 km an hour, and the third km at 6 km an hour.

Solution: Clearly, the harmonic mean is the proper average.

$$N = 3$$

$$\text{H.M.} = \frac{3}{\frac{1}{10} + \frac{1}{8} + \frac{1}{6}}$$

$$= \frac{3}{\frac{12 + 15 + 20}{120}}$$

$$= \frac{3}{\frac{47}{120}}$$

$$= \frac{3}{0.39}$$

$$= 7.6 \text{ km an hour}$$

Uses:

1. The use of H.M is very limited.
2. It is useful in averages involving time, rate & price.
3. It gives less weight to large items & more weight to small item.

Merits:

1. It is based on all the observations of the series.
2. It is suitable for algebric treatment.
3. It is rigidly defined.

Demerits:

1. It is difficult to compute.
2. Its values can not be computed when there are both positive & negative items.
3. It is not popular.

Find out the relation between A.M, G.M. and H.M.

[*C.U. (zoology) Part II (Hons)*: *2003*]

Solution :

- For any given set of observations, A.M. is greater than or equal to G.M. and G.M. is greater than or equal to H.M.

$$\text{A.M.} \geq \text{G.M.} \geq \text{H.M.}$$

They are equal, only when all observations are equal

$$\overline{X} = \text{G.M.} = \text{H.M.}$$

For example: We take two position items 6 and 6

$$\text{Mean (A.M.)} = \frac{6+6}{2} = 6$$

$$\text{G.M.} = \sqrt{6 \times 6} = 6$$

$$\text{H.M.} = \frac{2}{\frac{1}{6}+\frac{1}{6}} = \frac{2}{\frac{1}{3}} = 6$$

Thus $\overline{X}$ = G.M. = H.M.

- But if the size vary, mean (A.M.) will be greater than the geometric mean and geometric mean will be greater than the harmonic mean. This is because of the property of the geometric mean to give larger weight to smaller item and of the harmonic mean to give the largest weight to the smallest item.

$$\overline{X} > \text{G.M.} > \text{H.M.}$$

For example: We take two positive items 4 and 9

$$\text{Mean (A.M.)} = \frac{4+9}{2} = 6.5$$

$$\text{G.M.} = \sqrt{4 \times 9} = 6$$

$$\text{H.M.} = \frac{2}{\frac{1}{4}+\frac{1}{9}} = \frac{2}{\frac{13}{36}} = \frac{2 \times 36}{13} = 5.5$$

Thus A.M. > G.M. > H.M. or 6.5 > 6 > 5.5.

Which type of average would be suitable for the following?

(*a*) Size of the shoes sold at a shop.

(*b*) Marks of candidates obtained in an examination.

(*c*) Comparison of intelligence of students.

(*d*) Runs scored by a player in different matches.

(*e*) Per capita income in several countries

(*f*) Size of agricultural holdings

(*g*) Sale of shirts with collar size in cms 36, 37, 35, 36, 33, 36.

(*h*) Average sales for various years.

Ans: (*a*) Mode, (*b*) Median, (*c*) Median, (*d*) Mean, (*e*) Mean, (*f*) Mode, (*g*) Mode, (*h*) Mean

Comparison of Averages (Mean, Median & Mode)

Characteristics	*Mean*	*Median*	*Mode*
1. Computation	All items are involved in calculation of mean	All items are not involved in its calculation	All items are not involved in its calculation
2. Average	Calculated average	Positional average	Positional average
3. Arrangement of data	Does not require arraying	Arranging of the data in the series is essential	Arranging of the data in the series is essential
4. Effect of extreme values	Greatly affected	Does not affect	Does not affect
5. Sampling stability	Yes	No	No
6. Result	Only one mean	Only one median	One mode or more than one or no mode
7. Further mathematical treatment	Possible	No	No
8. Realiability	Most	Less	Less
9. Mathematical property	$\bar{X} = \frac{\sum X}{N}$	Size of $\left(\frac{N+1}{2}\right)$th item	Most common value
10. Application	Simple, easy to calculate & widely used (income & expenditure)	Easy to calculate. It is generally used in intelligence skill etc.	Easy to calculate weather forecast are based on mode. Useful in agriculture & socioeconomic survey
Relationship	**Mode** = 3 **Median** – 2 **Mean**		

CHAPTER

MEASURES OF VARIATION

Variability is essentially a normal character. The occurance of variability is a biological phenomenon. It is an important characteristic indicating the extent to which observations vary among themselves. There are three main types of variability viz. **Biological variability, Real variability** and **Experimental variability.**

Measures of Variability:

(*i*) Variability and its measurements are of fundamental importance in the Biological science.

(*ii*) The variability of a given set of observation will be zero, only when observations are equal. It takes positive when observations are unequal.

(*iii*) It help us to find out how individual observations are dispersed around the mean of a large series. They may also be called *measures of dispersion*, *variation* etc.

DISPERSION

A measure of dispersion is designed to state numerically the extent to which individual observations vary on the average.

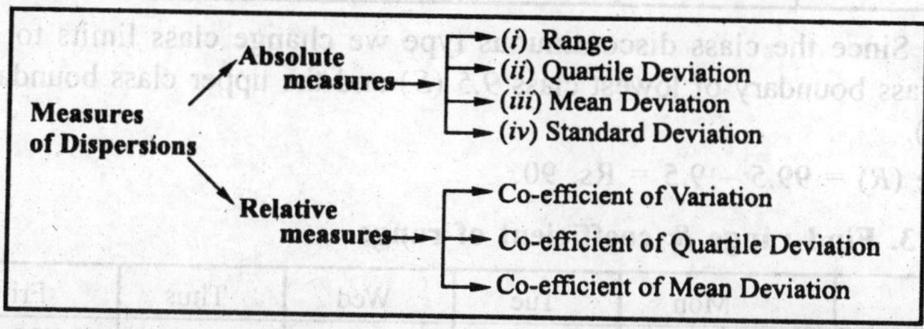

Absolute Measures of Dispersion:

(*i*) These are expressed in the same statistical unit in which the original data are given.

(*ii*) These values may be used to compare the variation in two sets of observations provided the variables are expressed in the same units and of the same average size.

(*iii*) Again when two sets of data are given in dissimilar units, the absolute measures of dispersion are not comparable.

Relative Measures of Dispersions:

(*i*) It is the ratio of a measure of absolute dispersion to an appropriate measure of central value and is expressed as pure number. These are independent of the units of measurement.

(*ii*) Relative measure may also be used to compare the relative accuracy of data.

Requisites of a Good Measures of Dispersions:

(*i*) It should be easily calculated.

(*ii*) It should be rigidly defined.

(*iii*) It should be based on all observations.
(*iv*) It should not be unduly affected by extreme items.
(*v*) It should be readily comprehensible.
(*vi*) It should have sampling stability.
(*vii*) It should be amenable to further algebraic treatment.

RANGE:

Range is the simplest measure of dispersion. It is the difference between the value of smallest item and the largest item included in the distribution.

Range (R) = Largest value (L) – Smallest value (S)

$$R = L - S$$

Co-efficient of Range: The relative measure corresponding to range is co-efficient of range.

$$= \frac{L - S}{L + S}$$

Example 1. Find the range of daily wages of 8 persons given below:

Rs. 10, Rs 11.50, Rs. 12, Rs. 21, Rs. 6.75, Rs. 18, Rs. 13, Rs. 20.

Solution: Here the largest values Rs. 20 & smallest value Rs. 6.75

$$R = 20 - 6.57 = 13.25.$$

Example 2. Find the range of the following:

Class (Rs.)	10–19	0–29	30–39	40–49	50–79	80–99
No. of persons	5	15	25	35	15	5

Solution: Since the class discontinuous type we change class limits to class boundaries & then lower class boundary of lowest class 9.5 (S) and the upper class boundary of the highest class = 99.5 (L).

The range (R) = 99.5 – 9.5 = Rs. 90.

Example 3. Find range & coefficient of range.

Days	Mon	Tue	Wed	Thus	Fri	Sat
Prices	20	21	23	16	25	22

Solution: Since $R = L - S$. Here $L = 25$, $S = 16$,

$$R = 25 - 16 = 9\ (R)$$

Coefficient of range $= \dfrac{L - S}{L + S} = \dfrac{25 - 16}{25 + 16} = \dfrac{9}{41} = 0.219.$

Merits & Demerits of Range:

Merits:

(*i*) It is easy to understand.
(*ii*) It is also simple to calculate.
(*iii*) It takes minimum time to calculate.

Demerits:

(*i*) It does not depend on all observations and is based on only the largest and smallest among them.

(*ii*) It is highly affected by extreme values.

(*iii*) It can not be calculated from frequency distribution with open end classes.

Uses:

1. **Quality Control:** It plays an important role in preparing control charts in the methods of statistical quality control.
2. **Weather Forecasts:** It is preferably used in determining the difference in minimum and maximum temperature for predicting the variation of temperature in a day.
3. **Estimating the Fluctuations in Prices:** It is useful in studying in the variations in prices of stocks & shares.

MEAN DEVIATION:

The mean deviation is also called average deviation. It is the average difference between the items in a distribution and the median and mean that series.

- **Mean Deviation**

  Mean deviation about the mean.

  Mean deviation about the median.

- **Coefficient of Mean Deviation:** It is the ratio of mean deviation to its arithmetic mean or median multiplied by 100.

$$\text{C.M.D.} = \frac{MD}{\text{Mean/Median}} \times 100$$

Calculation of Mean Deviation

A. Ungrouped data:

Procedure:

(*i*) Calculate mean or median.

(*ii*) Calculate the deviation from mean or median denoting by '*D*' and ignoring the sign positive (+) or negative (–).

(*iii*) '*n*' is the total number of items. Divide sum of the $\sum D$ by the total number of items.

$$\text{Mean deviation } (MD) = \frac{\sum |X - \bar{X}|}{n} = \frac{\sum |D|}{n}$$

(*a*) Mean Deviation about the mean. $MD = \frac{1}{N} \sum |X - \bar{X}| f$

(*b*) Mean Deviation about the median. $MD = \frac{1}{N} \sum f$ (*x*-median)

Example 1. Calculate mean deviation from the following:

X	10	11	12	13	14
F	3	12	18	12	3

Solution: Median deviation $= \frac{\sum fD}{N} = \frac{36}{48} = 0.75$

Example 2. Find the mean deviation of the following: 13, 84, 68, 24, 96, 139, 84, 27, about the median.

Solution: Since there are even number observations viz 8, the median is the average of the two middle most observations.

Let us arrange the data:

13, 24, 27, 68, 84, 84, 96, 139

$$\text{Median} = \frac{68+84}{2} = \frac{152}{2} = 76.$$

X	X – median = D
13	76 – 13 = 63
24	76 – 24 = 52
27	76 – 27 = 49
68	76 – 68 = 8
84	84 – 76 = 8
84	84 – 76 = 8
96	96 – 76 = 20
139	139 – 76 = 63
N = 8	$\sum D = 271$

Mean Deviation about the median

$$= MD = \frac{1}{N}\sum |x - \text{median}|$$

$$= \frac{1}{8} \times 271 = 33.88$$

B. Grouped data:

(a) Discrete Series:

(*i*) Calculate the mean or median.

(*ii*) Calculate the deviation from the mean & median ignoring the signs and denote them by $|D|$.

(*iii*) Multiply these deviation by respective frequencies & obtain the total $\sum f|D|$.

(*iv*) Divide the total $\left(\sum f|D|\right)$ by the number of observations giving the required value of Mean Deviation.

$$MD = \frac{\sum f|D|}{\sum f} = \frac{1}{N}\sum f|D|$$

$$D = |X - \bar{X}|$$

$$\sum f = N.$$

Example 1. Find the mean deviation of the following series:

X	10	11	12	13	14
Frequency	3	12	18	12	13

[*C.A. 63*]

Solution:

X	f	fx	$\lvert X - \bar{X} \rvert = D$	$f \lvert X - \bar{X} \rvert$
10	3	30	10 − 12 = 2	6
11	12	132	11 − 12 = 1	12
12	18	216	12 − 12 = 0	0
13	12	156	13 − 12 = 1	12
14	3	42	14 − 12 = 2	6
	$\sum f = 48$	576		36

$$MD = \frac{1}{48} \times 36 \qquad \sum f|D| = 36 \quad N = 48$$

$$= \frac{3}{4} = 0.75$$

(b) Continuous Series:

(*i*) Calculate mean or median of the series.

(*ii*) Take the deviation of the items from the mean or median, ignoring signs and denote them by $|D|$.

Here mean deviation are taken from the mid value of each class.

(*iii*) Multiply the deviation by the respective frequencies and obtain the total.

(*iv*) Divide the total by sum of the total number of observations.

Example 1. Calculate mean deviation from the mean from the following data.

Marks	0–10	10–20	20–30	30–40	40–50	Total
No. of students	5	8	15	16	6	50

Solution:

Class interval	Mid value (X)	Frequency (f)	fx	$\lvert D \rvert = \lvert X - 27 \rvert$	fD
0–10	5	5	25	5 − 27 = 22	110
10–20	15	8	120	15 − 27 = 12	96
20–30	25	15	375	25 − 27 = 2	30
30–40	35	16	560	35 − 27 = 8	128
40–50	45	6	270	45 − 27 = 18	108
		$\sum f = 50$	$\sum fx = 1350$		$\sum fD = 472$

$$\bar{X} = \frac{\sum fd}{N} = \frac{1350}{50} = 27$$

$$MD = \frac{\sum f|D|}{\sum f} = \frac{472}{50} = 9.44.$$

Merits & Demerits of Mean Deviation:

Merits:

(*i*) The outstanding advantage of the mean deviation is its relative simplicity.

(*ii*) It is simple to understand and easy to compute.

(*iii*) It is less affected by the value of extreme items as compared to standard deviation.

Demerits:

(*i*) In mean deviation the signs of all deviations are taken as positive and therefore it is not suitable for further algebric treatment.

(*ii*) This method may not yield accurate result.

(*iii*) It is yielding best result while taken from median, but median is not a satisfactory average where variability is too high.

(*iv*) It is rarely used in social science.

• **Quartile Deviation:** It is defined as half the difference between the upper and lower quartiles.

$$\text{Quartile Deviation} = \frac{Q_3 - Q_1}{2}$$

(*i*) The difference $Q_3 - Q_1$ being the distance between the two quartiles. This may be called *interquartile range* and half of this is "Semi interquartile Range".

(*ii*) Thus the name "Semi interquartile Range" itself gives the definition of "Quartile Deviation".

• **Co-efficient of Quartile Deviation:** It is the ratio of quartile deviation to it's median multiplied by 100 (hundred).

$$\text{So Coefficient of Quartile Deviation} = \frac{\text{Quartile Deviation}}{\text{Median}} \times 100$$

Example 1. From the following frequency distribution calculate the value of quartile (Q_1) median (Q_2) and upper quartile (Q_3):

Marks in Mathematics	10–19	20–29	30–39	40–49	50–59	60–69	Total
Frequency	8	11	15	17	12	7	70

Solution: $Q_1 = \frac{N}{4}, \quad Q_2 = \frac{N}{2}, \quad Q_3 = \frac{3N}{4}.$

Class interval	*Frequency*	*Class boundary*	*Cumulative frequency*
		9.5	0
10–19	8	19.5	8 $\frac{N}{4} = (17.5)$
20–29	11	$Q_1 \rightarrow$ 29.5	19
30–39	15	39.5	34 $\frac{2N}{4} = \frac{N}{2} = (35)$
40–49	17	$Q_2 \rightarrow$ 49.5	51
50–59	12	59.5	63 $\frac{3N}{4} = (52.5)$
60–69	7	$Q_3 \rightarrow$ 69.5	70 = N

$$Q_1 = L_1 + \frac{\frac{N}{4} - F_1}{f_1} \times i$$

L_1 = Lower boundary of quartiles

F_1 = Cumulative frequency

f_1 = frequency of quartile class

i = Width of class interval

$$Q_1 = 19.5 + \frac{17.5 - 8}{11} \times 10$$

$$= 19.5 + \frac{95}{11} = 19.5 + 8.6 = 28.1$$

$$Q_2 = 39.5 + \frac{35 - 34}{17} \times 10 = 39.5 + \frac{10}{17} = 39.5 + 0.58 = 40.08 = 40$$

$$Q_3 = 49.5 + \frac{52.5 - 51}{12} \times 10$$

$$= 49.5 + \frac{1.5 \times 10}{12} = 49.5 + 1.25 = 50.75 = 51$$

$Q_1 = 28$

$Q_2 = 40$

$Q_3 = 51.$

Example 2. From the following frequency distribution, calculate the value of the quartile deviation and the Coefficient of Quartile deviation:

X	Y
300–399	14
400–499	46
500–599	58
600–699	76
700–799	68
800–899	62
900–999	48
1000–1099	22
1100–1199	6

Solution: Quartile $(Q_1) = \frac{N}{4}$ $Q_2 = \frac{2N}{4}$ or $\frac{N}{2}$ and $Q_3 = \frac{3N}{4}$

	Class boundary	*Cumulative frequency (less than)*	
	299.5	0	
	399.5	14	
$Q_1 \rightarrow$	499.5	60	$\leftarrow 100 = \frac{N}{4}$
	599.5	118	
	699.5	194	
$Q_2 \rightarrow$			$\leftarrow 200 = \frac{N}{2}$
	799.5	262	
$Q_3 \rightarrow$			$\leftarrow 300 = \frac{3N}{4}$
	899.5	324	
	999.5	372	
	1099.5	394	
	1199.5	400	

$$\frac{N}{4} = \frac{400}{4} = 100 \qquad \frac{3}{4}N = 3 \times 100 = 300$$

$$Q_1 = L_1 + \frac{\frac{N}{4} - F_1}{f_1} \times h_1$$

$$Q_1 = 499.5 + \frac{100 - 60}{58} \times 100$$

$$= 499.5 + \frac{40}{58} \times 100 = 499.5 + 68.97 = 568.47$$

$$Q_3 = L_3 + \frac{\frac{3N}{4} - F_3}{f_1} \times h_3$$

$$Q_3 = 799.5 + \frac{300 - 262}{62} \times 100$$

$$= 799.5 + \frac{38}{62} \times 100 = 799.5 + 61.29 = 860.79$$

Quartile Deviation $= \frac{Q_3 - Q_1}{2} = \frac{860.79 - 568.47}{2} = \frac{292.32}{2} = 146.16$

Median (Q_2) $= L_2 + \frac{\frac{N}{2} - F_2}{f_2} \times h_1$

$$Q_2 = 699.5 + \frac{200 - 194}{68} \times 100$$

$$Q_2 = 699.5 + \frac{6}{68} \times 100 = 699.5 + 8.82 = 708.32$$

Coefficient Quartile Deviation $= \frac{\text{Quartile deviation}}{\text{Median } (Q_2)} \times 100$

$$= \frac{146.16}{708.32} \times 100 = 20.63\%.$$

Example 3. Calculate Quartile Deviation and Coefficient of Quartile Deviation from the following frequency distribution:

Class interval	10–15	15–20	20–25	25–30	30–40	40–50	50–60	60–70
Frequency	4	12	16	22	10	8	6	4 = 82

Solution: In order to calculate Quartile Deviation & its coefficient, we have to find Q_1, Q_2 & Q_3 *i.e.*, the value of variables corresponding to cumulative frequency $\frac{N}{4}$, $\frac{N}{2}$ and $\frac{3N}{4}$ total frequency *i.e.*, $N = 82 \quad \therefore \frac{N}{4} = 20.5 \quad \frac{N}{2} = 41 \quad \frac{3N}{4} = 61.5$.

Class interval	*Frequency*	*Class boundary*	*Cumulative frequency*
10–15	4	10	0
15–20	12	15	4
20–25	16	20	16
		$Q_1 \rightarrow$	$\leftarrow 20.5 = \frac{N}{4}$
25–30	22	25	32
		$Q_2 \rightarrow$	$\leftarrow 41 = \frac{N}{2}$
30–40	10	30	54
		$Q_3 \rightarrow$	$\leftarrow 61.5 = \frac{3N}{4}$
40–50	8	40	64
50–60	6	50	72
60–70	4	60	78
		70	82 = N

i = Class interval

$$Q_1 = L_1 + \frac{\frac{N}{4} - F_1}{f} \times i \qquad \frac{N}{4} = 20.5$$

$$Q_1 = 20 + \frac{20.5 - 16}{16} \times 5$$

$$Q_1 = 20 + \frac{4.5}{16} \times 5$$

F_1 = cumulative frequency corresponding to L_1

$$= 20 + \frac{22.5}{16} = 20 + 1.4 = 21.4$$

f = frequency of this class

$$Q_2 = 4 + \frac{\frac{N}{2} - F_1}{f} \times i$$

$$= 25 + \frac{41 - 32}{22} \times 10 = 25 + \frac{90}{22} = 25 + 4.0 = 29.0$$

$$Q_3 = 30 + \frac{61.5 - 54}{10} \times 10 = 30 + 7.5 = 37.5$$

$$Q = \text{(Quartile Deviation)} = \frac{Q_3 - Q_1}{2} = \frac{37.5 - 21.4}{2} = \frac{16.1}{2} = 8.0$$

$$\text{Coefficient of Quartile Deviation} = \frac{Q}{Q_2} \times 100 = \frac{8}{29} \times 100 = 27.58\%.$$

Example 4. Find the quartile deviation of the following distribution:

Class interval	40–45	45–50	50–55	55–60	60–65	65–70
Frequency	10	22	28	20	12	8

Solution:

Class interval	Frequency	Class boundary	Cumulative frequency
		40	0
40–45	10	45	10
45–50	22	$Q_1 \rightarrow$ 50	32 $\leftarrow 25 = \frac{N}{4}$
50–65	28	55	60
55–60	20	$Q_3 \rightarrow$ 60	80 $\leftarrow 75 = \frac{3N}{4}$
60–65	12	65	92
65–70	8	70	100 $= N$

$$Q_1 = L_1 + \frac{\frac{N}{4} - F}{f} \times i$$

$$\frac{N}{4} = 25 \quad F = 10$$

$$L_1 = 45 \quad f = 28$$

$$i = 5$$

$$= 45 + \frac{25 - 10}{22} \times 5$$

$$= 45 + \frac{15}{22} \times 5 = 45 + 3.4 = 48.4$$

$$Q_3 = 55 + \frac{75 - 60}{20} \times 5 = 55 + 3.75 = 58.75$$

$$\text{Quartile Deviation } (Q) = \frac{Q_3 - Q_1}{2} = \frac{58.75 - 48.4}{2}$$

$$= \frac{10.35}{2} = 5.175.$$

Standard Deviation:

Karl Pearson 1893 introduced the concept of standard deviation. It is most important and widely used measure of studying dispersion.

(*i*) It is the best measure of dispersion because it possesses almost all the requisites of a good measures dispersions.

(*ii*) It is best on all observations. Even if one of the observation is changed, *SD* also changes.

(*iii*) *S.D.* is least affected by the fluctuations of sampling.

(*iv*) The unique property of *S.D.* is that it is amenable to algebraic treatment.

(*v*) It is also known as "**root means squared deviation**" for the reason that it is the square root of the mean of the squared deviation from the arithmetic mean.

$$\left[\sigma = \sqrt{\frac{\sum (X - \bar{X})^2}{N}}\right].$$

(*vi*) A small standard deviation means a high degree of uniformity of the observation as well as homogeneity of a distribution on the other hand, a large *S.D.* means just opposite. Hence *S.D.* is extremely useful in judging the representation mass of the mean.

Standard Deviation (S.D.):

It is the square-root of the arithmetic mean squares of deviation from arithmetic mean. In short *S.D.* may be defined as "Root Mean Square Deviation from Mean".

It is usually denoted by the Greek small letter **sigma (σ)**.

Explanation:

(*i*) If $x_1\ x_2, \ldots x_n$ be a set of observations and $\bar{x}$ their A.M (arithmetic mean).

(*ii*) Deviation from mean $= (x_1 - \bar{x}), (x_2 - \bar{x}) \ldots (x_n - \bar{x})$.

(*iii*) Square deviation from mean $= (x_1 - \bar{x})^2\ (x_2 - \bar{x})^2 \ldots (x_n - \bar{x})^2$.

(*iv*) Mean-Square Deviation from mean *i.e.*,

$$= \frac{[(x_1 - \bar{x})^2 + (x_2 - \bar{x})^2 + \ldots (X_n - \bar{x})^2]}{n}$$

$$= \frac{\sum (x - \bar{x})^2}{n}$$

Root Mean Square Deviation from Mean *i.e.*, Standard Deviation $(\sigma) = \sqrt{\frac{1}{n} \Sigma (x - \bar{x})^2}$.

Coefficient of Standard Deviation:

It is ratio of the standard deviation to its arithmetic mean *i.e.*,

Coefficient of standard deviation $= \frac{\sigma}{\bar{X}}$.

Calculation of Standard Deviation

A. Simple Series:

Procedure:

(*i*) Calculate the mean.

(*ii*) Find the difference of each observations from the mean.

(*iii*) Square the differences of observations from the mean.

(*iv*) Add the square values to get the sum of the squares.

(*v*) Divide by the number of observations.

$$S.D.\ (\sigma) = \sqrt{\frac{\sum (X - \bar{X})^2}{n}} = \sqrt{\frac{d^2}{n}}$$

where $d = (X - \bar{X})$

X = Value of the variable

$\bar{X}$ = Arithmetic mean

n = Total number of observations.

Example 1. Find the standard deviation for the following distribution:

11, 12, 13, 14, 15, 16, 17, 18, 19, 20, 21

Solution: Here arithmetic mean

$$\bar{X} = \frac{11 + 12 + 13 + 14 + 15 + 16 + 17 + 18 + 19 + 20 + 21}{11}$$

$$\bar{X} = \frac{176}{11} = 16$$

Let us prepare the table to calculate *S.D.*

X	$X - \bar{X} = d$	$(X - \bar{X})^2 = D^2$
11	11 – 16 = –5	25
12	12 – 16 = –4	16
13	13 – 16 = –3	9
14	14 – 16 = –2	4
15	15 – 16 = –1	1
16	16 – 16 = 0	0
17	17 – 16 = +1	1
18	18 – 16 = +2	4
19	19 – 16 = +3	9
20	20 – 16 = +4	16
21	21 – 16 = +5	25
$n = 11$		$\sum D^2 = 110$

$$S.D.\ (\sigma) = \sqrt{\frac{d^2}{n}}$$

$$= \sqrt{\frac{110}{11}} = \sqrt{10}$$

$$= 3.16.$$

Short Cut Method:

(*i*) This method is applied to calculate the standard deviation when the arithmetic mean of the data comes out to be a fraction.

(*ii*) In this case it very difficult & tedious to find the deviations of all observations from the mean.

Here $S.D.\ (\sigma) = \sqrt{\frac{\sum d^2}{n} - \left(\frac{\sum d}{n}\right)^2}$ When $d = X - A$

A = Assumed mean

n = Total number of observations

Example 1. Find the standard deviation of the following data:

48, 43, 65, 57, 31, 60, 37, 48, 59, 78.

Solution: Let us prepare the following table in order to calculate value of *S.D.*

Value (X)	$d = (X - A).\ (A = 57)$	d^2
48	48 – 57 = –9	81
43	43 – 57 = –14	196
65	65 – 57 = +8	64
57	57 – 57 = 0	0
31	31 – 57 = –26	676
60	60 – 57 = +3	9

Value (X)	d = (X – A), (A = 57)	d^2
37	37 – 57 = –20	400
48	48 – 57 = –9	81
59	59 – 57 = +2	4
78	78 – 57 = +21	441
	–78 + 34 = –44	1952

Here assumed mean = 57

$$\sum d^2 = 1952 \quad \sum d = -44 \quad n = 10$$

$$S.D. = \sqrt{\frac{\sum d^2}{n} - \left(\frac{\sum d}{n}\right)^2} = \sqrt{\frac{1952}{10} - \left(\frac{-44}{10}\right)^2} = \sqrt{195.2 - 19.36}$$

$$= \sqrt{175.84} = 13.26$$

(b) Standard Deviation (Grouped Data) or Discrete Series.

(i) Direct method:

$$S.D. (\sigma) = \sqrt{\frac{\sum f\,(X - \bar{X})^2}{n}}$$

$\bar{X}$ = Arithmetic mean

f = Frequency

n = Number of item.

(ii) Short cut method: When mean has a fractional value then the following formula is used.

$$\sigma = \sqrt{\frac{\sum fd^2}{n} - \left(\frac{\sum fd}{n}\right)^2}$$

where $d = X - A$

A = assumed mean

$n = \sum f$.

Example 1. Find the mean and standard deviation of the following data:

Size of item	10	11	12	13	14	15	16
Frequency	2	7	11	15	10	4	1

Solution: Let us prepare the following table:

Size of the item	*Frequency*	*d = X – A (13)*	*fd*	*fd^2*
10	2	10 – 13 = –3	–6	18
11	7	11 – 13 = –2	–14	28
12	11	12 – 13 = –1	–11	11
13	15	13 – 13 = 0	0	0
14	10	14 – 13 = +1	10	10
15	4	15 – 13 = +2	8	16
16	1	16 – 13 = +3	3	9
	$\sum f = 50$		$\sum fd = +21 - 31 = -10$	$\sum fd^2 = 92$

Now $\bar{X} = A + \frac{\sum fd}{n} = 13 - \frac{10}{50} = 13 - .2 = 12.8$

$\therefore$ $\bar{X} = 12.8$

$$S.D. = \sqrt{\frac{\sum fd^2}{n} - \left(\frac{\sum fd}{n}\right)^2}$$

$$= \sqrt{\frac{92}{50} - \left(\frac{-10}{50}\right)^2}$$

$$= \sqrt{1.84 - (.2)^2}$$

$$= \sqrt{1.84 - .04}$$

$$= \sqrt{1.80}$$

$$= 1.342$$

Standard Deviation in Continuous Series:

(*a*) Direct method:

$$S.D.(\sigma) = \sqrt{\frac{\sum f(X - \bar{X})^2}{n}}$$

where X = Mid value

$\bar{X}$ = A.M.

f = Frequency

$\sum f = N$ Total frequency.

(*b*) Short cut method:

Procedure:

(*i*) For class interval, find mid values of (X) of each class.

(*ii*) Let 'A' be any assumed mean of X.

(*iii*) Compute $d = X - A$ and divide it by the common factor (i) *i.e.*, $d' = \frac{X - A}{i} = \frac{d}{i}$

(*iv*) Apply the formula

$$\sigma = \sqrt{\frac{\sum fd^2}{n} - \left(\frac{\sum fd}{n}\right)^2} \times i$$

where $d = \frac{X - A}{i}$

A = assumed mean

N = total frequency

i = class width.

Example 1. Calculate *S.D.* for the following distribution:

Profits (Rs.)	20–30	30–40	40–50	50–60	60–70	70–80	80–90	90–100
No. of Company	30	58	62	85	112	70	57	26

Solution:

Profits	Mid value X	No. of company (f)	$d = \frac{X-A}{C}$	fd	fd^2
20–30	25	30	25 – 65/10 = –4	–120	480
30–40	35	58	35 – 65/10 = –3	–174	522
40–50	45	62	45 – 65/10 = –2	–124	248
50–60	55	85	55 – 65/10 = –1	–85	85
60–70	65	112	65 – 65/10 = 0	0	0
70–80	75	70	75 – 65/10 = +1	70	70
80–90	85	57	85 – 65/10 = +2	114	228
90–100	95	26	95 – 65/10 = +3	78	234
		$\sum f = 500$		–503 + 262 = –241	1867

$$\sigma = \sqrt{\frac{\sum fd^2}{n} - \left(\frac{\sum fd}{n}\right)^2} \times i$$

$$= \sqrt{3.734 - (-.482)^2} \times 10$$

$$= \sqrt{\frac{1867}{500} - \left(\frac{-241}{500}\right)^2} \times 10$$

$$= \sqrt{3.734 - .232} \times 10$$

$$= \sqrt{3.502} \times 10$$

$$= 1.871 \times 10 = 18.71$$

- **Variance:** The square of standard deviation is called *variance* and is denoted by σ^2
  Variance = $(S.D.)^2 = \sigma^2$
- **Coefficient of Variance:** It is the product of the coefficient of standard deviation × 100

  Coefficient of variance $= \frac{\text{S.D.}}{\text{AM}} \times 100$ *i.e.,* $\frac{\text{Sigma}}{\text{A.M.}} \times 100.$

Example 2. Calculate the mean, median, S.D. variance and covariance of the following data:

Height in inches	95–105	105–115	115–125	125–135	135–145
No. of children	19	23	36	70	52

Solution: Let us take assumed mean 130

Class interval	Mid value X	No. of children	Cf	$\frac{X-A}{C} = d$	fd	fd^2
95–105	100	19	19	$\frac{100-130}{10} = -3$	–57	19 × 9 = 171
105–115	110	23	42	$\frac{110-130}{10} = -2$	–46	23 × 4 = 92
115–125	120	36	78	$\frac{120-130}{10} = -1$	–36	36 × 1 = 36

Class interval	Mid value X	No. of children	Cf	$\frac{X-A}{C}=d$	fd	fd^2
125–135	130	70	148	$\frac{130-130}{10}=0$	0	70 × 0 = 0
135–145	140	52	200	$\frac{140-130}{10}=+1$	+52	52 × 1 = 52
		N = 200			−139 + 52 = −87	= 351

$$\text{Mean} = A + \frac{\sum fd}{N} \times i$$

$$= 130 + \frac{-87}{200} \times 10$$

$$= 130 - 4.35 = 125.65$$

$$S.D.(\sigma) = \sqrt{\frac{\sum fd^2}{N} - \left(\frac{\sum fd}{N}\right)^2} \times i$$

$$= \sqrt{\frac{351}{200} - \left(\frac{-87}{200}\right)^2} \times 10$$

$$= \sqrt{1.75 - .189} \times 10$$

$$= \sqrt{1.56} \times 10$$

$$= 1.2489 \times 10 = 12.489$$

$$\text{Median} = L_1 + \frac{\frac{N}{2} - C}{f_m} \times i$$

$$\text{Median class } 125\text{–}135 = 125 + \frac{\frac{200}{2} - 78}{70} \times 10$$

$$= 125 + \frac{100 - 78}{70} \times 10$$

$$= 125 + \frac{22}{7} = 125 + 3.14$$

$$= 128.14$$

$$\text{Variance} = \sigma^2 = (12.489)^2$$

$$= 155.97$$

$$= 156$$

$$C.V. = \frac{12.48}{125.65} \times 100$$

$$= \frac{1248.00}{125.65} - \frac{124800}{12565} = 9.93.$$

- **Comparison between Mean Deviation & Standard Deviation:**

Mean Deviation	Standard deviation
1. Deviations are calculated from mean, median and mode.	1. Deviations are calculated only from mean.
2. Algebraic signs are ignored while calculating mean deviations.	2. Algebraic signs are taken into account.
3. It is simple to calculate.	3. It is difficult to calculate.
4. It lacks mathematical properties because algebraic signs are ignored.	4. It is mathematically sound because algebraic signs are taken into account.

COVARIANCE:

It is a measure of statistical association between two variables (X and Y) where the average product of the simultaneous deviation of the variables from their respective mean.

Features:

(*i*) Variance describes the tendency of a set of measurements (variables) to vary *i.e.*,

$$\sigma^2 = \frac{\sum(X-\bar{X})^2}{n-1} \quad \text{or} \quad \sigma^2 = \frac{\sum(X-\bar{X})^2}{n}$$

It may be defined as the square of standard deviation.
The covariance describes the tendency of pairs of numbers (variables) to vary together (co-vary).

(*ii*) Therefore the covariance is analogous to the variance but it involves the simultaneous deviation from the means of both the variables (X, Y)

$$COV_{XY} = \frac{\sum(X-\bar{X})(Y-\bar{Y})}{n}$$

The covariance indicates the joint variations of the two variables *i.e.*, covariance may be looked upon as *conjoint variation*.

Steps:

(*i*) Find out the mean of the two series *i.e.*, X and Y.

(*ii*) Take deviations of the two series from X and Y and denote dx and dy.

(*iii*) Square the deviations and get the total of the respective squares of deviation of X and Y and denote $\sum dx^2$ and $\sum dy^2$.

(*iv*) Multiply the deviations of X and Y and get total (This is covariance).

(*v*) Substitute the values of $\sum xy$, $\sum x^2$ and $\sum y^2$ in the formula.

Calculation:

(*i*) Calculation of covariance is similar to the calculation of variance except that the square deviation term $(X-\bar{X})^2$ is replaced with the product of the deviations of the pairs of measurements from their respective means $(X-\bar{X})(Y-\bar{Y})$.

(*ii*) Let two quantitative variables X and Y be observed for n units of observation.
Let $X_1, X_2, ..., X_n$ be the observed values of X and $Y_1, Y_2 ... Y_n$ be the observed values of Y.

Then $\bar{X} = \frac{n_1 + n_2 ... x_n}{n}$ $\bar{Y} = \frac{y_1 + y_2 ... y_n}{n}$ and the covariance of X and Y written as $COV(X, Y)$.

$$COV(X, Y) = \frac{(X_1-\bar{X})(Y_1-\bar{Y}) + (X_2-\bar{X})(Y_2-\bar{Y}) ... (X_n-\bar{X})(Y_n-\bar{Y})}{n}$$

$$= \frac{1}{n} \sum (X - \bar{X})(Y - \bar{Y}) \qquad \text{...(1)}$$

Expanding the expansion of the right

$$= \frac{1}{n} \sum (X - \bar{X})(Y - \bar{Y}) = \frac{1}{n} \sum XY - \frac{\bar{X}\sum Y}{n} - \bar{Y}\frac{\sum X}{n} + \frac{1}{n}\sum \bar{X}\bar{Y}$$

$$= \frac{1}{n} \sum XY - \bar{X}\bar{Y} - \bar{Y}\bar{X} + \frac{1}{n} n\bar{X}\bar{Y}$$

$$= \frac{1}{n} \sum XY - \bar{X}\bar{Y} - \bar{X}\bar{Y} + \bar{X}\bar{Y}$$

$$= \frac{1}{n} \sum XY - \bar{X}\bar{Y}$$

$$= \frac{\sum XY}{n} - \frac{\sum X}{n}\frac{\sum Y}{n} = \frac{\sum XY}{n} - \left(\frac{\sum Y}{n}\right)\left(\frac{\sum Y}{n}\right) \qquad \text{...(2)}$$

(*iii*) We know that variance is always positive but co-variance may be positive, negative or zero

$$\text{Variance of } X = \sigma_x^2 = \frac{1}{n} \sum (X - \bar{X})^2 = \frac{1}{n} \sum (X - \bar{X})(X - \bar{X})$$

$$\text{Variance of } Y = \sigma_y^2 = \frac{1}{n} \sum (Y - \bar{Y})^2 = \frac{1}{n} \sum (X - \bar{X})(Y - \bar{Y})$$

$$\text{Covariance of } X \text{ and } Y = \frac{1}{n} \sum (X - \bar{X})(Y - \bar{Y})$$

$$\text{Variance of } X = \frac{\sum x^2}{n} - \left(\frac{\sum x}{n}\right)^2 = \frac{\sum xx}{n} - \left(\frac{\sum x}{n}\right)\left(\frac{\sum x}{n}\right)$$

$$\text{Variance of } Y = \frac{\sum y^2}{n} - \left(\frac{\sum y}{n}\right)^2 = \frac{\sum yy}{n} - \left(\frac{\sum y}{n}\right)\left(\frac{\sum y}{n}\right)$$

$$\text{Covariance of } X \text{ and } Y = \frac{\sum XY}{n} - \left(\frac{\sum x}{n}\right)\left(\frac{\sum y}{n}\right)$$

1. Find the covariance of the following pairs of observations: (1, 3), (2, 2), (3, 5), (5, 4), (4, 6).

Solution: $\sum X = 1 + 2 + 3 + 5 + 4 = 15 \qquad \sum Y = 3 + 2 + 5 + 4 + 6 = 20$

$$\bar{X} = \frac{15}{5} = 3 \qquad \bar{Y} = \frac{20}{5} = 4$$

$n = 5$, $\sum XY = 1 \times 3 + 2 \times 2 + 3 \times 5 + 5 \times 4 + 4 \times 6 = 3 + 4 + 15 + 20 + 24 = 66$

$$\text{Hence covariance} = \frac{\sum XY}{n} - \left(\frac{\sum X}{n}\right)\left(\frac{\sum Y}{n}\right) = \frac{66}{5} - 3 \times 4 = 13.2 - 12 = 1.2$$

$$= \frac{66}{5} - 3 \times 4 = 13.2 - 12 = 1.2.$$

2. Find the covariance of the following pairs of observations: (1, 10), (2, 9), (3, 8), (4, 8), (5, 6), (6, 12), (7, 4), (8, 3), (9, 18), (10, 12).

Solution: $\sum X = 1 + 2 + 3 + 4 + 5 + 6 + 7 + 8 + 9 + 10 = 55$

$\sum Y = 10 + 9 + 8 + 8 + 6 + 12 + 4 + 3 + 18 + 2 = 80$

$n = 10$ therefore $\bar{X} = \frac{55}{10} \quad \bar{Y} = \frac{80}{10} = 9$

$$\sum XY = 1 \times 10 + 2 \times 9 + 3 \times 8 + 4 \times 8 + 5 \times 6 + 6 \times 12 + 7 \times 4 + 8 \times 3 + 9 \times 18 + 10 \times 22$$

$$= 10 + 18 + 24 + 32 + 30 + 72 + 28 + 24 + 162 - 20$$

$$= 420$$

Hence covariance $= \frac{420}{10} - \left(\frac{55}{10}\right)\left(\frac{80}{10}\right)$

$$= 42 - 5.5 \times 8$$

$$= 42 - 44.0$$

$$= -2.$$

Combined Standard Deviation

Combined mean $\bar{X} = \frac{n_1 \bar{X}_1 + n_2 \bar{X}_2}{n_1 + n_2}$

Let the standard deviation of these series be σ_1 and σ_2 respectively. Then their combined standard deviation *i.e.* σ or σ_{12} is given by the formula

$$\sigma \text{ or } \sigma_{12} \text{ (sigma)} = \sqrt{\frac{n_1 \sigma_1^2 + n_2 \sigma_2^2 + n_1 d_1^2 + n_2 d_2^2}{n_1 + n_2}}$$

$$= \sqrt{\frac{n_1 (\sigma_1^2 + d_1^2) + n_2 (\sigma_2^2 + d_2^2)}{n_1 + n_2}}$$

$$\begin{bmatrix} d_1 = (\bar{X}_1 - \bar{X}) \\ d_2 = (\bar{X}_2 - \bar{X}) \\ \bar{X} = \text{combined mean.} \end{bmatrix}$$

Example 1. The mean length of 500 and 600 carp fishes are 186 mm and 175 mm respectively. The corresponding standard deviations are 9 and 10 respectively. The length variablities are studied. Find the combined mean and variance of the combined sample.

Solution: $n_1 = 500 \quad \bar{X}_1 = 186 \quad \sigma_1 = 9$

$n_2 = 600 \quad \bar{X}_2 = 175 \quad \sigma_2 = 10$

$$\text{Combined mean } (\bar{X}) = \frac{n_1 \bar{X}_1 + n_2 \bar{X}_2}{n_1 + n_2} = \frac{500 \times 186 + 600 \times 175}{500 + 600}$$

$$= \frac{93000 + 105000}{1100} = \frac{198000}{1100} = 180 \text{ mm}$$

Combined (S.D.) $\sigma_{12} = \sqrt{\dfrac{n_1(\sigma_2^1 + d_1^2) + n_2(\sigma_2^2 + d_2^2)}{n_1 + n_2}}$

$$d_1 = \bar{X}_1 - \bar{X} = 186 - 180 = 6$$

$$d_2 = \bar{X}_2 - \bar{X} = 175 - 180 = -5$$

$$= \sqrt{\frac{500\,(9^2 + 6^2) + 600\,(10^2 + (-5)^2}{500 + 600}}$$

$$= \sqrt{\frac{500\,(81 + 36) + 600\,(100 + 25)}{1100}}$$

$$= \sqrt{\frac{500 \times 117 + 600 \times 125}{1100}} = \sqrt{\frac{58500 + 75000}{1100}} = \sqrt{\frac{133500}{1100}} = \sqrt{121.36}$$

∴ Variance in length of the combined sample is

$$\sigma^2 = (\sqrt{12136})^2 = 121.36.$$

Example 2. The mean weights of 50 and 100 crabs are 54.4 gram and 50.3 gram respectively and the standard deviations are 8 and 7. Obtain the mean and standard deviation of the sample of 150 (100+50) obtained by combining the two samples.

Solution:

	Sample 1	*Sample 2*	*Combined*
• Number of observation	$n_1 = 50$	$n_2 = 100$	$N = 100 + 50 = 150$
• Mean	$\bar{X}_1 = 54.4$	$\bar{X}_2 = 50.3$	$\bar{X} = ?$
• Standard deviation	$\sigma_1 = 8$	$\sigma_2 = 7$	$\sigma = ?$

Combined mean *i.e.*, $\bar{X} = \dfrac{n_1 \bar{X}_1 + n_2 \bar{X}_2}{n_1 + n_2} = \dfrac{50 \times 54.4 + 100 \times 50.3}{50 + 100}$

$$= \frac{2720 + 5030}{150} = \frac{7750}{150} = 51.666 = 51.7$$

$$S.D.\,(\sigma) = \sqrt{\frac{n_1 \sigma_1^2 + n_2 \sigma_2^2 + n_1 d_1^2 + n_2 d_2^2}{n_1 + n_2}}$$

$$d_1 = \bar{X}_1 - \bar{X} = 54.4 - 51.7 = 2.7$$

$$d_2 = \bar{X}_2 - \bar{X} = 50.3 - 51.7 = -1.4$$

$$\sigma^2 = \frac{50 \times (8)^2 + 100 \times (7)^2 + 50 \times (2.7)^2 + 100 \times (-1.4)^2}{150}$$

$$150 \times \sigma^2 = 50 \times 64 + 100 \times 49 + 50 \times 7.29 + 100 \times 1.96$$

$$150\sigma^2 = 3200 + 4900 + 364.50 + 196$$

$$150\sigma^2 = 8660.5$$

$$\sigma^2 = \frac{866.05}{150} = 57.73$$

$$\sigma = \sqrt{57.73} = 7.598 = 7.6.$$

Example 3. The first of two samples has 100 items with mean 15 and *S.D.* 3. If the whole group has 250 items with mean 15.6 and *S.D.* = $\sqrt{13.44}$. Find the *S.D.* of the second group.

Solution: Mean and *S.D.* of Combined Sample

First sample	*Second sample*	*Combined sample*
No. of observation = n_1 = 100	n_2 = 150	250
Mean = $\bar{X}_1 = 15$	$\bar{X}_2 = ?$	$\bar{X} = 15.6$
Standard deviation = $\sigma_1 = 3$	$\sigma_2 = ?$	$\sqrt{13.44}$

$$\bar{X} = \frac{n_1 \bar{X}_1 + n_2 \bar{X}_2}{n_1 + n_2}$$

$$15.6 = \frac{100 \times 15 + 150\bar{X}^2}{250}$$

$$3900 = 1500 + 150\bar{X}^2$$

$$\text{or} \quad 150\bar{X}_2 = 3900 - 1500$$

$$\bar{X}_2 = \frac{2400}{150} = 16$$

$$d_1 = \bar{X}_1 - \bar{X} = 15 - 15.6 = -0.6$$

$$d_2\, \bar{X}_2 - \bar{X} = 16 - 15.6 = 0.4$$

$$\sigma_2 = \frac{n_1 \sigma_1^2 + n_2 \sigma_2^2 + n_1 d_1^2 + n_2 d_2^2}{n_1 + n_2}$$

$$13.44 = \frac{100 \times (3)^2 + 150 \times \sigma_2^2 + 100\,(-0.6)^2 + 150 \times (0.4)^2}{250}$$

$$\text{or} \quad 250 \times 13.44 = 100 \times 9 + 150\sigma_2^2 + 3600 + 150 \times .16$$

$$\text{or} \quad 3360 = 900 + 150\sigma_2^2 + 36 + 24$$

$$150\sigma_2^2 = 3360 - (900 + 36 + 24)$$

$$150\sigma_2^2 = 3360 - 960 = 2400$$

$$\sigma_2^2 = \frac{2400}{150} = 16$$

$$\sigma_2 = \sqrt{16} = 4$$

The standard deviation of the second group is 4.

Example 4. The mean and S.D. of a sample of 100 observations was calculated as 40 and 5.1 respectively by a student who took by mistake 50 instead of 40 for one observation. Calculate the correct mean and *S.D.*

Solution: Using mean $\frac{\sum X}{N}$ and *S.D.* $(\sigma) = \sqrt{\frac{\sum X^2}{N} - \left(\frac{\sum X}{N}\right)^2}$

$$40 = \frac{\sum X}{100} \;.....(1) \qquad S.D. = 5.1 = \sqrt{\frac{\sum X^2}{100} - \left(\frac{\sum X}{100}\right)^2} \;.......(2)$$

or $$\sum X = 40 \times 100 = 4000$$

The correct value of $$\sum X = 4000 - 50 + 40 = 4000 - 10 = 3990$$

$$\therefore \quad \text{Correct mean} = \frac{\text{Correct value of} \sum X}{100} = \frac{3990}{100} = 39.9$$

From the equation (2), $$(5.1)^2 = \frac{\sum X^2}{100} - \left(\frac{\sum X}{100}\right)^2$$

$$26.01 = \frac{\sum X^2}{100} - \left(\frac{4000}{100}\right)^2 = \frac{\sum X^2}{100} - 1600$$

or $$26.01 + 1600 = \frac{\sum X^2}{100}$$

or $$\sum X^2 = 100\,(26.01 + 1600) = 2601 + 160000 = 162601$$

$$\therefore \text{ Correct value of } \sum X^2 = 162601 - (50)^2 + (40)^2$$
$$= 162601 - 2500 + 1600$$
$$= 162601 - 9000$$
$$= 161701$$

$$\therefore \quad \text{Correct } S.D. = \sqrt{\frac{\text{Correct} \sum X^2}{100} - \left(\frac{\text{Correct} \sum X}{100}\right)^2}$$

$$= \sqrt{\frac{161701}{100} - \left(\frac{3990}{100}\right)^2}$$

$$= \sqrt{1617.01 - 1592.01}$$

$$= \sqrt{25}$$

$$= 5$$

Standard Error of Mean (SE – $\bar{X}$)

The sampling distribution of any statistic will have its own mean, standard deviation etc. The sample estimates of statistic will differ from population parameter.

The difference or deviation between the value of statistic of a particular sample and the corresponding population parameter is known as *sampling error* or *standard error*. Thus standard error is a measure of variation of the mean.

- **Calculation of Standard Error Mean:** Standard error of mean is the standard deviation (S.D.) of the sample divided by the square root of the number of observation of the sample.

(*i*) $$SE - (\bar{X}) = \frac{S.D}{\sqrt{n}}$$

$SE - \bar{X}$ = standard error of Mean.

S. D. = Standard Deviation.

n = size of the sample

(*ii*) If the population have same standard deviation.

$$S\,E \text{ of } (\bar{X}_1 - \bar{X}_2) = S.D.\sqrt{\frac{1}{n_1} + \frac{1}{n_2}}$$

$\bar{X}_1$ & $\bar{X}_2$ = Population

n_1 & n_2 = [size of the sample]

(*iii*) If two random samples with $\bar{X}_1, \sigma_1, n_1$ & $\bar{X}_2, \sigma_2, n_2$ are drawn from different populations.

$$S\,E \text{ of difference between mean} = \sqrt{\frac{\sigma_1^2}{n_1} + \frac{\sigma_2^2}{n_2}}$$

σ_1 = S. D.

n_1 = size of the sample.

- **Factors controlling S.E.:**
  1. **The sample size:** Increase the size of the sample decrease S.E.
  2. **Standard deviation:** The values of SE varies directly with the size of S.D.
  3. **The nature of statistic:** Example—mean, variance, etc.
  4. The mathematical form of the sampling distribution.
- **Uses:**

  (*i*) The standard error of the mean is used to measure of the extent of sampling error in the mean.

  (*ii*) It helps to determine whether the population is drawn from a known population or not.

  (*iii*) It helps to calculate the size of the sample.

- **Problem 1: Find out *SE* of the mean when the unbiased S. D. of sample of 36 cases is 23.61.**

Solution:

$$S.E. \text{ of } \bar{X} = \frac{S.D.}{\sqrt{n}}$$

$S.D. = 23.61$

$n = 36$

$$= \frac{23.61}{\sqrt{36}} = \frac{23.61}{6} = 3.935$$

- **Problem 2: Systolic blood pressures of 566 male were recorded & the unbiased S.D. are 13.05 mm. Calculate S.E. of the mean.**

Solution:

$$S.E. \text{ of } \bar{X} = \frac{S.D.}{\sqrt{n}}$$

$S.D. = 13.05$

$n = 566$

$$= \frac{13.05}{\sqrt{566}} = \frac{13.05}{23.79} = 0.548 = 0.55$$

- **Problem 3 : A sample of 100 days student yield examination results as under S.D 14.8 and other sample of 200 evening students yields examination result as under S.D. 17.9. Calculate S.E. of mean.**

Solution:

$$\text{S.E. } (\overline{X}_1 - \overline{X}_2) = \sqrt{\frac{\sigma_1^2}{n_1} + \frac{\sigma_2^2}{n_2}}$$

$$= \sqrt{\frac{(14.8)^2}{100} + \frac{(17.9)^2}{200}}$$

X_1 = Day student

X_2 = Evening student

$$= \sqrt{\frac{219.04}{100} + \frac{320.41}{200}}$$

$$= \sqrt{2.19 + 1.60} = \sqrt{3.79} = 1.9467$$

$n_1 = 100, \sigma_1 = 14.8$

$n_2 = 200, \sigma_2 = 17.9$

- **Problem 4 : Calculate standard error of mean from the following data, showing the amount paid by 100 shopping stalls in Calcutta on the occasion of Durga Puja.**

Mid value (Rs):	39	49	59	69	79	89	99
No. of shopping stall:	2	3	11	20	32	25	7

Solution:

$$\text{S. E. } \overline{X} = \frac{\sigma}{\sqrt{N}}$$

Mid Value (X)	f	$\frac{X - A(69)}{i} = d$	fd	fd^2
39	2	$\frac{39-69}{10} = -3$	–6	18
49	3	$\frac{49-69}{10} = -2$	–6	12
59	11	$\frac{59-69}{10} = -1$	–11	11
69	20	$\frac{69-69}{10} = 0$	0	0
79	32	$\frac{79-69}{10} = +1$	+32	32
89	25	$\frac{89-69}{10} = +2$	+50	100
99	7	$\frac{99-69}{10} = +3$	+21	63
	$N = 100$		$\Sigma fd = +80$	$\Sigma fd^2 = 236$

$$\text{S. D. } (\sigma) = \sqrt{\frac{\sum fd^2}{N} - \left(\frac{\sum fd}{N}\right)^2} \times i$$

$$= \sqrt{\frac{236}{100} - \left(\frac{80}{100}\right)^2} \times 10 = \sqrt{2.36 - (0.8)^2} \times 10$$

$$= \sqrt{2.36 - 0.64} \times 10 = \sqrt{1.72} \times 10 = 1.311 \times 10 = 13.11$$

$$S.E. - \overline{X} = \frac{13.11}{\sqrt{100}} = \frac{13.11}{10} = 1.311$$

Merits and Demerits Standard Deviations

- **Merits:**

I. It is based on all the observations.

II. It is rigidly defined.

III. It is less affected by fluctuations of sampling as compared to other measures of dispersion.

IV. It is extremely used in correlation.

- **Demerits**

I. It is difficult to compute unlike other measures of dispersions.

II. It is not simple to understand.

III. It gives more weightage to extreme values.

- **Uses of standard Deviation**

I. It summaries the deviations of a large distribution from mean.

II. It indicates whether the variation of difference of an individual from the mean is by chance. *i.e.*, natural or real due to some special reasons.

III. It also helps in finding the standard error which determines whether the difference between means of two similar samples is by chance or real.

IV. It also helps in finding the suitable size of sample for valid conclusion.

- **The Standard Deviation (σ) may be used**
  - When arithmetic mean is used for the central tendency.
  - When the statistics having the greater stability is sought.
  - When the correlation coefficient (*r*) & other statistics which depend on the standard deviation are to be computed.
  - The standard deviation is used in preference to other deviation as it is a neat method of removing the negatives.

Example 1. Compute the median & mode of the following distribution of tracheal ventilation scores (ml per minute) of a sample of beetle.

Class interval:	61–65	66–70	71–75	76–80	81–85
Frequencies:	12	25	45	30	8

[Vidyasagar Univ. M.Sc (Zoology) 2000]

Solution:

Class intervals Score limit	*Class interval (true limit)*	*Frequencies (f)*	*Cumulative frequencies (cf)*
61 – 65	60.5 – 65.5	12	12
66 – 70	65.5 – 70.5	25	37
71 – 75	70.5 – 75.5	45	82
76 – 80	75.5 – 80.5	30	112
81 – 85	80.5 – 85.5	8	120

$$\frac{N}{2} = \frac{120}{2} = 60$$

It $\left(\frac{N}{2}\right)$ lies between cumulative frequency 82 and 112, hence the median class is 70.5 – 75.5.

L_1 = Lower limit of the median class *i.e.*, 70.5

f_m = frequency of the median class = 45

C = Cumulative frequency of the class preceding class = 37

i = width of the class interval = 5

$$\text{Median } (M) = L_1 + \frac{\frac{N}{2} - C}{fm} \times i$$

$$= 70.5 + \frac{60 - 37}{45} \times 5$$

$$= 70.5 + \frac{23}{45_9} \times 5 = 70.5 + 2.56$$

$$= 73.06$$

$$\text{Mode} = L_1 = 70.5, \quad fm = 45, \quad f_1 = 25, \quad f_2 = 30, \quad i = 5$$

$$\text{Mode} = L_1 + \frac{fm - f_1}{2f_m - f_2 - f_1} \times i$$

$$= 70.5 + \frac{45 - 25}{2 \times 45 - 30 - 25} \times 5$$

$$= 70.5 + \frac{20}{90 - 55} \times 5$$

$$= 70.5 + \frac{20}{35_7} \times 5 = 70.5 + 2.86$$

$$= 73.36.$$

Example 2. Compute the mean standard deviation and coefficient of variation of the following distribution of the body weights (grams) of a sample of animals.

Class intervals:	101–105	106–110	111–115	116–120	121–125
Frequencies:	6	22	40	25	7

[Vidyasagar Univ. M.Sc. (Zoology)–2001]

Solution:

Class	Mid value	Frequency (f)	Cumulative frequency (cf)	$\frac{X-A}{i}=d$	fd	fd^2
101–105	103	6	6	$\frac{103-133}{5}=-2$	12	24
106–110	108	22	28	$\frac{108-113}{5}=-1$	22	22
111–115	113	40	68	$\frac{113-113}{5}=0$	0	0
116–120	118	25	93	$\frac{118-113}{5}=+1$	25	25
121–125	123	7	100	$\frac{123-113}{5}=+2$	14	28
			N = 100		+39–34=5	99

Let us take assumed mean = 113

$$\text{Mean} = A+\frac{\sum fd}{N}\times c = 113+\frac{5}{\cancel{100}_{20}}\times \cancel{5} = 113+.25 = 113.25$$

$$\text{Standard deviation (S.D.)} = \sqrt{\frac{\sum fd^2}{N}-\left(\frac{\sum fd}{N}\right)^2}\times i$$

$$= \sqrt{\frac{99}{100}-\left(\frac{5}{100}\right)^2}\times 5 = \sqrt{.99-.0025}\times 5$$

$$= \sqrt{.9875}\times 5 = .99\times 5 = 4.95$$

$$\text{Coefficient of variation} = \frac{\text{S.D.}}{\text{A.M.}}\times 100$$

$$= \frac{4.95}{113.25}\times 100 = \frac{495}{113.25} = 4.37$$

Example 3. Work out the mean & the standard error of the mean of the following frequency distribution of blood sugar scores (mg per deciliter) of a sample of goats.

Class intervals:	51–55	56–60	61–65	66–70	71–75	76–80
Frequencies:	7	15	30	25	14	9

[*Vidyasagar Univ. M.Sc. (Zoology)–2002*]

Solution: Let us take assumed mean = 63

Class interval	Frequencies (f)	Mid Value	Cumulative frequency (cf)	$\frac{X-A}{i} = d$	fd	fd^2
51–55	7	53	7	$\frac{53-63}{5} = -2$	–14	28
56–60	15	58	22	$\frac{58-63}{5} = -1$	–15	15
61–65	30	63	52	$\frac{63-63}{5} = 0$	0	0
66–70	25	68	77	$\frac{68-63}{5} = +1$	25	25
71–75	14	73	91	$\frac{73-63}{5} = +2$	28	56
76–80	9	78	100	$\frac{78-63}{5} = +3$	27	81
	100				+80–29=51	205

$$\text{Mean} = A + \frac{\sum fd}{N} \times i$$

$$\text{Mean} = 63 + \frac{51}{\cancel{100}_{20}} \times \cancel{5} = 63 + 2.55 = 65.55$$

$$\sum fd = 51$$

$$i = 5$$

$$N = 100$$

$$\text{S.D.} = \sqrt{\frac{\sum fd^2}{N} - \left(\frac{\sum fd}{N}\right)^2} \times i$$

$$\sqrt{2.05 - 2601} \times 5$$

$$= \sqrt{\frac{205}{100} - \left(\frac{51}{100}\right)^2} \times 5$$

$$= \sqrt{2.05 - .2601} \times 5 = \sqrt{1.79} \times 5 = 1.34 \times 5 = 6.70$$

$$\text{Standard Error of mean} = \frac{S.D.}{\sqrt{N}} = \frac{6.7}{\sqrt{100}} = \frac{6.7}{10} = 0.67$$

Example 4. The following table shows a distribution of bristle number in *Drosophila*. Calculate the mean, variance & standard deviation.

Bristle number	*Number of individuals*
1	1
2	4
3	7
4	31
5	56
6	17
7	4

[C.U. B.Sc. (Botany Hon's. Part II) 1998, 2 008]

Solution: Let us take assumed mean = 4

Bristle Number X	*Number of individual (f)*	*d = X – A (4)*	*fd*	*fd²*
1	1	1 – 4 = –3	–3	9
2	4	2 – 4 = –2	–8	16
3	7	3 – 4 = –1	–7	7
4	31	4 – 4 = 0	0	0
5	56	5 – 4 = +1	+56	56
6	17	6 – 4 = +2	+34	68
7	4	7 – 4 = +3	+12	36
	$\sum f = N = 120$		$\sum fd = 120 - 18 = 84$	$\sum fd^2 = 192$

$$\text{Mean} = \bar{X} = A + \frac{\sum fd}{N}$$

$$= 4 + \frac{84}{120} = 4 + .7 = 4.7$$

$$\text{S.D. } (\sigma) = \sqrt{\frac{\sum fd^2}{N} - \left(\frac{\sum fd}{N}\right)^2} = \sqrt{\frac{192}{120} - \left(\frac{84}{120}\right)^2} = \sqrt{1.6 - (.7)^2}$$

$$= \sqrt{1.6 - .49} = \sqrt{1.11} = 1.053$$

$$\text{Variance} = (\text{S.D})^2 = \left(\sqrt{1.11}\right)^2 = 1.11$$

Example 5. According to height, 200 jute plants can be grouped as

Frequency:	10	30	75	50	30	5
Class value:	60	62	64	66	68	70

Calculate the mean height & the mean deviation.

[C.U. (Bot. Hons.) 1998]

Solution: Let us assumed mean 64.

Class value (X)	Frequency (f)	$\frac{X-A}{i}=d$	fd	$X - \bar{X} = \lvert D \rvert$	$f\lvert D \rvert$
60	10	$\frac{60-64}{2}=-2$	–20	60–64.75 = –4.75	47.50
62	30	$\frac{62-64}{2}=-1$	–30	62–64.75 = 2.75	82.50
64	75	$\frac{64-64}{2}=-0$	0	64–64.75 = 0.75	56.25
66	50	$\frac{66-64}{2}=+1$	+50	66–64.75 = +1.25	62.50
68	30	$\frac{68-64}{2}=+2$	+60	68–64.75 = +3.25	97.50
70	5	$\frac{70-64}{2}=+3$	+15	70–64.75 = +5.25	26.25
	$\sum f = N = 200$		$\sum fd = 75$		$\sum fD = 372.5$

$$\text{Mean}\left(\bar{X}\right) = A + \frac{\sum fd}{N} \times i$$

$$= 64 + \frac{75}{200} \times 2$$

$$= 64 + .75$$

$$= 64.75$$

$$\text{Mean deviation} = \frac{\sum fD}{N} = \frac{372.5}{75} = 4.966$$

$$= 4.97$$

Example 6. The following measurements of head width & wing length were made on a series steamer ducks.

Specimen	Head width (cm)	Wing length (cm)
1	2.75	30.3
2	3.20	36.2
3	2.86	31.4
4	3.24	35.7
5	3.16	33.4
6	3.32	34.8
7	2.52	27.2
8	4.16	52.7

Calculate the mean & standard deviation of head width & length for these birds.

[C.U. (Bot. Hon.) 1999]

Solution: Let us take assumed mean for head width 3.16

	Head width (cm) X	d = X – A (3.16)	d^2
1	2.75	2.75 – 3.16 = –0.41	0.1681
2	3.20	3.20 – 3.16 = +0.04	0.0016
3	2.86	2.86 – 3.16 = –0.30	0.0909
4	3.24	3.24 – 3.16 = +0.08	0.0064
5	3.16	3.16 – 3.16 = 0	0
6	3.32	3.32 – 3.16 = +0.16	0.0256
7	2.52	2.52 – 3.16 = –0.64	0.4096
8	4.16	4.16 – 3.16 = 1.00	1.0000
		$\sum 0 = 1.28 - 1.35$	$\sum d^2 = 1.7013$

$$\text{Mean}\left(\bar{X}\right) = A + \frac{\sum d}{N} = 3.16 + \frac{(-0.07)}{8} = 3.16 - .0088 = 3.1512 = 3.15$$

$$\text{S.D.} = \sqrt{\frac{\sum d^2}{N} - \left(\frac{\sum d}{N}\right)^2} = \sqrt{\frac{1.7013}{8} - \left(\frac{-0.07}{8}\right)^2} = \sqrt{0.212 - \frac{.0049}{64}}$$

$$= \sqrt{0.212 - .000076} = \sqrt{.2119} = \sqrt{.212} = .46$$

Solution: Let us take assumed mean = 35.7 (*b*) Wing length.

Wing length cm	d = X – A (35.7)	d^2
30.3	30.3 – 35.7 = -54	29.16
36.2	36.2 – 35.7 = +0.5	0.25
31.4	31.4 – 35.7 = –4.3	18.49
35.7	35.7 – 35.7 = 0	0
33.4	33.4 – 35.7 = –2.3	5.29
34.8	34.8 – 35.7 = –0.9	0.81
27.2	27.2 – 35.7 = –8.5	72.25
52.7	52.7 – 35.7 = +17.0	289.00
	$\sum d = +17.5 - 21.4 = -3.9$	$\sum d^2 = 415.25$

$$\text{Mean}\left(\bar{X}\right) = A + \frac{\sum d}{N} = 35.7 - \frac{3.9}{8} = 35.7 - .488 = 35.21$$

$$\text{S.D.}\ (\sigma) = \sqrt{\frac{\sum fd^2}{N} - \left(\frac{\sum fd}{N}\right)^2} = \sqrt{\frac{415.25}{8} - \left(\frac{-3.9}{8}\right)^2}$$

$$= \sqrt{\frac{415.25}{8} - (-0.488)^2} = \sqrt{51.906 - 0.238} = \sqrt{51.67}$$

$$= 7.18$$

Example 7. A sample of 20 plants from a population was measured in inches as follows: 18 , 21, 20, 23, 20, 21, 22, 20, 20, 19, 17, 21, 20, 22, 20, 21, 20, 22, 19 and 23. Calculate mean and standard deviation.

[C.U. (Bot. Hon.) 2000]

Solution: Let as take assumed mean = 20

Height of plants (X)	*Frequency (f)*	*d = X – A (20)*	*fd*	*fd^2*
17	1	17 – 20 = –3	–3	9
18	1	18 – 20 = –2	–2	4
19	2	19 – 20 = –1	–2	2
20	7	20 – 20 = 0	0	0
21	4	21 – 20 = +1	+4	4
22	3	22 – 20 = +2	+6	12
23	2	23 – 20 = +3	+6	18
	$\sum f = N = 20$		$\sum fd = +16 - 7 = +9$	$\sum fd^2 = 49$

$$\text{Mean}\left(\bar{X}\right) = A + \frac{\sum fd}{N} = 20 + \frac{9}{20} = 20 + 0.45 = 20.45$$

$$\text{S.D.} = \sqrt{\frac{\sum fd^2}{N} - \left(\frac{\sum fd}{N}\right)^2} = \sqrt{\frac{49}{20} - \left(\frac{9}{20}\right)^2} = \sqrt{2.45 - (.45)^2}$$

$$= \sqrt{2.45 - 0.2025} = \sqrt{2.247} = 1.49$$

Example 8. There are 9 boys and 9 girls in a class. Their heights are given below.

Boys (Height in cm):	170	175	165	162	168	170	172	163	161
Girls (Height in cm):	152	165	155	161	168	160	155	154	152

Calculate variance of height for the boys and girls separately.

[C.U. (Bot. Hon.) 2000]

Solution: Let us take assumed mean 168 (cm)

Sl. No.	*Boys Height (cm)*	*d = X – A (168)*	*d^2*
1	170	170 – 168 = +2	4
2	175	175 – 168 = +7	49
3	165	165 – 168 = –3	09
4	162	162 – 168 = –6	36
5	168	168 – 168 = 0	0
6	170	170 – 168 = +2	04
7	172	172 – 168 = +4	16
8	163	163 – 168 = –5	25
9	161	161 – 168 = –7	49

$\sum d = 15 - 21 = -6 \sum d^2 = 192$

$$\text{Mean}(\bar{X}) = A + \frac{\sum d}{N} = 168 - \frac{6}{9} = 168 - .67 = 167.33$$

$$\text{S.D.} = \sqrt{\frac{\sum d^2}{N} - \left(\frac{\sum d}{N}\right)^2}$$

$$= \sqrt{\frac{192}{9} - \left(\frac{-6}{9}\right)^2}$$

$$\left[\begin{array}{c}\sum d = -6 \\ \sum d^2 = 192 \\ N = 9\end{array}\right]$$

$$= \sqrt{21.34 - .449}$$

$$= \sqrt{20.89} = 4.57$$

$$\text{Variance} = (\text{S.D})^2 = 20.89.$$

Let us take assumed mean 168

	Height (girl cm)	d = X – A (168)	d^2
1	152	152 – 168 = –16	256
2	165	165 – 168 = –3	09
3	155	155 – 168 = –13	169
4	161	161 – 168 = –7	49
5	168	168 – 168 = 0	0
6	160	160 – 168 = –8	64
7	155	155 – 168 = –13	169
8	154	154 – 168 = –14	196
9	152	152 – 168 = –16	256
		$\sum d = -90$	$\sum d^2 = 1168$

$$\text{Mean}(\bar{X}) = A + \frac{\sum d}{N} = 168 - \frac{90}{9} = 168 - 10 = 158$$

$$\text{S.D.} = \sqrt{\frac{\sum d^2}{N} - \left(\frac{\sum d}{N}\right)^2} = \sqrt{\frac{1168}{9} - \left(\frac{-90}{9}\right)^2}$$

$$= \sqrt{129.78 - 100} = \sqrt{29.78} = 5.46$$

$$\text{Variance} = (\text{S.D})^2 = (5.46)^2 = 29.78$$

Example 9. Following data relate to increase in dry weight of the pods of a plant after a particular treatment. Calculate the mean standard deviation and standard error from the following distribution.

Observation:	1	2	3	4	5	6	7	8	9	10
Increase in dry weight:	4.25	4.20	4.15	3.35	3.25	4.70	3.25	3.75	3.70	3.90

[C.U. (Bot. Hon.)2001]

Solution:

Let us take assumed mean 3.25

Observation	Dry weight X	d = X – A	d^2
1	4.25	4.25 – 3.25 = 1.00	1.00
2	4.20	4.20 – 3.25 = 0.95	0.9025
3	4.15	4.15 – 3.25 = 0.90	0.81
4	3.35	3.35 – 3.25 = 0.10	0.01
5	3.25	3.25 – 3.25 = 0	0.00
6	4.70	4.70 – 3.25 = 1.45	2.1025
7	3.25	3.25 – 3.25 = 0	0.00
8	3.75	3.75 – 3.25 = 0.50	0.25
9	3.70	3.70 – 3.25 = 0.45	0.2025
10	3.90	3.90 – 3.25 = 0.65	0.4225
		$\sum d = 6.00$	$\sum d^2 = 5.7000$

$$\text{Mean}\left(\bar{X}\right) = A + \frac{\sum d}{N} = 3.25 + \frac{6}{10} = 3.25 + 0.6 = 3.85$$

$$\text{Standard deviation } (\sigma) = \sqrt{\frac{\sum d^2}{N} - \left(\frac{\sum d}{N}\right)^2}$$

$$= \sqrt{\frac{5.7}{10} - \left(\frac{6}{10}\right)^2} = \sqrt{0.57 - 0.36} = \sqrt{0.21} = 0.46$$

$$\text{Standard Error} = \frac{S.D}{\sqrt{N}} = \frac{0.46}{\sqrt{10}} = \frac{0.46}{3.16} = 0.1455$$

$$= 0.146$$

Example 10. Grain lengths of two varieties of rice are given below. Calculate the mean, standard error & Co-efficient of variation of grain length of two varieties.

[C.U. (Bot. Hon.) 2002]

Variety A		Variety B	
Grain length (mm)	No. of grain	Grain length (mm)	No. of grain
9 – 11	3	9 – 11	0
12 – 14	5	12 – 14	8
15 – 17	9	15 – 17	8
18 – 20	3	18 – 20	4

Solution: Variety A

Let us take assumed mean = 13

Grain length	Mid value	No. of grain	$d = \frac{X - A(13)}{i}$	fd	fd^2
9 – 11	10	3	$\frac{10-13}{3} = -1$	–3	3
12 – 14	13	5	$\frac{13-13}{3} = 0$	0	0
15 – 17	16	9	$\frac{16-13}{3} = +1$	9	9
18 – 20	19	3	$\frac{19-13}{3} = +2$	6	12
		$\Sigma f = N = 20$		$\Sigma fd = 12$	$\Sigma fd^2 = 24$

$$\text{S.D. } (\sigma) = \sqrt{\frac{\sum fd^2}{N} - \left(\frac{\sum fd}{N}\right)^2} \times i$$

$$\text{Mean}\left(\bar{X}\right) = A + \frac{\sum fd}{N} \times i$$

$$= 13 + \frac{12}{20} \times 3$$

$$= 13 + 1.8$$

$$= 14.8 \text{ mm.}$$

$$S.D(\sigma) = \sqrt{\frac{24}{20} - \left(\frac{12}{20}\right)^2} \times 3$$

$$\sigma = \sqrt{1.2 - (0.6)^2} \times 3$$

$$\sigma = \sqrt{1.2 - 0.36}$$

$$\sigma = \sqrt{0.84} \times 3$$

$$\sigma = 0.92 \times 3 = 2.76$$

$$\text{Co-efficient of variance } (CV) = \frac{S.D.}{A.M.} \times 100$$

$$= \frac{2.76}{14.8} \times 100 = 18.6$$

$$\text{Standard error} = \frac{SD(\sigma)}{\sqrt{N}} = \frac{2.76}{\sqrt{20}}$$

$$= \frac{2.76}{4.47} = 0.617$$

$$= 0.62$$

Let us take assumed mean = 10

Variety B

Grain length (mm) (X)	Mid value	No. of grain f	$d = \frac{X - A(16)}{i}$	fd	fd^2
9 – 11	10	0	$\frac{10-16}{3} = -2$	0	0
12 – 14	13	8	$\frac{13-16}{3} = -1$	–8	8
15 – 17	16	8	$\frac{16-16}{3} = +0$	0	0
18 – 20	19	4	$\frac{19-16}{3} = +1$	4	4
		$\Sigma f = N = 20$		$\Sigma fd = -4$	$\Sigma fd^2 = 12$

$$\text{Mean}\left(\bar{X}\right) = A + \frac{\sum fd}{N} \times i$$

$$= 16 + \frac{-4}{20} \times 3$$

$$= 16 - \frac{3}{5}$$

$$= 16 - 0.6$$

$$= 15.4$$

$$\text{S.D. } (\sigma) = \sqrt{\frac{\sum fd^2}{N} - \left(\frac{\sum fd}{N}\right)^2} \times i$$

$$= \sqrt{\frac{12}{20} - \left(\frac{-4}{20}\right)^2} \times 3$$

$$= \sqrt{0.6 - 0.04} \times 3$$

$$= \sqrt{0.56} \times 3$$

$$= 0.748 \times 3$$

$$= 2.24$$

$$\text{Co-efficient of variance } (C.V) = \frac{S.D.}{A.M.} \times 100$$

$$= \frac{2.24}{15.4} \times 100$$

$$= 0.1454 \times 100 = 14.5400$$

$$= 14.54$$

$$\text{Standard Error} = \frac{S.D.}{\sqrt{N}} = \frac{2.24}{\sqrt{20}} = \frac{2.24}{4.47} = 0.501$$

Example 11. There are 200 rice plants in an experimental plot. The plants range from 28 to 36 cm in height, which may be grouped in 5 classes. Example. 28, 30, 32, 34, 36 cm showing distribution frequency 32, 80, 55, 28 & 5 respectively.

Calculate the mean deviation of height of these plants.

[C.U. (Bot. Hon.) 2003]

Solution:

Let us take assumed mean = 30

Height of plants cm (X)	Frequency (f)	$\frac{X-A}{i}=d$	fd	$\lvert X-\bar{X}\rvert=\lvert D\rvert$	$f\lvert D\rvert$
28	32	$\frac{28-30}{2}=-1$	−32	28 − 30.94 = 2.94	94.08
30	80	$\frac{30-30}{2}=0$	0	30 − 30.94 = 0.94	75.20
32	55	$\frac{32-30}{2}=+1$	+55	32 − 30.94 = +1.06	58.30
34	28	$\frac{34-30}{2}=+2$	+56	34 − 30.94 = +5.06	85.68
36	5	$\frac{36-30}{2}=+3$	+15	36 − 30.94 = +3.06	25.30
	$\sum f = N = 200$		$\sum fd = 126 - 32 = 94$		$\sum f\lvert D\rvert = 338.56$

$$\text{Mean}\left(\bar{X}\right) = A + \frac{\sum fd}{N}\times i$$

$$= 30 + \frac{94}{200}\times 2$$

$$= 30 + 0.94$$

$$= 30.94$$

Mean deviation (M.D)

$$= \frac{\sum fD}{N} = \frac{338.56}{200}$$

$$= 1.69$$

Example 12. From a field of garden pea plants, random sampling of 13 plants is made. Their height in cms is as follows: 161, 183, 177, 157, 181, 176, 180, 162, 163, 174, 179, 169, 187. Calculate the (*i*) variance & (*ii*) standard deviation. *[C.U.(Bot. Hon.) 2004]*

Solution:

	X	$X-\bar{X}=d$	$(X-\bar{X})^2=d^2$
1	161	161 − 173 = −12	144
2	183	183 − 173 = +10	100
3	177	177 − 173 = +4	16
4	157	157 − 173 = −16	256
5	181	181 − 173 = +8	64
6	176	176 − 173 = +3	09
7	180	180 − 173 = +7	49
8	162	162 − 173 = −11	121
9	163	163 − 173 = −10	100
10	174	174 − 173 = +1	01
11	179	179 − 173 = +6	36
12	169	169 − 173 = −4	16
13	187	187 − 173 = +14	196
	2249		1108

$$N = 13$$

$$\bar{X} = \frac{2249}{13} = 173$$

$$\text{S.D} = \sqrt{\frac{\sum (X - \bar{X})^2}{N}}$$

$$= \sqrt{\frac{1108}{13}} = \sqrt{85.23} = 9.23$$

$$\text{Variance} = (\text{S.D.})^2 = (9.23)^2 = 85.19$$

Example 13. Compute mean & S.D of following distribution.

Scores:	60–64	55–59	50–54	45–49	40–44	35–39	30–34	25–29	20–24
Frequency:	2	2	4	6	9	12	7	5	3

Solution:

Scores (x)	*Mid value (m)*	*Frequency (f)*	*fm*	$d = (m - \bar{x})$	d^2	fd^2
60–64	62	2	124	22.5	506.25	1012.5
55–59	57	2	114	17.5	306.25	612.5
50–54	52	4	208	12.5	156.25	625
45–49	47	6	282	7.5	56.25	337.5
40–44	42	9	378	2.5	6.25	56.25
35–39	37	12	444	–2.5	6.25	75
30–34	32	7	224	–7.5	56.25	393.75
25–29	27	5	135	–12.5	156.25	781.25
20–24	22	3	66	–17.5	306.25	918.75
		$\sum f = 50$	$\sum fm = 1975$		$\sum fd^2 = 4812.5$	

Now Mean value $(\bar{x})$ is $= \frac{\sum fm}{N} = \frac{1975}{50} = 39.5$ [or taking assumed mean 42]

$$\therefore \quad \text{S.D.} = \sqrt{\frac{\sum fd^2}{N}} = \sqrt{\frac{4812.5}{50}} = \sqrt{96.25} = 9.81$$

Example 14. Calculate the mean, the median & the mode of the frequency distribution.

Class limits:	130–134	135–139	140–144	145–149	150–154	155–159	160–164
Frequency:	5	15	28	24	17	10	1

[M. Sc. (Zoology) C.U., 2002]

Solution:

Class limits	Mid value (x)	Class boundaries	f	fx	cf
130–134	132	129.5–134.5	5	660	5
135–139	137	134.5–139.5	15	2055	20
140–144	142	139.5–144.5	28	3975	48
					$\frac{N}{2} = 50$
145–149	147	144.5–149.5	24	3528	72
150–154	152	149.5–154.5	17	2584	89
155–159	157	154.5–159.5	10	1570	99
160–164	162	159.5–164.5	1	162	100
			$\sum f = 100$	$\sum fx = 14534$	

Now $\text{Mean}(\bar{x}) = \frac{\sum fx}{\sum f} = \frac{14534}{100} = 145.34$ [or taking assumed mean 147]

$$\text{Median} = L + \frac{N/2 - C}{fm} \times i$$

Here N/2 = 100/2 = 50

$L = 144.5$

$fm = 24$

$C = 48$

$i = 5$

$$\therefore \quad \text{Median} = 144.5 + \frac{50 - 48}{24} \times 5$$

$$= 144.5 + \frac{2}{24} \times 5$$

$$= 144.5 + .4166$$

$$= 144.9166 \text{ (Ans)}$$

$$\text{Mode} = L + \frac{fm - f_1}{2fm - f_1 - f_2} \times i$$

The maximum frequency of the frequency distribution is 28.

So, Modal class – 140 – 144

So $L = 139.5$, $fm = 28$, $f_1 = 15$, $f_2 = 24$, $i = 5$

$$\text{Mode} = 139.5 + \frac{28 - 15}{(2 \times 28) - 15 - 24} \times 5$$

$$= 139.5 + \frac{13}{56 - 15 - 24} \times 5$$

$$= 139.5 + \frac{13}{17} \times 5$$

$$= 139.5 + 3.82$$

$$= 143.32 \text{ (Ans).}$$

Example 15. Find the standard deviation & coefficient of variation from the following table giving wages of 230 persons.

Wages (Rs):	70–80	80–90	90–100	100–110	110–120	120–130	130–140	140–150
No. of persons:	12	18	35	42	50	45	20	8

Solution:

Wages	*Mid value (x)*	*f*	*fx*	$d = (x - \bar{x})$	d^2	fd^2
70–80	75	12	900	–35.43	1255.29	15063.42
80–90	85	18	1530	–25.43	646.69	11640.33
90–100	95	35	3325	–15.43	238.1	8333
100–110	105	42	4410	–5.43	29.49	1238.36
110–120	115	50	5750	4.57	20.9	1044.25
120–130	125	45	5625	14.57	212.28	9552.8
130–140	135	20	2700	24.57	603.68	12073.7
140–150	145	8	1160	34.57	1195	9560.7
		$\sum f = 230$	$\sum fx = 25400$			$\sum fd^2 = 68506.56$

$$\text{Mean} = \frac{\sum fx}{\sum f} = \frac{25400}{230}$$

[or taking assumed mean 115]

$$= 110.43$$

$$\text{S.D.} = \sqrt{\frac{\sum fd^2}{N}} = \sqrt{\frac{68506.56}{230}}$$

$$= \sqrt{297.85}$$

$$= 17.26$$

$$\text{\& Coefficient of variance} = \frac{S.D.}{A.M} \times 100$$

$$= \frac{17.26}{110.43} \times 100$$

$$= 15.63$$

Example 16. Calculate the mean, median, mode, S.D coefficient of Variation & S.E of mean from the following data.

Class Interval:	10–19	20–29	30–39	40–49	50–59	60–69	70–79
Frequency:	5	19	10	13	4	4	2

[M. Sc. (Zoology) C.U., 1995]

Solution:

Class interval	Mid value (x)	Class boundaries	f	cf	fx	$d^2 = (x - \bar{x})$	fd^2
10–19	14.5	9.5–19.5	5	5	72.5	488.41	2442.05
20–29	24.5	19.5–29.5	19	24	465.5	14641	2781.8
30–39	34.5	29.5–39.5	10	34	345	4.41	44.1
40–49	44.5	39.5–49.5	13	47	578.5	62.41	811.33
50–59	54.5	49.5–59.5	4	51	218	320.41	1281.64
60–69	64.5	59.5–69.5	4	55	258	778.41	3113.64
70–79	74.5	69.5–79.5	2	57	149	1436.41	2872.82
			N = 57	$\sum fx = 2086.5$			$\sum fd^2 = 13347.38$

Mean $\bar{x} = \frac{\sum fx}{N} = \frac{2086.5}{57} = 36.61$ [or taking assumed mean 44.5]

$$\text{Median} = L_1 + \frac{N/2 - C}{fm} \times i \qquad N/2 = 57/2 = 28.5$$

$$= 29.5 + \frac{28.5 - 24}{10} \times 10$$

$$= 29.5 + \frac{4.5}{10} \times 10$$

$$= 34$$

$$\text{Mode} = L_1 + \frac{fm - f_1}{2fm - f_1 - f_2} \times i$$

$$= 19.5 + \frac{19 - 5}{(2 \times 19) - 5 - 10} \times 10$$

$$= 19.5 + \frac{14}{23} \times 10$$

$$= 19.5 + 140/23$$

$$= 19.5 + 6.08$$

$$= 25.58$$

$$\text{S.D.} = \sqrt{\frac{\sum fd^2}{N}} = \sqrt{\frac{13347.38}{57}} = \sqrt{234.16} = 15.3$$

$$\text{Coefficient of variance} = \frac{SD}{AM} \times 100$$

$$= \frac{15.3}{36.61} \times 100$$

$$= 41.79$$

$$S.E. = SD \times \sqrt{\frac{1}{N}}$$

$$= 15.3 \times \sqrt{\frac{1}{57}}$$

$$= 15.3 \times .1324$$

$$= 2.02572$$

Example 17. Compute the variance, SD, the coefficient of variance & the coefficient of dispersion of the following frequency distribution of interorbital width (m.m) of a sample of 100 pigeons. *[M. Sc (C.U.) 2000]*

Class interval:	11–13	14–16	17–19	20–22	23–25
Frequency:	8	20	40	25	7

Solution:

Class interval	*Mid value (x)*	*f*	*fx*	*d = (x − x̄)*	d^2	fd^2
11–13	12	8	96	–6.09	37.1	296.7
14–16	15	20	300	–3.09	9.55	190.96
17–19	18	40	720	–.09	.0081	.324
20–22	21	25	525	2.91	8.47	211.7
23–25	24	7	168	5.91	34.92=35	244.49
		$\Sigma f = 100$	$\Sigma fx = 1809$		$\Sigma fd^2 = 944.174$	

$$\text{Mean}(\bar{x}) = \frac{\sum fx}{\sum f} = \frac{1809}{100} = 18.09$$

[or taking assumed mean 18]

$$\text{S.D.} = \sqrt{\frac{\sum fd^2}{N}} = \sqrt{\frac{944.174}{100}} = \sqrt{9.44174}$$

$$= 3.07$$

$$\text{Variance} = (\text{S.D.})^2 = 9.4249$$

$$\therefore \quad \text{Coefficient of variance} = \frac{S.D.}{A.M.} \times 100$$

$$= \frac{3.07}{18.09} \times 100$$

$$= 16.97$$

Example 18. Compute mean & S.D from the following frequency distribution.

Scores:	20–22	23–25	26–28	29–31	32–34	35–37	38–40
Frequency:	2	5	7	13	8	4	1

$N = 40$

Solution:

Score	Mid value (x)	f	fx	d = (x − x̄)	d^2	fd^2
20–22	21	2	42	−3.7	75.69	151.38
23–25	24	5	120	−5.7	32.49	162.46
26–28	27	7	189	−2.7	7.29	51.03
29–31	30	13	390	.3	.09	1.17
32–34	33	8	264	3.3	10.89	87.12
35–37	36	4	144	6.3	39.69	158.76
38–40	39	1	39	9.3	86.49	86.49
		$\sum f = 40$	$\sum fx = 1188$			$\sum fd^2 = 698.41$

$$\text{Mean} = \bar{x} = \frac{\sum fx}{\sum f} = \frac{1188}{40} = 29.7$$

[or taking assumed mean 33]

$$\text{S.D.} = \sigma = \sqrt{\frac{\sum fd^2}{N}} = \sqrt{\frac{698.41}{40}} = \sqrt{17.46} = 4.18$$

Example 19. Compute the mean & median from the following frequency distribution.

Scores:	20–24	25–29	30–34	35–39	40–44	45–49
Frequency:	7	9	12	6	4	2

$N = 40$

Solution:

Scores	Mid value (x)	Class boundaries	f	fx	cf
20–24	22	19.5–24.5	7	154	7
25–29	27	24.5–29.5	9	243	16
30–34	32	29.5–34.5	12	384	28
35–39	37	34.5–39.5	6	222	34
40–44	42	39.5–44.5	4	168	38
45–49	47	44.5–49.5	2	94	40
		$\sum f = N = 40$		$\sum fx = 1265$	

$$\text{Mean } (\bar{x}) = \frac{\sum fx}{\sum f} = \frac{1265}{40} = 31.625$$

[or taking assumed mean 32]

$$\text{Median} = L_1 + \frac{N/2 - C}{fm} \times i$$

Here $N/2 = 40/2 = 20$

$L_1 = 29.5,\ fm = 12,\ C = 16,\ i = 5$

$\therefore$ $$\text{Median} = 29.5 + \frac{20 - 16}{12} \times 5$$

$$= 29.5 + \frac{4}{12} \times 5$$

$$= 29.5 + 1.667$$

$$= 31.167$$

Example 20. Compute the mean, median & S.D. from the following distribution.

Scores:	10–19	20–29	30–39	40–49	50–59	60–69	70–79	80–89	90–99
Frequency:	2	5	3	5	8	12	25	30	10

$N = 100$

Solution:

Scores	*Mid value (x)*	*Class boundaries*	*f*	*fx*	*c.f*	*d = (x − x̄)*	*d²*	*fd²*
10–19	14.5	9.5–19.5	2	29	2	–55.8	3113.64	6227.28
20–29	24.5	19.5–29.5	5	122.5	7	–45.8	2097.64	10488.2
30–39	34.5	29.5–39.5	3	103.5	10	–35.8	1281.64	3844.92
40–49	44.5	39.5–49.5	5	222.5	15	–25.8	665.64	3328.2
50–59	54.5	49.5–59.5	8	436	23	–15.8	249.64	1997.12
60–69	64.5	59.5–69.5	12	774	35	–5.8	33.64	403.68
70–79	74.5	69.5–79.5	25	1862.5	60	4.2	17.64	441
80–89	84.5	79.5–89.5	30	2535	90	14.2	201.64	6049.2
90–99	94.5	89.5–99.5	10	945	100	24.2	585.64	5856.4
			$n = 100$	$\sum fx = 7030$				$\sum fd^2 = 38636$

$$\text{Mean } (\bar{x}) = \frac{\sum fx}{N} = \frac{7030}{100} = 70.30$$

[or taking assumed mean 74.5]

$$\text{Median} = L_1 + \frac{N/2 - C}{fm} \times i$$

$$= 69.5 + \frac{50 - 35}{25} \times 10$$

$$= 69.5 + \frac{150}{25}$$

$$\text{Median} = 69.5 + 6$$

$$= 75.5$$

$$\text{S.D.} = \sigma = \sqrt{\frac{\sum fd^2}{N}} = \sqrt{\frac{38636}{100}} = \sqrt{386.36} = 19.656$$

Example 21. Compute the mean, median & S.D. of the following distribution.

Class Interval:	40–44	45–49	50–54	55–59	60–64	65–69	70–74	75–79	80–84
Frequency:	2	4	2	6	8	10	12	20	16

$N = 80$

Solution:

Class interval	*Mid values*	*Class boundaries*	*f*	*fx*	*cf*	*d* $(x - \bar{x} = d)$	d^2	fd^2
40–44	42	39.5–44.5	2	84	2	–28	784	1568
45–49	47	44.5–49.5	4	188	6	–23	529	2116
50–54	52	49.5–54.5	2	104	8	–18	324	648
55–59	57	54.5–59.5	6	342	14	–13	169	1014
60–64	62	59.5–64.5	8	496	22	–8	64	512
65–69	67	64.5–69.5	10	670	32	–3	9	90
70–74	72	69.5–74.5	12	864	44	2	4	48
75–79	77	74.5–79.5	20	1540	64	7	49	980
80–84	82	79.5–84.5	16	1312	80	12	144	2304
			$N = 80$	$\Sigma fx = 5600$				$\Sigma fd^2 = 9280$

$$\text{Now Mean } (\bar{x}) = \frac{\sum fx}{N} = \frac{5600}{80} = 70$$

[or taking assumed mean 72]

$$\text{Median} = L + \frac{N/2 - C}{fm} \times i$$

$$= 69.5 + \frac{40 - 32}{12} \times 5$$

$$= 69.5 + \frac{8 \times 5}{12}$$

$$= 69.5 + 3.33 = 72.83$$

$$\text{S.D. } (\sigma) = \sqrt{\frac{\sum fd^2}{N}} = \sqrt{\frac{9280}{80}} = \sqrt{166} = 10.77.$$

6

CHAPTER

THEORETICAL DISTRIBUTION

A. NORMAL DISTRIBUTION:

The normal distribution was first discovered in 1733 by English mathematician **De Movire**. He obtained this continuous distribution as a limiting case of binomial distribution. Normal distribution is also known as **Gaussian distribution** named after **Karl Friedrich Gauss**, who used this normal curve to describe the theory of accidental errors of measurements involved in calculation of orbits of heavenly bodies. Now a days normal probabilities model is one of the most important probabilities models in statistical analysis.

- **Definition:** Normal distribution is a continuous probability distribution which is bell shaped, unimodal and symmetrical.
- **Normal Curve:** The normal distribution of a variable when represented graphically, takes the shape of a symmetrical curve known as **Normal Curve**. This curve is asymptotic to x-axis (base line) on either sides.

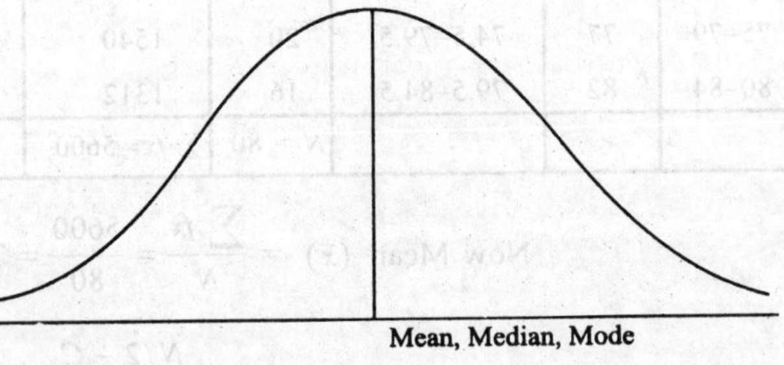

Fig. 6.1 Normal Curve

Properties of the Normal Distribution and Normal Probability Curve:

1. The normal distribution has two parameters viz. mean (μ) and standard deviation (σ).

2. The normal curve is bell shaped and symmetrical about the line $x = \mu$.

3. It has only one mode occurring at μ *i.e.*, it is unimodal.

4. The value of mean, median, mode will coincide at the centre ($x = \mu$) because the distribution is symmetrical and single peaked.

$$\textbf{Mean} = \textbf{Median} = \textbf{Mode} = \mu$$

5. The normal curve has asymptotic tails *i.e.*, progressively nearing the abscissa or x-axis.

6. The range is unlimited, infinite in both directions but as the distance from μ increases, the curve approaches the horizontal axis more and more closely and never touches the horizontal axis (X).

7. It has two points of inflections, the points where the change in curvature occurs at a distance σ on either side of mean. The points of inflection of the normal curve are at $x = \mu \pm \sigma$. The curve changes from concave to convex and vice versa.

8. The quartiles *i.e.*, the first (Q_1) and third (Q_3) are equidistant from μ.

$$Q_3 - \mu = \mu - Q_1$$

$$\text{Quartile deviation} = \frac{Q_3 - Q_1}{2} = 0.67458$$

$$Q_1 = \mu - 0.6745$$

$$Q_3 = \mu + 0.6745$$

Thus the range of the normal distribution from –0.6745 to +0.6745.

9. The normal distribution is bilaterally symmetrical so, it is free from *skewness*, its *coefficient* of *skewness* amounts to *zero*.

Skewness = 0   Kurtosis = 0

10. The maximum ordinate (y) lies at the mean *i.e.*, at $x = \mu$. It's value is $y = \frac{1}{\sigma\sqrt{2x}}$. When S.D increases the maximum ordinate decreases.

11. Since mean = median = μ, the coordinate at $x = \mu$ (or $Z = 0$) divides the whole area into two equal parts. The area right to the ordinate and left to the ordinate at $x = \mu$ (or $Z = 0$) is 0.5.

12. To fix the curve or the mathematical equation is completely determined if the mean (μ) and standard deviation (σ) of the variable is completely known.

13. Change in the value of mean (μ) and standard deviation (σ) or in both will alter the shape of the curve. But the curve will always remain in symmetrical about the maximum ordinate.

14. The total area under the normal curve is equal to unity. The maximum ordinate is at the mean (μ) and at various standard deviation (σ) the distances are in a fixed proportion to the ordinate at the mean.

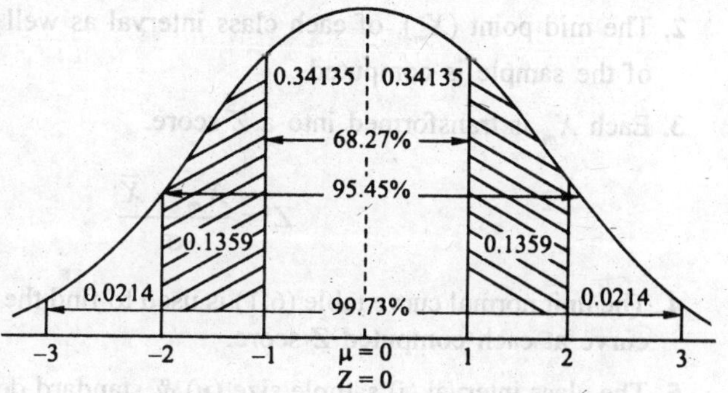

Fig. 6.2 Area under standard Normal Curve

The following table gives the area under the normal probability curve for some important value of Z.

Table 6.1

Distance from the mean ordinate in terms of ± σ	*Area under the curve*
Z = ±0.6745	50% = 0.50
Z = ±1.0	68.27% = 0.6827
Z = ±1.96	95% = 0.95
Z = ±2.00	95.45% = 0.9545
Z = ±2.58	99% = 0.99
Z = ±3.0	99.73% = 0.9973

Uses or Application of Normal Distribution:

1. It is used in sampling theory.
2. It can be used to approximate the binomial and Poisson distribution.
3. It has considerable application in statistical quality control.

4. It is widely used in testing statistical hypothesis and test significance in which it is always assumed that the population from which the samples have been drawn should have normal distribution.

5. It has many mathematical properties which make it popular and comparatively easy to manipulate for the use in social natural sciences.

- The standard normal distribution is also known as unit normal distribution or ***Z* distribution**.
- The standard normal curve helps us to find the areas within two assigned limits under the curve.

• Computation of Best Fitting Normal Distribution:

It is that normal distribution which fits best with an observed distribution and has the same mean ($\bar{X}$), the same standard deviation (σ) and the same sample size (*n*) as the latter.

1. The observed scores are arranged in a continuous frequency distribution with equal size of class intervals (*i*).

2. The mid point (X_m), of each class interval as well as mean ($\bar{X}$) and standard deviation (σ) of the sample is computed.

3. Each X_m is transformed into a *Z* score.

$$Z = \frac{X_m - \bar{X}}{\sigma}$$

[σ = sigma/S.D]

4. The unit normal curve table (6.1) is used to find the height *y* of the ordinate of the unit normal curve at each computed *Z* score.

5. The class interval (*i*) sample size (*n*) & standard deviation (σ) are each equal to 1.00 in case of the unit normal curve.

6. Using *i*, *n* & σ (sigma), the ordinate *Y* of the best fitting normal curve, corresponding to each recorded *y* of the unit normal curve is then computed.

$$Y = y\frac{in}{\sigma}$$

7. Each *Y* so computed gives the expected frequency of (*fe*) of the best fitting normal distribution for the class interval whose X_m corresponds to *Z* score for that *Y*. Thus computed value of *Y* constitute the distribution of *fe* values for the best fitting normal distribution.

8. Each *Y* score may be graphically plotted against the X_m of the corresponding class interval for drawing the best fitting normal curve.

Example 1. Compute the best fitting normal distribution for the following blood glucose concentration of 80 patients of Midnapure district hospital.

Class interval:	100–109	110–119	120–129	130–139	140–149	150–159	160–169
Frequency:	6	11	10	17	16	13	7

- Compute the mean, and the SD of the given data.

Solution:

Let us take assumed mean 134.5

Class interval	Mid value (X_m)	Frequency (f)	Assumed mean A = 134.5 $X_m - A$	$\frac{X_m - A}{C} = d$	fd	fd^2
100–109	104.5	6	104.5–134.5 = –30	–3	–18	54
110–119	114.5	11	114.5–134.5 = –20	–2	–22	44
120–129	124.5	10	124.5–134.5 = –10	–1	–10	10
130–139	134.5	17	134.5–134.5 = 0	0	0	0
140–149	144.5	16	144.5–134.5 = +10	+1	+16	16
150–159	154.5	13	154.5–134.5 = +20	+2	+26	52
160–169	164.5	7	164.5–134.5 = +30	+3	+21	63
		$\Sigma f = 80$			$\Sigma fd = 63 - 50 = 13$	239

$$\text{Mean} = A + \frac{\sum fd}{N} \times i$$

$$X = 134.5 + \frac{13}{80} \times 10 = 134.5 + 1.625 \qquad N = 80$$

$$i = 10$$

$$= 136.125 \qquad \sum fd = 13$$

$$= 136.1 \qquad \sum fd^2 = 239$$

$$\text{Standard deviation (S.D)} = \sqrt{\frac{\sum fd^2}{N} - \left(\frac{\sum fd}{N}\right)^2} \times i$$

$$= \sqrt{\frac{239}{80} - \left(\frac{13}{80}\right)^2} \times i$$

$$= \sqrt{2.9875 - (1.63)^2} \times 10 = \sqrt{2.988 - 0.02656} \times 10$$

$$= 10\sqrt{2.988} - 0.0266 = 10\sqrt{2.96} = 10 \times 1.72 = 17.2$$

(*i*) The deviation of each X_m from $\bar{X}$ is then transformed into *Z* score which entered in the following table.

- **Computation of the best-fitting normal distribution for blood sugar data.**

Class interval	Frequency	Mid value	$X_m - \bar{X}$	$Z = \frac{X_m - \bar{X}}{\sigma(sd)}$	y	Y
100–109	6	104.5	104.5–136.1 = –31.6	$\frac{-31.6}{17.2} = -1.84$	0.0734	3.4
110–119	11	114.5	114.5–136.1 = –21.6	$\frac{-21.6}{17.2} = -1.26$	0.1804	8.4
120–129	10	124.5	124.5–136.1 = –11.6	$\frac{-11.6}{17.2} = -0.67$	0.3187	14.8
130–139	17	134.5	134.5–136.1 = –1.6	$\frac{-1.6}{17.2} = -0.09$	0.3973	18.5

Class interval	Frequency	Mid value	$X_m - \bar{X}$	$Z = \frac{X_m - \bar{X}}{\sigma(sd)}$	y	Y
140–149	16	144.5	144.5–136.1 = +8.4	$\frac{8.4}{17.2} = 0.49$	0.3538	16.5
150–159	13	154.5	154.5–136.1 = +18.4	$\frac{18.4}{17.2} = 1.07$	0.2251	10.5
160–169	7	164.5	164.5–136.1 = +28.4	$\frac{28.4}{17.2} = 1.65$	0.1023	4.8
	$\sum f = 80$					

(*ii*) Neglecting the algebraic sign of the computed Z score, the height y of the ordinate at each Z score is then recorded from unit normal curve table (6.1).

Example: Z score of –1.26, $y = 0.1804$

(*iii*) The height Y of the ordinate of the best fitting normal distribution is computed for each Z score by multiplying its y score with $\frac{in}{\sigma}$ (i = class interval, n = total frequency, σ = Standard deviation) for example

$$Y = y\frac{in}{\sigma} = 0.1804 \times \frac{10 \times 80}{17.2} = 0.1804 \times 46.51 = 8.39 = 8.4$$

Example 2. Workout the best fitting normal distribution for the following observed frequency distribution of body weight scores (kg) of a 120 people of Arambagh subdivisional town (Hooghly).

Class interval:	41–47	48–54	55–61	62–68	69–75	76–82	83–89
Frequency:	5	15	25	43	21	10	1

Compute the mean & SD of the given data.

Solution: Let us take assumed mean 65

Class interval	Mid value (X_m)	Frequency (f)	Assumed mean A = 4.5 X_m – A	$\frac{X_m - A}{i} = d$	fd	fd^2
41–47	44	5	44–65 = –21	$\frac{-21}{7} = -3$	–15	45
48–54	51	15	51–65 = –14	$\frac{-14}{7} = -2$	–30	60
55–61	58	25	58–65 = –7	$\frac{-7}{7} = -1$	–25	25
62–68	65	43	65–65 = 0	0	0	0
69–75	72	21	72–65 = +7	$\frac{7}{7} = +1$	+21	21
76–82	79	10	79–65 = +14	$\frac{14}{7} = +2$	+20	40
83–89	86	1	86–65 = +21	$\frac{21}{7} = +3$	+3	9
		$\sum f = 120$			$\sum fd$ = +44 – 70 = –26	$\sum fd^2 = 200$

$N = 120 \quad \sum fd = -26 \quad \sum fd^2 = 200 \quad i = 7$

$$\overline{X} = 65 + \frac{-26}{120} \times 7$$

$$= 65 - 1.5 = 63.5$$

$$\text{Standard deviation } (\sigma) = \sqrt{\frac{\sum fd^2}{N} - \left(\frac{\sum fd}{N}\right)^2} \times i$$

$$= \sqrt{\frac{200}{120} - \left(\frac{-26}{120}\right)^2} \times 7 = \sqrt{1.67 - (-0.22)^2} \times 7$$

$$= 7\sqrt{1.67 - 0.048} = 7\sqrt{0.952} = 7 \times 0.98 = 6.86 = 6.8$$

1. The deviation of X_m from $\overline{X}$ is then transformed into Z score, which entered in the following table.

Computation of the best-fitting normal distribution of body weight.

Class interval	*Mid value*	*Frequency*	$X_m - \overline{X}$	$Z = \frac{X_m - \overline{X}}{\sigma}$	*y*	*Y*
41–47	44	5	44–63.5 = –19.5	$\frac{-19.5}{6.8} = -2.87$	0.0065	0.8
48–54	51	15	51–63.5 = –12.5	$\frac{-12.5}{6.8} = -1.84$	0.0734	9.0
55–61	58	25	58–63.5 = –5.5	$\frac{-5.5}{6.8} = -0.80$	0.2897	35.8
62–68	65	43	65–63.5 = 1.5	$\frac{1.5}{6.8} = +0.22$	0.3894	48.0
69–75	72	21	72–63.5 = 8.5	$\frac{8.5}{6.8} = +1.25$	0.1826	22.6
76–82	79	10	79–63.5 = 15.5	$\frac{15.5}{6.8} = +2.28$	0.0297	3.7
83–89	86	1	86–63.5 = 22.5	$\frac{22.5}{6.8} = +3.30$	0.0017	0.2
		$\sum f = 120$				

2. Neglecting the algebraic sign of the computed Z score the height of y of the ordinate at each Z score is then recorded from unit normal curve table (6.1).

3. The height Y of the ordinate of the best fitting normal distribution is computed for each Z score by multiplayer its y score with $\frac{in}{\sigma}$. For example;

$$\frac{in}{\sigma} = \frac{7 \times 120}{6.8} = 123.5$$

$$Y = y\frac{in}{\sigma} = 0.0065 \times 123.5 = 0.80275 = 0.8$$

Example 3. Work out the best fitting normal distribution for body weight (scores in kg) of 255 people (adult & child) of a tribal community of Ranchi.

Class interval:	2–6	7–11	12–16	17–21	22–26	27–31	32–36	37–41	42–46	47–51	52–56
Body weight:	1	11	33	53	63	54	21	6	6	5	2

Solution: We have to compute the mean $\overline{X}$ and SD (σ) of the above sample (population). Computing the mean & the SD of the given data.

Let us take assumed mean 29.

Class interval	*Mid value* (X_m)	*Frequency* (*f*)	*Assumed mean* $A = 29,\ X_m - A$	$\frac{Xm - A}{c} = d$	*fd*	fd^2
2–6	4	1	4–29 = –25	–5	–5	25
7–11	9	11	9–29 = –20	–4	–44	176
12–16	14	33	14–29 = –15	–3	–99	297
17–21	19	53	19–29 = –10	–2	–106	212
22–26	24	63	24–29 = –5	–1	–63	63
27–31	29	54	29–29 = 0	0	0	0
32–36	34	21	34–29 = +5	+1	21	21
37–41	39	6	39–29 =+10	+2	12	24
42–46	44	6	44–29 = +15	+3	18	54
47–51	49	5	49–29 = +20	+4	20	80
52–56	54	2	54–29 = +25	+5	10	50
		$\Sigma f = 255$			$\Sigma fd = -317 + 81 = -236$	1002

$\sum f = N = 255,\qquad i = 5\qquad \sum fd = -236\qquad \sum fd^2 = 1002$

(*i*) Mean $(\overline{X}) = 29 - \dfrac{236}{255} \times 5$

$$= 29 - \frac{1180}{255} = 29 - 4.6 = 24.4$$

$$\text{Standard deviation } (\sigma) = \sqrt{\frac{1002}{255} - \left(\frac{-236}{255}\right)^2} \times 5$$

$$= \sqrt{3.929 - (-0.925)^2} \times 5 = 5\sqrt{3.929 - 0.855} = 5\sqrt{3.074}$$

$$= 5 \times 1.75 = 8.75$$

(*ii*) The deviation between each X_m and $\overline{X}$ is then transformed into Z score which is entered in the table. (Table 6.1).

Example: Class interval 47–51 $\quad x_m = \dfrac{47 + 51}{2} = 49$

$$X_m - \overline{X} = 49 - 24.4 = +14.6$$

$$Z = \frac{14.6}{8.75} = 1.66$$

(*iii*) Neglecting the algebraic sign of computed Z score the height of the y of the ordinate at each Z score is then recorded from the unit normal curve table.

Example: For the Z score of 1.66, $y = 0.4515$.

(*iv*) The height Y of the ordinate of the best fitting normal distribution is computed for each Z score by multiplying it y score with in/σ.

For example: $i = 5$, $n = 255$, $\sigma = 8.75$

$$\therefore \frac{i \times n}{\sigma} = \frac{5 \times 255}{8.75} = \frac{1275}{8} = 145.7$$

$$y = 0.4515$$

$$Y = y \times 145.7 = 0.4515 \times 145.7 = 65.78 = 65.8$$

***Computation* of the best fitting normal *distribution*:**

Class interval	Mid value X_m	Frequency (f)	$X_m - \bar{X}$ (24.4)	$Z = \frac{X_m - \bar{X}}{\sigma}$	y	Y (fe)
2–6	4	1	4 – 24.4 = –20.4	$\frac{-20.4}{8.75} = -2.33$	.0264	3.8
7–11	9	11	9 – 24.4 = –15.4	$\frac{-15.4}{8.75} = -1.76$	.0848	12.4
12–16	14	33	14 – 24.4 = –10.4	$\frac{-10.4}{8.75} = -1.18$	.1989	28.9
17–21	19	53	19 – 24.4 = –5.4	$\frac{-5.4}{8.75} = -0.62$	.3292	47.9
22–26	24	63	24 – 24.4 = –0.4	$\frac{0.4}{8.75} = -0.05$	.3984	58.0
27–31	29	54	29 – 24.4 = +4.6	$\frac{4.6}{8.75} = +0.53$	.3467	50.5
32–36	34	21	34 – 24.4 = +9.6	$\frac{9.6}{8.75} = +1.09$	2203	32.0
37–41	39	6	39 – 24.4 = +14.6	$\frac{14.6}{8.75} = +1.66$	.1006	14.7
42–46	44	6	44 – 24.4 = +19.6	$\frac{19.6}{8.75} = +2.24$	.0325	4.7
47–51	49	5	49 – 24.4 = +24.6	$\frac{24.6}{8.75} = +2.81$	.0077	1.1
52–56	54	2	54 – 24.4 = +29.6	$\frac{29.6}{8.75} = +3.38$	.0013	0.2

***Comparison* of *normal & t distribution*:**

Normal distribution	t distribution
(*i*) Theoretical & continuous probabilities distribution.	(*i*) Theoretical & continuous probabilities distribution.
(*ii*) Unimodal & bilaterally symmetrical.	(*ii*) Unimodal & bilaterally symmetrical.
(*iii*) No Skewness.	(*iii*) Non-skewed
(*iv*) Mean of the distribution amounts to 0.00 which is the Z score for μ.	(*iv*) Mean of the distribution amounts to 0.00 which is the t score for μ.

B. PERMUTATION & COMBINATION:

Before explaining the theory of probability, it is essential to understand the concept of permutation and combination. In simple language permutation means *arrangement* and combination means groups or *selection*.

Permutation: Each of the different arrangements which can be made by taking some or all of a given number of things at a time is called a permutation.

The number of permutation of n things taken r at a time is denoted by $^n p_r$ [P = Permutation].

Combination: Each of the different groups (or selection) which can be made by taking some or all of a given number of things at a time (irrespective of the order) is called a combination.

The number of combination of n things taken r at a time is meant the number of groups of r things which can be formed from n things and is denoted by $^n c_r$ [c = combination].

Difference between a Permutation and a Combination:

Permutation	Combination
(*i*) It is an arrangement of objects in a definite order with reference to its placement.	(*i*) It is an arrangement of objects irrespective of the order.
(*ii*) Concerns selection as well as order.	(*ii*) Concerns only the selection.
(*iii*) Ordering is essential $^n p_r = \frac{\lfloor n}{\lfloor n-r}$.	(*iii*) Ordering of the selecting object is immeterial $^n c_r = \frac{\lfloor n}{\lfloor (n-r) \lfloor r}$.

Comparison:

Number of object (n)	Taking at a time (r)	Combination (C)	Permutation (P)
AB	2	AB	AB, BA
ABC	3	ABC	ABC, ACB, BCA, BAC, CAB, CBA

Example 1. In how many ways can 5 PG students of zoology occupy 3 vacant seats?

Solution: Total number of ways

$$^N p_r = {^5 p_3} = \frac{\lfloor 5}{\lfloor 3 \lfloor 5-3} = \frac{\lfloor 5}{\lfloor 2}$$

$$= \frac{\lfloor 5}{\lfloor 2} = 5.4.3 = 60$$

Example 2. In a family 4 brothers and 3 sisters are to be seated in one row.

(*a*) In how many ways can they be seated if all the sisters sit together?

(*b*) In how many ways can they be seated if all the sisters can not sit together?

Solution:

(*a*) (*i*) Taking 3 sisters as a unit, we have to arrange 4 + 1 *i.e.*, 5 different units taking 5 at a time in a row.

(*ii*) This is done by $^5 p_5 = 5 \times 4 \times 3 \times 2 \times 1 = 120$ ways.

(*iii*) In each of 120 ways 3 sister can be arranged among themselves $^3 p_3 = 3 \times 2 \times 1 = 6$ ways. Hence the required number of ways = 6 × 120 = 720.

(b) (i) Sisters can not sit together. Therefore they are placed between the two brothers or they may be placed at two ends of the row.

(ii) Place in between two brothers is one & among 4 brothers it will be 3 and 2 place at two ends. Total number of place 3 + 2 = 5. Sisters can be arranged in $^5p_3$ ways = 5 × 4 × 3 = 60 ways.

(iii) 4 Brothers are arranged in $\lfloor 4$ ways = 4 × 3 × 2 × 1 = 24.

Hence required number of ways = 60 × 24 = 1440.

Example 3. In how many ways can 7 students of microbiology of R. K. Mission Vidyamandira be seated at a round table?

Solution:

(i) 7 students occupy their seats at a round table; so we shall have to arrange the 7 students with respect to the table and not with respect to themselves.

(ii) Hence the arrangement is equivalent to the linear permutation.

∴ The required number of permutation = Number of arrangements of 7 students in 7 places

i.e., $\lfloor 7$ = 7.6.5.4.3.2.1 = 5040.

Example 4. In how many ways 6 students of Genetics form a ring?

Solution:

(i) In this case, we are concerned with the relative position of the students. Therefore we fix the position of one students and arrange the other students among themselves.

(ii) The remaining (6–1) = 5 students can be arranged altogether in $\lfloor 5$ ways.

(iii) Hence the required number of ways to form a ring = 5.4.3.2 = 120.

Example 5. In how many ways 6 boys and 4 girls be arranged in a straight line so that no two girls are ever together?

Solution:

(i) First we fix the position of 6 boys. They are arranged in $\lfloor 6$ ways = 6 × 5 × 4 × 3 × 2 = 720.

(ii) 4 girls are placed in between the boys *i.e.*, one girl is in between two boys. Among 6 boys it will 5 places & two sides *i.e.*, 5 + 2 = 7 places.

(iii) Girls are arranged $^7p_4$ = 7 × 6 × 5 × 4 = 840 ∴ Required total ways = 840 × 720 = 604800.

Example 6. There are 6 books on Zoology, 3 on Botany and 2 on Biochemistry. In how many ways can there be placed on a shelf if the books on the same subject are to be together?

Solution:

(i) Since the books on the same subject are to be together, let us consider the 6 books Zoology as 1 unit, 3 books on botany as another unit and the 2 books on Biochemistry as a different unit.

(ii) Therefore we have to arrange 3 different units. This can be done $^3p_3$ ways 3 × 2 × 1 = 6 ways.

(iii) But the 6 books on Zoology remaining together can be arranged among themselves in $^6p_6$ ways *i.e.*, 6.5.4.3.2.1 = 720 ways.

(iv) Similarly 3 books on Botany can be arranged among themselves in $^3p_3$ ways 3 × 2 × 1 = 6 ways.

(*v*) Another 2 books of Biochemistry can be arranged among themselves in $^2p_2$ ways 2×1 = 2 ways.

Hence the required number of ways $6 \times 720 \times 6 \times 2 = 51,840$ ways.

Example 7. In how many ways can 6 boys and 4 girls be arranged in a round table so that two girls are never together?

Solution:

(*i*) Let the boys first take up their seats. They can sit in $\lfloor 6-1 = \lfloor 5$ ways $= 5 \times 4 \times 3 \times 2 = 120$ ways.

(*ii*) When they have been seated there remain 6 places for the girls each between two boys.

(*iii*) Therefore the girls can sit in $^6p_4$ ways $= 6 \times 5 \times 4 \times 3 = 360$ ways.

Hence the required number of ways $= 360 \times 120 = 43200$ ways.

Example 8. In how many ways can 4 students be selected from among 9 students of PG studies in Zoology of Serampore college. How many times will a particular student be always selected?

Solution:

(*a*) The number of ways in which 4 students can be selected among 9 students.

$$^9c_4 = \frac{9 \times 8 \times 7 \times 6}{4 \times 3 \times 2 \times 1} = 126 \text{ ways.}$$

(*b*) The number of ways in which a particular student is always be selected *i.e.*, 4–1 = 3 & total students = 9–1 = 8.

$$^8c_3 = \frac{8 \times 7 \times 6}{3 \times 2 \times 1} = 56 \text{ ways.}$$

Example 9. Prama has 6 friends. In how many ways may she invite one or more of them in her birthday celebration?

Solution: Prama may invite one, two, three .... six of her friends at a time in her birthday celebration.

∴ The required number of ways

$$= {}^6c_1 + {}^6c_2 + {}^6c_3 + {}^6c_4 + {}^6c_5 + {}^6c_6$$

$$= \frac{6\lfloor 5}{1.\lfloor 5} + \frac{6.5\lfloor 4}{2.\lfloor 4} + \frac{6.5.4\lfloor 3}{3.2\lfloor 3} + \frac{6.5.\lfloor 4}{2.\lfloor 4} + \frac{6.\lfloor 5}{\lfloor 5} + \frac{6}{\lfloor 6.\lfloor 0}$$

$$= 6 + 15 + 20 + 15 + 6 + 1$$

$$= 63$$

or $2^6 - 1 = 64 - 1 = 63$ ways.

Example 10. A question paper of Biostatistics contains 5 questions each question has alternative question. In how many ways a student can select one or more questions?

Solution:

(*i*) One or more questions may be selected. One student may take three type decisions for each question. (*a*) Selection of question (*b*) Selection of alternative question (*c*) Without selection of the question.

(*ii*) Therefore selections for 2 questions are $3 \times 3 = 3^2$ ways. In case of five questions it will be 3^5 ways.

(*iii*) But he/she may not select question. Therefore this decision should be subtracted from the total selection of questions. The required number of selection = $3^5 – 1$ = (3 × 3 × 3 × 3 × 3–1) = 243–3 = 242.

Example 11. In how many ways can 10 examination papers be arranged so that the best and the worst papers may never come together?

Solution:

(*i*) Clearly 10 examination papers can be arranged among themselves in $\underline{|10}$ ways.

(*ii*) Now considering the best and worst papers as a single examination papers we get all in 10–1 = 9 papers. Therefore these 9 examination papers can be arranged among themselves in $\underline{|9}$ ways and for each such arrangement the best and worst papers can arrange among themselves in $\underline{|2}$ ways.

(*iii*) Hence the number of arrangement of 10 examination papers in which the best and worst papers always come together $= \underline{|9} \times \underline{|2} = 2\underline{|9}$.

(*iv*) Therefore the required number of arrangement in which the best & worst papers never come together.

$$\underline{|10} - 2\underline{|9} = 10\underline{|9} \quad 2\underline{|9} = 8\underline{|9} \text{ ways.}$$

C. BINOMIAL DISTRIBUTION:

This distribution was discovered by Swiss mathematician **James Bernoulli** (1654–1705). Binomial distribution is derived from a process known as *Bernoulli trial.*

It is a discrete probability distribution which is obtained when the probability p of the happening of an event is same in all the trials and there are only two events in each trial.

A binomial probability distribution is a distribution of probabilities of random occurrences of different combinations of cases from the two classes of dichotomous variable in a sample drawn from such population.

It is a theoretical probability distribution because it can be worked out theoretically using the series of terms of the binomial equation.

• Conditions under which Binomial Distribution is used:

1. The variable should be discrete *i.e.*, the values of X should be 1, 2, 3, 4 or 5 etc. and never 1.5, 2.1, 3.47 etc.
2. A dichotomy exist *i.e.*, the happening of an event has two possible outcomes *i.e.*, either success or failure.
3. The number of trials 'n' should be finite & small.
4. The trials or events must be independent *i.e.*, happening of one event or trial must not affect the happening of other events.
5. The trials or events must be repeated under identical conditions.

• Properties of Binomial Distribution:

1. It is a discrete probability distribution.
2. Binomial distribution has two parameters p or q, the probability of success or failure and n the number of trials. The parameter n is always integer.
3. The mean (μ), SD (σ). variance (σ^2) and the coefficient of dispersion (C.D) of a binomial

distribution of cases of the class having the proportion p in the population are obtained from the sample size (n) and the proportions (p and q) of the cases in the two classes.

- Mean $(\mu) = np$
- Standard deviation $(\sigma) = \sqrt{npq}$
- Variance $(\sigma^2) = npq$
- Coefficient of dispersion (CD) $= \dfrac{npq}{np} = q$

4. Skewness and kurtosis of binomial distribution depends on the proportion of *p and q* in the population.

   (*a*) Skewness $(\gamma_1) = \dfrac{(q-p)^2}{npq}$

   (*b*) Kurtosis $(\gamma_2) = 3 + \dfrac{1-6\,pq}{npq}$

5. Binomial distribution is symmetrical of $p = q = 0.5$.
6. It is positively skewed if $p < 0.5$ and it is negatively skewed if $p > 0.5$.
7. The binomial coefficients are given by the **Pascal's triangle**.

• Computation of Binomial probabilities:

A. Using Binomial Expansion:

n = total number of events/trials.

p = probabilities of occurrence of success.

q = probabilities of occurrence of failure.

$p + q = 1$ or $q = 1 - p$

The general expression of Binomial expansion

$$(p+q)^n = p^n + np^{n-1}q + \frac{n(n-1)}{1\times 2}p^{n-2}q^2 + \frac{n(n-1)(n-2)}{1\times 2\times 3}p^{2-3}q^3 + \ldots + \frac{n(n-1)(n-2)\times 3\times 2}{1\times 2\times 3\times \ldots\ldots \times (n-1)}pq^{n-1} + q^n$$

The probabilities distribution of binomial expansion for r success in n trials is given by

$$P(r) = {}^n c_r\, q^{n-r}\, p^r$$

It has mean $= np$ and variance $= npq$.

B. Using Bernoulli expansion:

n = total number of events

p = classes

q = classes

X = number of cases in p classes

$n\text{–}X$ = number of cases in q classes

Probability of $P(X)$ is expressed by Bernoulli expansion

$$P(X) = \frac{\underline{|n}\, p^x q^{n-x}}{\underline{|X}\,\underline{|n-X}} = \frac{[n(n-1)(n-2)\ldots..\times 3\times 2\times 1]p^x q^{n-x}}{[X(X-1)(X-2)\ldots..\times 2\times 1][n-X\ldots..\times 2\times 1]}$$

C. Single Terms of the expansion:

(*i*) The coefficients of the binomial expansion represent the number of ways in which the conditions of each term may be satisfied.

(*ii*) The number of combination (*C*) of *n* different things taken *k* at a time is expressed by

$$n^{C}k = \frac{\lfloor n}{\lfloor k \lfloor n-k}$$

D. Multimonial Distribution:

(*i*) The binomial distribution may be regarded as generalization of Binomial distribution for accomodating any number of variables.

(*ii*) When there are more than two mutually exclusive outcomes of a trial, the observation leads to **multimonial distribution.**

(*iii*) Suppose $E_1, E_2 \ldots\ldots E_k$ are *k* mutually exclusive and exhaustive outcomes of an event/trial with respective probabilities $p_1, p_2 \ldots\ldots p_k$ will occur $k_1, k_2 \ldots\ldots k_n$ times respectively is

$$C = \frac{\lfloor N}{\lfloor k_1 \lfloor k_2 \ldots \lfloor k_n} p_1^{k_1} p_2^{k_2} \ldots p_n^{k_n}$$

Where $k_1 + k_2 + \ldots k_n = N$

C = the number of permutation of the events $E_1, E_2 \ldots E_k$.

• Mean, variance and standard deviation of Binomial Distribution:

(*i*) The probability distribution of binomial distribution for *r* success in *n* events/trials is given by

$$p(r) = {}^{n}C_r \, q^{n-r} \, p^r$$

(*ii*) It has $mean\,(\mu) = np$

$Variance\,(\sigma^2) = npq$

Standard deviation $(\sigma) = \sqrt{npq}$

n = number of independent events
p = probability of success
q = probability of failures

• Expression of Binomial Theorem:

The expression of binomial theorem is $(p + q)^n = 1$ where *p* and *q* are respective probabilities of two alternative outcomes and *n* is the number of trials. The binomial equation can be expanded for binomial expressions as follows:

n	*Binomial equation*	*Expanded binomial expression*
1	$(p + q)^1$	$p + q$
2	$(p + q)^2$	$p^2 + 2pq + q^2$
3	$(p + q)^3$	$p^3 + 3p^2q + 3pq^2 + q^3$
4	$(p + q)^4$	$p^4 + 4p^3q + 6p^2q^2 + 4pq^3 + q^4$
5	$(p + q)^5$	$p^5 + 5p^4q + 10p^3q^2 + 10p^2q^3 + 5pq^4 + q^5$
6	$(p + q)^6$	$p^6 + 6p^5q + 15p^4q^2 + 20p^3q^3 + 15p^2q^4 + 6pq^5 + q^6$
7	$(p + q)^7$	$p^7 + 7p^6q + 21p^5q^2 + 35p^4q^3 + 35p^3q^4 + 21p^4q^5 + 7pq^6 + q^7$

4 $(q + p)^n \times q + p = 1$

- ***Short cut method:***

(*i*) A simple formula based on factorial (!) binomial theorem can be used to calculate the probability by short cut method.

(*ii*) If the probability of an event (*X*) is p and an alternative (*Y*) is *q*, then probability in *n* trials that event *X* will occur s times and *Y* will occur *t* times is

$$P = \frac{\underline{|n}}{\underline{|s}\,\underline{|t}}\, p^s q^t$$

(*iii*) n = total number of events

s = the number of times outcome (*X*) occurs

t = the number of times outcome (*Y*) occurs

Therefore $n = s + t$

- ***Pascal's Triangle:***

The expression of binomial theorem is $(a + b)^n = 1$. For each value of n, the binomial expression can be expanded.

Pascal's triangle is a triangular arrays of numbers made up of the coefficients of the binomial expansion.

The coefficients denote the number of ways by which a particular combination of events can occur.

The first and last number in each row equals 1 and the others equal the sum of the adjacent numbers in the row immediately above.

Sl. No.	*Number of sample n*	*Binomial coefficients in the expansion $(p + q)^n$*	*Sum*
1	$n = 0\ (p + q)^0$	1	1
2	$n = 1\ (p + q)^1$	1 1	2
3	$n = 2\ (p + q)^2$	1 2 1	4
4	$n = 3\ (p + q)^3$	1 3 3 1	8
5	$n = 4\ (p + q)^4$	1 4 6 4 1	16
6	$n = 5\ (p + q)^5$	1 5 10 10 5 1	32
7	$n = 6\ (p + q)^6$	1 6 15 20 15 6 1	64
8	$n = 7\ (p + q)^7$	1 7 21 35 35 21 7 1	128

* All numbers other than 1's are equal to the sum of the two numbers directly above them.

When $n = 5$ the 4th term in this line is 10 which is obtained by adding together the 3rd term and 4th term of the line $n = 4$ (*i.e.*, 10 = 6 + 4).

Example 1. Calculate the probability of getting head three times when a coin is tossed 5 times.

Solution:

(*i*) When a coin is tossed, there are only two outcomes either head or tail.

(*ii*) Therefore it shows binomial distribution. If the probability of getting head is *p* and of tail is *q*. The probability of getting head 3 times can be calculated in the following way.

$$n = 5,\ X = 3,\ p = \frac{1}{2} = 0.5,\ q = \frac{1}{2} = 0.5$$

$$P(X) = \frac{\lfloor n\, p^X q^{n-X}}{\lfloor X \lfloor n-X} = \frac{\lfloor 5 (0.5)^3 (0.5)^{5-3}}{\lfloor 3 \lfloor 5-3} = \frac{5.4.3.2 \times (0.5)^3 \times (0.5)^2}{3.2.2} = 5 \times (0.5)^5$$

$$= 10 \times 0.5 \times 0.5 \times 0.5 \times 0.5 \times 0.5$$

$$= 5 \times 0.03125$$

$$= 0.3125$$

Example 2. Calculate the probability of having two male and female children in a family of four children by applying binomial theorem.

Solution: Total number of children = 2 + 2 = 4

Male = 2 and female = 2

(*i*) Initial probabilities of male child (p) $= \frac{2}{4} = \frac{1}{2} = 0.5$

(*ii*) Initial probabilities of female child (q) $= \frac{2}{4} = \frac{1}{2} = 0.5$

According to binomial equation:

$$(p + q)^4 = p^4 + 4p^3q + 6p^2q^2 + 4pq^3 + q^4$$

$$p = 6p^2q^2$$

$$= 6.(0.5)^2(0.5)^2$$

$$= 6 \times 0.5 \times 0.5 \times 0.5 \times 0.5$$

$$= 6 \times 0.0625 = 0.375$$

Thus the probabilities in a families of four children having two boys & two girls is 0.375.

Example 4. Rayna has three children, what is the probability that two are boys and one is a girl?

Solution:

(*i*) For any one child probability of boy is $\frac{1}{2}$ and girl is also $\frac{1}{2}$.

(*ii*) Each child is produced independently. Thus the probabilities of 2 boys and 1 girl is as follows.

(*iii*) Total number of children $n = 2 + 1 = 3$

p (boy) $= \frac{1}{2}$, q (girl) $= \frac{1}{2}$

$s = 2$, $t = 1$

$$p = \frac{\lfloor n}{\lfloor s \lfloor t} \times (p)^s (q)^t$$

$$= \frac{\lfloor 3}{\lfloor 2 \lfloor 1} \times \left(\frac{1}{2}\right)^2 \left(\frac{1}{2}\right)^1 = \frac{3.\lfloor 2}{1.\lfloor 2} \times \frac{1}{4} \times \frac{1}{2} = \frac{3}{8}$$

$$(p + q)^3 = p^3 + 3p^2q + 3pq^2 + q^3$$

$$p = 3p^2q = 3.\left(\frac{1}{2}\right)^2 \frac{1}{2} = 3\frac{1}{4}.\frac{1}{2} = \frac{3}{8}$$

Example 5. What are the probabilities that a family (parents are heterozygous) of five will consist of blue eyed and 2 brown eyed children (in any order) (Blue is dominant over brown).

Solution: Total members $(n) = 3 + 2 = 5$

Probabilities of blue eyes $(p) = \frac{3}{4}$, brown eyes $(q) = \frac{1}{4}$

Blue eyes $(s) = 3$, brown eyes$(t) = 2$

$$p = \frac{\lfloor n}{\lfloor s \lfloor t} \times p^s q^t = \frac{\lfloor 5}{\lfloor 3 \lfloor 2} \times \left(\frac{3}{4}\right)^3 \times \left(\frac{1}{4}\right)^2$$

$$= \frac{5.4\lfloor 3}{2.1\lfloor 3} \times \frac{3}{4} \times \frac{3}{4} \times \frac{3}{4} \times \frac{1}{4} \times \frac{1}{4}$$

$$= \frac{10 \times 27}{64} \times \frac{1}{16} = \frac{270}{1024} = 0.2636 = 0.264$$

Example 6. What is the probability of having (*a*) three boys and one girl in a family of four children or (*b*) three girls and one boy in a family of 4 children? Apply binomial theorem?

Solution:

(*a*) Total members $(N) = 4$ Boys $(p) = 3$ Girl $(q) = 1$

Probabilities $(p) = \frac{1}{2} = 0.5$ $(q) = \frac{1}{2} = 0.5$

$$(p + q)^4 = p^4 + 4p^3q + 6p^2q^2 + 4pq^3 + q^4$$

As three girls & one boy, therefore probabilities

$$p = 4p^3q$$
$$= 4.(0.5)^2 \times 0.5 = 4 \times 0.5 \times 0.5 \times 0.5 \times 0.5$$
$$= 4 \times 0.0625 = 0.25$$

(*b*) Three girls & one boy

$$q = 4.pq^3$$
$$= 4 \times 0.5 \times 0.5 \times 0.5 \times 0.5$$
$$= 4 \times 0.0625 = 0.25$$

Example 7. A normal woman whose father is haemophilic marries a normal man and they have four children 2 boys and 2 girls. What is the probability that

(*a*) All four of the children will be haemophilic?

(*b*) None of the children will be haemophilic?

(*c*) Exactly two children will be haemophilic?

(*d*) Only one of the children will be haemophilic.

Solution:

(*i*) Normal woman is carrier because her father is haemophilic so her genotype is $X_c^+ X_c$.

(*ii*) She marries normal man *i.e.*, $X_c^+ Y$

$$X_c^+ X_c \times X_c^+ Y$$

$$F_1 = X_c^+ X_c , X_c^+ Y , X_c^+ X_c , X_c Y$$

Normal woman ($X_c^+ X_c^+$) & carrier woman ($X_c^+ X_c$)

Normal male $X_c^+ Y$

Haemophilic male $X_c Y$

Genotype	*Probabilities*
$X_c^+ X_c^+$	$\frac{1}{4}$
$X_c^+ X_c$	$\frac{1}{4}$
$X_c^+ Y$	$\frac{1}{4}$
$X_c Y$	$\frac{1}{4}$

(*a*) It is not possible that all four would have haemophilic.

(*b*) The probabilities that male child not having haemophilia $\frac{1}{2}$ & female child have haemophilia *i.e.*, $\frac{1}{2} \times \frac{1}{2} = \frac{1}{4}$ that none of the children are haemophilic.

(*c*) Two children will be haemophilic (*c*) $\frac{1}{4} \times \frac{1}{4} = \frac{1}{16}$

(*d*) Considering on the males, the probability, $\frac{1}{2} \times \frac{1}{2} = \frac{1}{4}$ for neither, therefore the remainder would be probability for one child is with haemophilia *i.e.*,

$$1 - \frac{1}{4} - \frac{1}{4} = 1 - \left(\frac{1+1}{4}\right) = 1 - \frac{2}{4}$$

$$= 1 - \frac{1}{2} = \frac{1}{2}$$

Example 8. If two parents, both heterozygous carriers of the autosomal recessive gene causing cystic fibrosis have five children. What is the probability that exactly three will be normal?

Solution: The probability of having a normal child during each pregnancy is

$$P_A = \text{normal} = \frac{3}{4}$$

The probability of having an affected offspring is

$$P_a = \text{(affected)} = \frac{1}{4}$$

$N = 5$, s (normal) $= 3$, t (affected) $= 2$

$$p = \frac{\lfloor n}{\lfloor s \lfloor t} p^s q^t = \frac{\lfloor 5}{\lfloor 3 \lfloor 2} p^3 q^2 = \frac{\lfloor 5}{\lfloor 3 \lfloor 2}\left(\frac{3}{4}\right)^3\left(\frac{1}{4}\right)^2$$

$$= \frac{5.4.3.2}{3.2.2}\left(\frac{3}{4}\right)^3\left(\frac{1}{4}\right)^2$$

$$= 10 \times \frac{27}{64} \times \frac{1}{16} = \frac{270}{1024} = 0.26$$

Example 9. Phenylketonuria, a metabolic disease in humans is caused by a recessive allele *p*. If two heterozygous carriers of the allele marry and plan a family of five children, (*a*) what is the chance that all their children will be unaffected? (*b*) What is the chance that four children will be unaffected and one affected with phenylketonuria? (*c*) What is the chance that at least three children will be unaffected? (*d*) What is the chance that the first child will be an unaffected girl?

Solution:

(*i*) Mating between two heterozygous result the probability of unaffected child is $\frac{3}{4}$ and affected child is $\frac{1}{4}$.

(*ii*) The probabilities for any one child born *i.e.*, boy $= \frac{1}{2} = 0.5$ and girl $= \frac{1}{2} = 0.5$.

(*a*) Calculation for that five children will be unaffected. For each child the probabilities of unaffected child is $\frac{3}{4}$. For 5 children it will be $\left(\frac{3}{4}\right)^5 = \frac{243}{1024} = 0.237$.

(*b*) To calculate the chance of 4 children will be unaffected and 1 is affected.

$$p = \frac{\lfloor N}{\lfloor s \lfloor t} p^s q^t = \frac{\lfloor 5}{\lfloor 4 \lfloor 1}\left(\frac{3}{4}\right)^4\left(\frac{1}{4}\right)^1 = \frac{5.\lfloor 4}{\lfloor 4.1} \times \frac{81}{256} \times \frac{1}{4} = 5 \times \frac{81}{1024} = \frac{405}{1024} = 0.396$$

(*c*) To calculate that at least three children will be unaffected. Sum the first three terms of the binomial distribution.

Event	Binomial Formula	Probabilities
5 unaffected, 0 affected	$\frac{\lfloor 5}{\lfloor 5}\left(\frac{3}{4}\right)^5\left(\frac{1}{4}\right)^0 = \frac{243}{1024}$	0.237
4 unaffected, 1 affected	$\frac{5.\lfloor 4}{\lfloor 4 \lfloor 1} \times \left(\frac{3}{4}\right)^4\left(\frac{1}{4}\right)^1 = \frac{405}{1024}$	0.396
3 unaffected, 2 affected	$\frac{\lfloor 5}{\lfloor 3 \lfloor 2} \times \left(\frac{3}{4}\right)^3\left(\frac{1}{4}\right)^2 = \frac{5.4\lfloor 3}{\lfloor 3.2} \times \frac{27}{64}.\frac{1}{16} = \frac{270}{1024}$	0.264
		0.897

(*d*) To calculate that the probabilities of first child will be unaffected girl.

$$p = \frac{3}{4} \times \frac{1}{2} = \frac{3}{8}$$

Example 10. Two hundred families with three children a population of Arambagh subdivision are sampled at random. How many families do we expect to have (*a*) no girls, (*b*) one girl (*c*) two girls? Assume the sex ratio to be 1 : 1.

Solution: (*i*) Probabilities for boys & girls $= \frac{1}{2}$. *g* for girls and *b* for boys. Now expand the binomial (*g* & *b*), $n = 3$.

$$(g + b)^3 = g^3 + 3g^2b + 3gb^2 + b^3$$

(*a*) No girls relates to b^3 term

$$\frac{1}{2} \times \frac{1}{2} \times \frac{1}{2} = \frac{1}{8} \quad = \frac{1}{8} \times 200 = 25 \text{ (200 = families)}$$

(*b*) One girl relates to $3gb^2$ term

$$3.\frac{1}{2}\left(\frac{1}{2}\right)^2 = 3\frac{1}{2}.\frac{1}{4} = \frac{3}{8} = \frac{3}{8} \times 200 = 75$$

(*c*) Two girls relate to $3g^2b$ term

$$3.\left(\frac{1}{2}\right)^2 \frac{1}{2} = 3.\frac{1}{4}.\frac{1}{2} = \frac{3}{8} = \frac{3}{8} \times 200 = 75$$

Example 11. A plant breeder has 45 different inbred strains of Brinjal plants. How many different hybrids can be obtained from a total 45 plants?

Solution: Hybrid has two genes.

$n = 45$ and $r = 2$

According to formula;

$${}_nC_r = \frac{\lfloor n}{\lfloor r \lfloor n - r} = \frac{\lfloor 45}{\lfloor 2 \lfloor 45 - 2} = \frac{45 \times 44 \times \lfloor 43}{2.1 \times 4 \times \lfloor 43}$$

$$= \frac{45 \times 44}{2} = 45 \times 22 = 990$$

Example 12. Consider a family with two children in Serampore subdivision where both parents are heterozygous for albinism. What proportion of these family would be expected to have (*a*) neither child with albinism, (*b*) one child with albinism, (*c*) both children with albinism?

Solution:

(*i*) Let the symbol '*a*' for albinism and '*A*' for normal.

(*ii*) Expand binomial expansion

$$(A + a)^2 = A^2 + 2Aa + a^2$$

(*iii*) Both parents are heterozygous therefore probabilities of normal $\frac{3}{4}$ and albino $\frac{1}{4}$.

(*a*) Two children in the family both are normal *i.e.*, $A^2 = \left(\frac{3}{4}\right)^2 = \frac{9}{16}$.

(*b*) Among the two children, one is with albinism *i.e.*, $2Aa = 2.\frac{3}{4}.\frac{1}{4} = \frac{6}{16}$.

(*c*) both children with albinism *i.e.*, $a^2 = \left(\frac{1}{4}\right)^2 = \frac{1}{16}$.

Example 13. Consider parents of a Sinha Roy family in which both of them heterozygous for a severe genetic syndrome, that is autosomal recessive. Of their six children five of them have this particular syndrome. How unlucky is this family?

Solution:

$n = 6,\ p = \frac{3}{4}$ and $q = \frac{1}{4}$ $\left[\begin{array}{l} p = \text{Normal} \\ q = \text{Autosomal recessive Syndrome} \end{array}\right]$

Probabilities for disease $(t) = 5$

Normal $(s) = 1$

Using $p = \frac{\lfloor n}{\lfloor s \lfloor t} p^1 q^5$

$$= \frac{\lfloor 6}{\lfloor 1 \lfloor 5}\left(\frac{3}{4}\right)^1\left(\frac{1}{5}\right)^5 = \frac{6.\lfloor 5}{1.\lfloor 5} \times \frac{3}{4} \times \frac{1}{1024}$$

$$= \frac{18}{4096} = 0.00439$$

$$= 0.0044$$

Example 14. A couple is heterozygous for albinism (*Aa*). What is the probability that (*a*) 4 out of 6 children born to them are normal? (*b*) 4 normal & 2 albino out of 6 children?

Solution:

Let '*a*' = allele for albinism

A = allele for normal skin colour

Generally heterozygous parents have $\frac{3}{4}$ normal children & $\frac{1}{4}$ albino children.

$p_A = \frac{3}{4},\ q(a) = \frac{1}{4}$ and $n = 6$

(*a*) Probabilities 4 children being normal

$$p_A = \left(\frac{3}{4}\right)^4 = \frac{3}{4} \times \frac{3}{4} \times \frac{3}{4} \times \frac{3}{4} = \frac{81}{256} = 0.316$$

(*b*) $n = 6$, $s = 4$ and $t = 2$ [4 normal & 2 albino]

$$p = \frac{\lfloor n}{\lfloor s \lfloor t} \times (p)^4 \times (q)^2 = \frac{\lfloor 6}{\lfloor 4 \lfloor 2}\left(\frac{3}{4}\right)^4 \times \left(\frac{1}{4}\right)^2$$

$$= \frac{6.5\lfloor 4}{2.\lfloor 4} \times \frac{3}{4} \times \frac{3}{4} \times \frac{3}{4} \times \frac{3}{4} \times \frac{1}{4} \times \frac{1}{4}$$

$$= 15 \times \frac{81}{256} \times \frac{1}{16} = \frac{1215}{4096} = 0.2966$$

$$= 0.297$$

Or through binomial expansion;

$$(p + q)^6 = p^6 + 6p^5q + 15p^4q^2 + 20p^3q^3 + 15p^2q^4 + 6pq^5 + q^6$$

$$p = 15p^4q^2 \qquad [\because 4 \text{ normal } \& \text{ 2 albino}]$$

$$= 15 \times \left(\frac{3}{4}\right)^4 \times \left(\frac{1}{4}\right)^2$$

$$= 15 \times \frac{81}{256} \times \frac{1}{16} = \frac{1215}{4096} = 0.2966 = 0.297$$

Example 15. Four babies were born in Uttarpara general hospital. (*a*) What was the chance that two will be boys and two girls? (*b*) What was the chance that all four would be girls?

Solution: The probabilities of boys & girls were $\frac{1}{2} = 0.5$

$p = \frac{1}{2}$ and $q = \frac{1}{2}$

(*a*) $n = 4$, $s = 2$ and $t = 2$ $\left[\begin{array}{l} s = \textit{for girls} \\ t = \textit{for boys} \end{array}\right]$

$$p = \frac{\lfloor n}{\lfloor s \lfloor t} \times p^s \times q^t = \frac{\lfloor 4}{\lfloor 2 \lfloor 2} \times \left(\frac{1}{2}\right)^2 \times \left(\frac{1}{2}\right)^2$$

$$= \frac{4.3.\lfloor 2}{2.1\lfloor 2} \times \frac{1}{4} \times \frac{1}{4} = 6 \times \frac{1}{4} \times \frac{1}{4} = \frac{3}{8}$$

Thus probability was $\frac{3}{8}$.

(*b*) $(p + q)^4 = p^4 + 4p^3q + 6p^2q^2 + 4pq^3 + q^4$

$$p = q^4 = \left(\frac{1}{2}\right)^4 = \frac{1}{8}$$

Example 16. In a family of 8 children, where both parents are heterozygous for albinism, what mathematical expression predicts the probability that six are normal & two are albinos?

Solution: As both the parents are heterozygous, the probabilities of normal is $\frac{3}{4}$ and albinos $\frac{1}{4}$.

i.e., $p = \frac{3}{4}$, $q = \frac{1}{4}$, $n = 8$, $s = 6$ and $t = 2$

$$p = \frac{\lfloor n}{\lfloor s \lfloor t} \times p^s q^t = \frac{\lfloor 8}{\lfloor 6 \lfloor 2} \times \left(\frac{3}{4}\right)^6 \times \left(\frac{1}{4}\right)^2$$

$$= \frac{8.7\lfloor 6}{2.1\lfloor 6} \times \left(\frac{3}{4}\right)^6 \times \left(\frac{1}{4}\right)^2 = 28 \times \left(\frac{3}{4}\right)^6 \times \left(\frac{1}{4}\right)^2$$

$$= 28 \times \frac{3 \times 3 \times 3 \times 3 \times 3 \times 3}{4 \times 4 \times 4 \times 4 \times 4 \times 4} \times \frac{1}{4} \times \frac{1}{4}$$

$$= \frac{7 \times 729}{16384} = \frac{5103}{16384} = 0.31146$$

Example 17. A multiple allelic system is known to consist of seven alleles. Assuming that this is a diploid species, how many different genotypes could exist in the population?

Solution:

Number of possible genotypes = Number of different allelic combination. (heterozygotes) + number of genotypes with two of the same allele (homozygotes)

$$= \frac{\underline{|n}}{\underline{|k}\,\underline{|n-k}} + n \qquad n = 7,\ k = 2 \text{ (heterozygotes)}$$

$$= \frac{\underline{|7}}{\underline{|2}\,\underline{|7-2}} + 7 = \frac{7.6\underline{|5}}{2.\underline{|5}} + 7 = 21 + 7 = 28 \text{ genotypes}$$

Example 18. The $M\,N$ blood types of humans are under the genetic control of a pair of codominant alleles. In a population of Patna city, the families of size six where both parents are blood type MN. What is the chance of finding 3 children type M, 2 of type MN and $1N$?

Solution:

Parents are MN *i.e.*, $MN \times MN$

$$F_1 = \frac{1}{4}MM,\ \frac{1}{2}MN\ \&\ \frac{1}{4}NN \text{ type.}$$

Let p_1 = The probabilities of child being type M *i.e.*, $\frac{1}{4}$.

p_2 = The probabilities of child being type MN *i.e.*, $\frac{1}{2}$.

p_3 = The probabilities of child being type NN *i.e.*, $\frac{1}{4}$.

Let k_1 = Required number of M child = 3

k_2 = Required number of MN child = 2

k_3 = Required number of N child = 1

$$(p_1 + p_2 + p_3)^N = \frac{\underline{|N}}{\underline{|k_1}\,\underline{|k_2}\,\underline{|k_3}} \times p_1^{k_1}\, p_2^{k_2}\, p_3^{k_3} \qquad N = \sum k_1 + k_2 + k_3 = 6$$

$$= \frac{\underline{|6}}{\underline{|3}\,\underline{|2}\,\underline{|1}} \times \left(\frac{1}{4}\right)^3 \times \left(\frac{1}{2}\right)^2 \times \left(\frac{1}{4}\right)^1 = \frac{6.5.4\underline{|3}}{2.1\underline{|3}} \times \frac{1}{64} \times \frac{1}{4} \times \frac{1}{4}$$

$$= \frac{15 \times 1}{256} = \frac{15}{256} = 0.05859 = 0.0586$$

Example 19. There are 64 beds in a garden and 3 seeds of a particular type of flower are sown in each bed. The probability of a flower being white is 1/4. Find the number of beds with 3, 2, 1 and 0 white flower.

Solution: The probability p of a white flower is $\frac{1}{4}$

So $q = 1 - p = 1 - \frac{1}{4} = \frac{3}{4}$

Here $n = 3$, $N = 64$ and $r = 0, 1, 2, 3$

$$f(r) = N^nc_rp^rq^{n-r}$$

from this formula we can get

I. Beds with zero white flower

$$f(0) = 64.^3c_0\left(\frac{1}{4}\right)^0\left(\frac{3}{4}\right)^{3-0} = 64.1.1.\frac{27}{64} = 27$$

II. Beds with one white flower

$$f(1) = 64.^3c_1\left(\frac{1}{4}\right)^1\left(\frac{3}{4}\right)^{3-1} = 64.3.\frac{1}{4}.\frac{9}{16} = 27$$

III. Beds with two white flowers

$$f(2) = 64.^3c_2\left(\frac{1}{4}\right)^2\left(\frac{3}{4}\right)^{3-2} = 64.3.\frac{1}{16}.\frac{3}{4} = 9$$

IV. Beds with three white flowers

$$f(3) = 64.^3c_3\left(\frac{1}{4}\right)^3\left(\frac{3}{4}\right)^{3-3} = 64.1.\frac{1}{64} = 1$$

Example 20. If two parents, both heterozygous carriers of the autosomal recessive gene causing cystic fibrosis, have five children. What is the probability that three will be normal?

Solution: Probability of having a normal child during each pregnancy is $\frac{3}{4}\left[Aa \times Aa = \frac{3}{4}(AA + Aa) + \frac{1}{4}(aa)\right]$ and affected child is $\frac{1}{4}$

i.e., $p = \frac{3}{4}$ (normal) $q = \frac{1}{4}$ (cystic fibrosis) $n = 5$ $r = 3$

$$f(r) = {}^nc_rp^rq^{n-r} = {}^5c_3\left(\frac{3}{4}\right)^3\left(\frac{1}{4}\right)^{5-3}$$

$$= \frac{5.4.3}{3.2}\frac{27}{64}\frac{1}{16}$$

$$= \frac{10.27}{1024} = \frac{270}{1024} = 0.26$$

The probability of three will be normal 0.26.

Example 21. Three coins are tossed. Find the probability of (*a*) 0 heads, 1 heads, 2 heads, 3 heads. (*b*) more than one head (*c*) at least one head.

Solution:

Probability of success (head) in single trial $= p = \frac{1}{2}$

Probability of failure (tail) in single trial $= q = 1 - \frac{1}{2} = \frac{1}{2}$

Number of independent trial $= n = 3$

(*a*) 0 heads $= f(0) = {}^3c_0\left(\frac{1}{2}\right)^0\left(\frac{1}{2}\right)^{3-0} = 1.1.\frac{1}{8} = \frac{1}{8}$

$$1 \text{ heads} = f(1) = {}^3c_1\left(\frac{1}{2}\right)^1\left(\frac{1}{2}\right)^{3-1} = 3.\frac{1}{2}.\frac{1}{4} = \frac{3}{8}$$

$$2 \text{ heads} = f(2) = {}^3c_2\left(\frac{1}{2}\right)^2\left(\frac{1}{2}\right)^{3-2} = \frac{3.2}{2.1}\frac{1}{4}\frac{1}{2} = \frac{3}{8}$$

$$3 \text{ heads} = f(2) = {}^3c_3\left(\frac{1}{2}\right)^3\left(\frac{1}{2}\right)^{3-3} = \frac{3.2}{3.2.1}.\frac{1}{8}.1 = \frac{1}{8}$$

(*b*) More than 1 success.

$$f(2) + f(3) = \frac{3}{8} + \frac{1}{8} = \frac{4}{8} = \frac{1}{2}$$

(*c*) At least one success *i.e.*,

$$1 - f(0) = 1 - \frac{1}{8} = \frac{7}{8}$$

D. POISSON DISTRIBUTION:

This type of distribution was derived by French mathematician **Simeon Denis Poisson** (1837) and is called Poisson distribution after the name of its discoverer.

It represent the probability distribution of discrete, random variables of rare events whose probability occurance is very small but the number of events/trials is very large & may approach infinity. It is basically applicable to such events where binomial expression formula can be used in determining theoretical probabilities.

Definition: It is a discrete probability distribution of rare events and the mean and variance is equal.

• Characteristics:

(*i*) The occurrences of events are independent and random.

(*ii*) It is a limited form of binomial distribution.

(*iii*) It may be expected in cases where the chance of any individual event being success is small.

(*iv*) Poisson distribution has single parameter, the mean of the distribution.

(*v*) Here mean and variance are equal.

(*vi*) Poisson distribution is a discrete probability distribution because it is a probability distribution of whole number (0, 1, 2, 3.....*n*) of events.

(*vii*) Poisson distribution is positively skewed and declines with the rise of value of mean.

(*viii*) Poisson distribution is *leptokurtic* which decreases with the increase of mean.

• Conditions under which Poisson Distribution is used:

(*i*) Random variables should be discrete.

(*ii*) It is applicable in those cases where the number of trials or observations (*n*) is very large but the probability of success is very small.

(*iii*) A dichotomy exist *i.e.*, the happening of the events must be divided into two classes–*viz*. Success or failure, occurrence or non occurrence.

(*iv*) It is independent *i.e.*, happening of one event does not affect the happening of other event.

(*v*) *P* should be small (close to zero).

• Computation of Poisson Distribution:

The probability distribution of a random variable of X is said to have Poisson distribution.

$$P(X) = \frac{e^{-m} m^x}{\lfloor x} \qquad [x = 0, 1, 2, 3....n]$$

P = Probability of success

x = variables 0, 1, 2....n

e = constant 2.7183 (base of natural logarithm)

m = finite positive constant and is known as parameter of poison distribution.

Number of success (X)	*0*	*1*	*2*	*3*	*r*	*n*	.....	*Total*
Probabilities $P(X)$	e^{-m}	$\frac{e^{-m} m}{\lfloor 1}$	$\frac{e^{-m} m^2}{\lfloor 2}$	$\frac{e^{-m} m^3}{\lfloor 3}$	$\frac{e^{-m} m^r}{\lfloor r}$	$\frac{e^{-m} m^n}{\lfloor n}$	.....	1

$p(0) + p(1) + p(2) + p(3) + p(r) + p(n) + ... \alpha$

$$= e^{-m} + \frac{e^{-m} m}{\lfloor 1} + \frac{e^{-m} m^2}{\lfloor 2} + \frac{e^{-m} m^3}{\lfloor 3} + \frac{e^{-m} m^r}{\lfloor r} + \frac{e^{-m} m^n}{\lfloor n} + ..\alpha$$

$$= e^{-m}\left[1 + \frac{m}{\lfloor 1} + \frac{m^2}{\lfloor 2} + \frac{m^3}{\lfloor 3} + \frac{m^r}{\lfloor r} + \frac{m^n}{\lfloor n} + ...\alpha\right]$$

$$= e^{-m} e^{m} = e^{0} 1 = 1$$

The Poisson distribution is a discrete distribution with a parameter m. The various constants are

- Mean $= m = p$
- Standard deviation $= \sqrt{m}$
- Skewness given by (β_1) $= \frac{1}{m}$
- Kurtosis given by (β_2) $= 3 + \frac{1}{m}$
- Variance $= m$

• Examples of Poisson Distribution:

1. Number of bacterial colonies in a given culture per unit area of microscopic slide.
2. Number of fish deaths in a tank in one week due to some pesticide treatment in water.
3. The emission of radio active particles (*alpha* = α).
4. The number of mistakes committed by a good typist per page.
5. The number of car passing through a certain road **(M.G. Road Calcutta)**.
6. The number of suicide or death by rare disease **(cancer or heart attack)** in any cities **(Calcutta)** hospital in one year.

Example 1. Suppose a Zoology book with 585 pages contains 43 typological errors. If these errors are randomly distributed throughout the book, what is the probability that 10 pages, selected at random, will be free from errors?

(Use $e^{-0.735} = 0.4795$)

Solution:

Here $n = 10$, Book has 585 pages

Typological errors 43 pages

Therefore probabilities $P = \frac{43}{585} = 0.0735$

$$\text{Mean } (m)\ np = 10 \times 0.0735 = 0.735$$

Poisson distribution $(Pr) = \frac{e^{-m} m^r}{\lfloor r} = \frac{-0.735 \times (0.735)^r}{\lfloor r}$

Probability zero error

$$P(0) = \frac{e^{-0.735} \times (0.735)^0}{\lfloor 0} = e^{-0.735} \times 1 = e^{-0.735} = 0.4795$$

Example 2. In a History book of 520 pages, 390 typological errors occur. Assuming Poisson law for the number of errors per page, find probabilities that random sample of 5 pages will contain no error.

Solution:

Here $n = 5$, Book has 520 pages

Typological errors 390 pages

Therefore probabilities $(P) = \frac{390}{520} = 0.75$

$$\text{Mean} = np = 5 \times 0.75 = 3.75$$

Using Poisson probabilities law

$$P(r) = \frac{e^{-m} m^r}{\lfloor r} = \frac{e^{-0.75} m^r}{\lfloor r}$$

Probabilities error zero, therefore

$$P(0) = \frac{e^{-0.75} (3.75)^0}{\lfloor 0} = e^{-0.75}\, 3.75$$

Example 3. Work out the Poisson probability of finding 9 mutant *Drosophila* in sample of 500 *Drosophila* flies collected from natural population of Uttarpara known to have 80 mutant per 10,000. Interpret your result ($\alpha = 0.05$).

Solution:

$N = 500$, $X = 9$

$$P(X) = \frac{\text{Number of mutants in a given population size}}{\text{Given population size}}$$

$$= \frac{80}{10{,}000} = 0.008$$

$$\text{Mean } (m)\ np = 500 \times 0.008 = 4$$

$$\text{Poisson probability} \quad (Px) = \frac{e^{-m}m^x}{\lfloor x} = \frac{e^{-m}(4)^9}{\lfloor 9} = \frac{e^{-m}.4.4.4.4.4.4.4.4.4}{9.8.7.6.5.4.3.2.1}$$

$$= \frac{e^{-m} \times 64 \times 64 \times 64}{72 \times 42 \times 20 \times 6} = \frac{e^{-m} \times 2048}{2835}$$

$$= e^{-m} \times 0.72239 = e^{-m} \times 0.723$$

$$= \frac{0.723}{(2.718)^4} = \frac{0.723}{7.3875 \times 7.3875} = \frac{0.723}{54.575}$$

$$= 0.01324 = 0.0133$$

Interpretation: $\alpha = 0.05 \quad p = 0.0133$

Since the computed poisson probabilities is 0.0133 lower than the chosen α 0.05 *i.e.*, $p < 0.05$. It is due to random sampling.

Example 4. In Arambagh general hospital, the mortalities rate for malignant malaria is 7 in 1000. What is the Poisson probability for just 2 on account of the disease in a group of 400 people. Given $e^{-2.8} = 0.06$.

Solution:

$$\text{Here} \quad P = \frac{\text{The number of mortality}}{\text{Total number of people}} = \frac{7}{1000} = 0.007$$

$m = np$, $n = 400$ and $r = 2 \quad \therefore m = 400 \times 0.007 = 2.8$

$$\text{Poisson probability } (P) = \frac{e^{-m} \times m^r}{\lfloor r} = \frac{(2.8)^2}{\lfloor 2 . e^m} = \frac{e^{-2.8} \times 2.8 \times 2.8}{2.1}$$

$$= 1.4 \times 2.8 \times e^{-2.8}$$

$$= 1.4 \times 2.8 \times 0.06 = 0.2352$$

$$= 23.52\%$$

Example 5. In R.G Kar medical college hospital receives patient at the rate of 3 patients per minute on average. What is the probability of receiving no patient in one minute interval? [Given $e = 2.718$].

Solution:

X = The number of patient received per minute & the poisson parameter $m = 3$

$m = n \times p = 3$

$$p(x) = \frac{m^r}{\lfloor r\, e^m} = 3 \qquad r = 0, 1, 2, 3\ldots\ldots\ldots$$

Probability of no patient *i.e.*, $r = 0$

$$p(x) = \frac{(3)^0}{\lfloor 0 .(2.718)^3} = \frac{1}{1 \times 20.079} = \frac{1}{20.079} = 0.049803 = 0.0498$$

Example 6. Radiation in mutational genetics, genes are often regarded as targets. Ionizing radiation transfers energy in discrete packages interacting with the genes. These interactions

(hits) cause mutation and are independent to each other, that is they follow a poisson distribution. If a malignant tissue is irradiated such that the result is a mean of two hits per target.

(a) How many targets will not be hit at all?

(b) How many targets will get hit at least once?

(c) How many targets will get hit less than three times?

(d) How many targets will get hit exactly six times?

(Given $e^{-m} = 0.135$)

Solution:

$$\text{Poisson equation } (Pr) = \frac{e^{-m} m^r}{\underline{|r}}$$

$$m = 2$$

(a) Target not hit at all. *i.e.*, $r = 0$

$$P(0) = \frac{e^{-m}(2)^0}{\underline{|0}} = \frac{.135 \times 1}{1} = 0.135 = 1.35\%$$

(b) Hit at least once *i.e.*, $1 - 0.135 = 0.865 = 865\%$

(c) $$P(0) = 0.135$$

$$P(1) = \frac{.135 \times (2)^1}{\underline{|1}} = 0.135 \times 2 = 0.270 = 2.7\%$$

$$P(2) = \frac{0.135 \times (2)^2}{\underline{|2}} = \frac{0.135 \times 4}{2} = 0.135 \times 2 = 0.270 = 2.7\%$$

Less than 3 times $0.135 + 0.270 + 0.270 = 0.675 = 67.5\%$

(d) Hit exactly 6 times

$$P(6) = \frac{0.135 \times (2)^6}{\underline{|6}} = \frac{0.135 \times 2 \times 2 \times 2 \times 2 \times 2 \times 2}{6.5.4.3.2}$$

$$\frac{0.135 \times 64}{30 \times 8 \times 3} = \frac{0.135 \times 8}{90} = \frac{1.08}{90} = 0.012 = 1.2\%$$

Example 7. In a Genetics research laboratory, radioactive emission occurs average one particle per minute. If emission continues for several hundred minutes during which time the particles are emitted randomly, in what proportion of these minutes would we expect the following? (Given $e^{-m} = e^{-1} = 0.36788$)

(a) Exactly one particles emitted?

(b) Exactly two particles per minute?

(c) More than two particles per minute ?

Solution:

$$\text{Poisson equation} \quad (Pr) = \frac{e^{-m} m^r}{\underline{|r}}$$

$m = 1$ $\quad e^{-1} = 0.36788 = 0.3678$

(*a*) $r = 1$ $\qquad P(1) = \dfrac{e^{-m} \times (1)^1}{\lfloor 1} = .3678 \times 1 = 0.368 = 36.8\%$

(*b*) $r = 2$ $\qquad P(2) = \dfrac{0.368 \times (1)^2}{\lfloor 2} = \dfrac{0.368 \times 1}{2} = 0.184 = 18.4\%$

(*c*) $r = 0$ $\qquad P(0) = \dfrac{0.368 \times (1)^0}{\lfloor 0} = 0.368 = 36.8\%$

More than two is therefore

$1 - (0.368 + 0.368 + 0.184) = 1 - 0.92 = 0.08 = 8\%$

Example 8. How many mammalian cells would be killed if an irradiation dose administered to a cell population was sufficient for an average of 5 lethal hits per target, when in fact only 2 hits are needed for lethality? [Given e^{-m} *i.e.*, $e^{-5} = .0067$]

Solution:

Poisson equation $\qquad (Pr) = \dfrac{e^{-m} m^r}{\lfloor r}$

$m = 5$

$$P(0) = \frac{e^{-m}(5)^0}{\lfloor 0} = \frac{0.0067 \times 1}{1} = 0.0067$$

$$P(1) = \frac{0.0067 \times (5)^1}{\lfloor 1} = 0.0067 \times 5 = 0.0335$$

Total $= 0.0067 + 0.0335 = 0.0402$

and $\qquad 1 - 0.0402 = 0.9598 = 0.96$

Example 9. In the following data are obtained from vector cytogenetics research laboratory of Serampore college.

Events (x_1)	*Frequency (f)*
0	20
1	26
2	16
3	4
4	2

(*a*) What is the mean?

(*b*) What is e^{-m}?

(*c*) What is $P(0)$?

Solution:

(*a*) Mean $(\bar{X}) = \dfrac{\sum fx}{\sum f} = \dfrac{0 + 26 + 32 + 12 + 8}{20 + 26 + 16 + 4 + 2} = \dfrac{78}{68} = 1.147 = 1.15$

(*b*) $(\bar{X}) = m = 1.15$

$e^{-1.15} = 0.32$

$$(c)\ P(0) = \frac{e^{-m}(1.15)^0}{\lfloor 0} = \frac{0.32 \times 1}{1} = 0.32$$

Example 10. A discrete random variable X follows Poisson distribution such that $Pr(X = 1) = Pr(X = 2)$. Find the mean and variance of the Poisson distribution.

Solution:

Here random variable X follows poisson distribution

Let m the parameter of X .....

$$P(X) = \frac{e^{-m} m^x}{\lfloor x} \qquad [x = 0, 1, 2, 3.....n]$$

$$\text{Here } Pr = (n = 1) \text{ i.e., } P(1) = \frac{e^{-m} \times m^1}{\lfloor 1} = e^{-m} m$$

$$(Pr) = n = 2 \text{ i.e., } P(2) = \frac{e^{-m}(m)^2}{\lfloor 2} = \frac{e^{-m} m^2}{2.1} = \frac{e^{-m} m^2}{2}$$

According to the problem $Pr\ (n = 1) = Pr\ (n = 2)$

$$e^{-m}.m = e^{-m}.\frac{m^2}{2}$$

$$m = \frac{m^2}{2} \text{ or } m = 2 \qquad [\because e^{-m} \neq 0\ m \neq 0]$$

Therefore mean of the Poisson distribution *i.e.*,
$m = 2$ and variance $= m = 2$

Example 11. If the mean of the Poisson distribution is 4 then find (*i*) S.D (σ), (*ii*) β_1, (*iii*) β_2, (*iv*) μ_3 & (*v*) μ_4.

Solution:

$m = 4$

(*i*) Variance $(\sigma)^2 = m = 4$

$$\text{S.D}(\sigma) = \sqrt{\sigma^2} = \sqrt{4} = 2$$

(*ii*) Skewness $(\beta_1) = \frac{1}{m} = \frac{1}{4} = 0.25$

(*iii*) Kurtosis $(\beta_2) = 3 + \frac{1}{m} = 3 + \frac{1}{4} = 3.25$

(*iv*) $\mu_3 = m = 4$

(*v*) $\mu_4 = m + 3m^2 = 4 + 3(4)^2 = 4 + 3.16 = 4 + 48 = 52$

Example: 12. Between the hours of 3 and 4 pm, the average number of phone calls per minute coming into the switch board of a pathological laboratory of Calcutta is 2.5. Find the probability that one particular minute there will be no phone call (*ii*) 4 calls during the same time.

(Given $e^{-2.5} = 0.0821$)

Solution:

(*i*) The random variable X is the number of telephone calls received during the period 3–4 pm. Since x is assumed to follow Poisson distribution, the parameter 'm' is equal to the mean of the distribution *i.e.*, $m = 25$.

(*ii*) Hence the probability of x call is

$$p(x) = \frac{e^{-m} m^x}{\lfloor x} \qquad [x = 0, 1, 2......]$$

The probability of no call *i.e.*, $x = 0$

$$P(0) = \frac{e^{-m}(2.5)^0}{\lfloor 0} = \frac{e^{-2.5}.1}{1} = e^{-2.5} = 0.0821$$

(*iii*) The probability of 4 calls

$$P(4) = \frac{e^{-m}(2.5)^4}{\lfloor 4} = \frac{e^{-2.5} \times 2.5 \times 2.5 \times 2.5 \times 2.5}{4.3.2.1}$$

$$= \frac{0.0821 \times 39.0625}{24} = \frac{3.2070312}{24} = 0.13362 = 0.1336$$

Example 13. Poisson distribution has a double mode $x = 1$ and $x = 2$. What is the probability that x will have one or the other of these two values?

Solution:

(*i*) If the Poisson distribution is bimodal then two modes are the points $x = m - 1$ and $x = m$ where m is the parameter of the Poisson distribution.

(*ii*) Since we are given that the two modes are at the points $x = 1$ and $x = 2$, we find that $m = 2$.

$$\therefore \qquad P(x = n) = \frac{e^{-m} m^x}{\lfloor x} = \frac{e^{-2}(2)^x}{\lfloor x} \qquad n = 0, 1, 2.......$$

$$P(x = 1) = \frac{e^{-2}(2)^1}{\lfloor 1} = e^{-2}.2$$

$$P(x = 2) = \frac{e^{-2}(2)^2}{\lfloor 2} = \frac{e^{-2}.4}{2 \times 1} = e^{-2}.2$$

Required probability $= P(x = 1) + P(x = 2) = 2e^{-2} + 2e^{-2}$

(*i*) Genes that are **closely linked** and located in the **same chromosome** are usually unlikely to experience more than one crossing over. Hence, small map distances are likely to reflect the actual amount of crossing over.

(*ii*) However, as the **gene distance increases**, the opportunity for multiple crossing over increases, tending to follow a **Poisson distribution**

(*iii*) Meiosis involving any finite number of crossing over per meiosis produces an observed recombination frequency (*RF*) of 50% among their products. Theoretically they can not give ***RF* more than 50%**. No matter how far apart two loci are linked.

(*iv*) The further **apart two genes** are, the greater the error of estimate of true linkage distance provided by the recombinant progeny.

(*v*) The proportion of meiosis in which **at least one crossing over** occurs is predicted by

$R.F = \frac{1}{2}(1 - e^{-\mu})$ where μ the actual mean number of zero crossing over events (*not observed number*).

(*vi*) If there are no zero crossing over events, then $R.F = \frac{1}{2}(1 - 0) = \frac{1}{2}$ because the expected zero class frequency is $= \frac{e^{-\mu}\mu^0}{\lfloor 0} = e^{-\mu}$. Given any observed *RF*, μ can be calculated by this above formula.

(*vii*) The following formula can be used to convert observed *RF* to most likely amount of crossing over (*for closely linked genes*) $RF^1 = \frac{\mu}{2}$ because when $\mu = 0.02$, $e^{-\mu} = 0.98$ and $RF = \frac{1}{2}(1 - 0.98) = \frac{1}{2}(0.02) = \frac{\mu}{2}$. Likewise when $\mu = 0.05$, $e^{-\mu} = 0.95$ and $RF = \frac{1}{2}(1 - 0.95) = \frac{1}{2}(0.05) = \frac{\mu}{2}$. Thus $RF^1 = \frac{\mu}{2}$.

(*viii*) *A* mean (μ) of one crossing over between two loci equates 50 map units.

Example 14. If two linked genes produce 33% recombinants, estimate the probable amount of crossing over that actually occurred between there loci, assuming that probabilities of 0, 1, 2....*n* exchange occur during meiosis according to a Poisson distribution.

Solution: 33% recombinants = 0.33

$$RF = \frac{1}{2}(1 - e^{-\mu})$$

$$0.33 = \frac{1}{2}(1 - e^{-\mu})$$

$0.66 = 1 - e^{-\mu}$ or $e^{-\mu} = 1 - 0.66 = 0.34$

(*i*) From the table we find that $e^{-0.34}$ lies between $\mu = 1$ (0.36788) and $\mu = 2$ (0.13534).

Thus $\frac{0.34}{0.36788} = 0.9242$.

(*ii*) 0.9242 corresponds to $\mu = 0.08$. Therefore $\mu = 1.08$ and $RF^{1} = \frac{1.08}{2} = 0.54$ or 54 map units.

(*iii*) The error of under estimate of actual crossing over $= \frac{54}{33} = 1.636 = 1.64$ or about 64%.

Example 15. If two linked genes produce 27.5% recombinants, estimate the probable amount of crossing over that actually occurred between these loci. Assuming the probabilities of 0, 1, 2.....*n* exchange occur during meioses according to Poisson distribution.

Solution: 27.5 % crossing over = 0.275

$$RF = \frac{1}{2}(1 - e^{-\mu})$$

$$0.275 = \frac{1}{2}(1 - e^{-\mu}) \text{ or } 0.55\phi = 1 - e^{-\mu} \text{ or } e^{-\mu} = 1 - 0.55 = 0.45$$

(*i*) From the table we find that $e^{-0.45}$ lies between $\mu = 1$ (0.36788) & $\mu = 0$ (1.0000).

Thus $\frac{0.45}{0.36788} = 1.223.$

7

CHAPTER

SKEWNESS KURTOSIS & MOMENTS

A distribution is said to be symmetrical when mean, median are coincide.

1. It has three parts left tail, right tail and middle part.
2. Right & left tail are of equal length.

• Skewness:

The word "skewness" is used to denote the "extent of a symmetry" in the data. When the frequency distribution is not symmetrical it is said to be skewed. The literally meaning of "skew ness" is "lack of symmetry". A symmetrical distribution has therefore zero skewness.

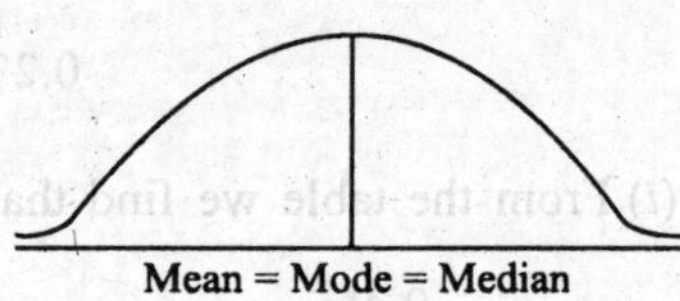

Fig. 7.1 Symmetrical Curve

• Features:

1. Skewness may be **Positive** or **Negative**.

   Positive Skewness:

   (*i*) If the curve of the distribution has longer tail toward the right *i.e.*, the higher values of the variable.

   (*ii*) *Mean > Median > Mode*

   (*a*) If the curve of the distribution has a longer tail towards the left *i.e.*, the lower values of the variable.

   (*b*) *Mean < Median < Mode*

2. Here mean median and mode are failed to coincide. Both median and mean are displaced from the mode toward the skewed tail.

   Mean > Median > Mode **(Positively skewed)**

   Mean < Median < Mode **(Negatively skewed)**

3. Here the first quartile is displaced toward the skewed tail. Therefore

   $Q_3 - Q_2 > Q_2 - Q_1$ **(Positively skewed)**

   $Q_2 - Q_1 > Q_3 - Q_2$ **(Negatively skewed)**

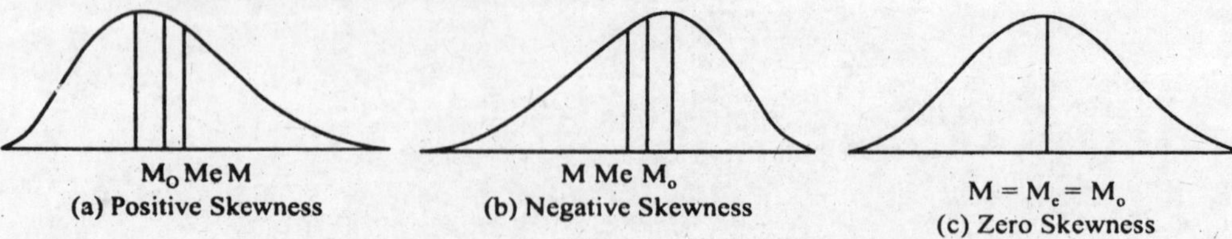

Fig. 7.2 Position of Mean (M), Median (M_e) and Mode (Mo) for different types of Skewness.

• MEASURES OF SKEWNESS:

1. The measures of symmetry are normally called measures of skewness.
2. The measures of skewness indicate not only the extent of skewness (in numerical expression) but also the direction *i.e.*, the manner in which the deviations are distributed.
3. The absolute measures are known as *measures* of *skewness*. It tells us the extent of asymmetry whether it is positive or negative.

Absolute skewness = Mean – Mode

Mean > Mode **(Positive skewness)**

Mode > Mean **(Negative skewness)**

4. The relative measures are known as *coefficient of skewness* and is given by $\beta_1 = \frac{\mu_3^2}{\mu_2^3}$

μ_3 = 3rd moment and μ_2 = 2nd moment. There are important measures of relative skewness

(*a*) Karl Pearson's coefficient of skewness.

$$(i)\ S_k = \frac{\text{Mean} - \text{Mode}}{\text{Standard Deviation}}$$

$$(ii)\ S_k = \frac{3(\text{Mean} - \text{Median})}{\text{Standard Deviation}}$$

(*b*) Bowley's quartile coefficient of skewness.

$$S_k = \frac{Q_3 - 2Q_2 + Q_1}{Q_3 - Q_1} \quad \begin{bmatrix} Q_1 = \text{First Quartile} \\ Q_3 = \text{Third Quartile} \end{bmatrix}$$

(*c*) Kelly's coefficient of skewness.

$$S_k = \frac{P_{90} + P_{10} - 2\,\text{Median}}{P_{90} - P_{10}} \quad \begin{bmatrix} P_{10} = 10^{th}\ \text{Percentiles} \\ P_{90} = 90^{th}\ \text{Percentiles} \end{bmatrix}$$

This method is rarely used.

• Kurtosis:

The expression of "Kurtosis" is used to describe the degree of peakedness of a frequency distribution compared to that of normal distribution.

• Features:

1. Kurtosis measure the peakedness of a normal curve.
2. Karl Pearson called it a "Measures of Convexity" of the curve. He introduced three broad patterns.
3. If Peakedness *viz.* Mesokurtic, Leptokurtic and Platykurtic.
   - The curve which is neither flat nor peaked is known as *mesokurtic i.e.*, normal curve.
   - The curve which has higher & sharper peaked and narrower body then the normal curve is called *Leptokurtic*.
   - The curve which is flatter at its centre, broader in the body & thinner at tails, than normal curve is called *Platykurtic*.

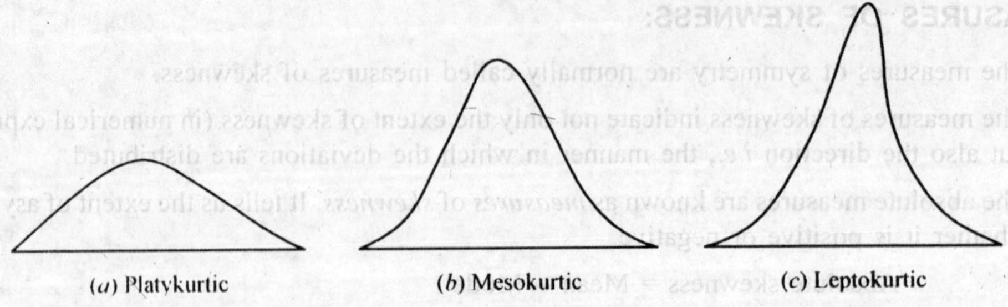

(*a*) Platykurtic (*b*) Mesokurtic (*c*) Leptokurtic

Fig. 7.3 Different types of Kurtosis.

• Measures of Kurtosis:

Measures of kurtosis of a frequency distribution are based upon the fourth moment about the mean of the distribution.

$$\beta_2 = \frac{\mu_4}{\mu_2^2} = \frac{\mu_4}{\sigma_4} \qquad \left[\begin{array}{l}\mu_4 = 4^{th} \text{ moment} \\ \mu_2 = 2^{nd} \text{ moment}\end{array}\right]$$

$\beta_2 = 3$ Mesokurtic
$\beta_2 > 3$ Leptokurtic
$\beta < 3$ Platykurtic

• Importance of Skewness:

1. It tells the direction and extent of asymmetry in a series.
2. It provides us an idea about the nature and degree of concentration of items.

Difference between Skewness and Dispersion

Skewness	*Dispersion*
1. It shows us the departure from symmetry *i.e.*, direction of variation.	1. It shows us the spread of individual values about the mean *i.e.*, central value.
2. It shows whether the concentration is in higher or lower value.	2. It shows the degree of variability.
3. It judges the differences between the central tendencies.	3. It judges the truthfulness of the central tendencies.
4. It is not an average but it is measured by the use of mean, median & mode.	4. It is a type of averages of deviation–average of the second order.

• Significance:

1. It tells us the extent to which a distribution is more peaked or more flat topped than normal curve.
2. It also denotes the shape of the top of a frequency curve.

• Moments:

It is generally used in physics, mechanics etc. But when it is applied in statistics, it describes the various characteristics of frequency distribution *viz.* central tendency, dispersions, skewness and kurtosis.

Moment can be defined as arithmetic mean of various powers of deviations taken from the mean of distribution.

• Role of Moments:

1. **First moment** (μ_1) of frequency distribution $= \frac{\sum X - \bar{X}}{N}$	It is always zero *i.e.*, $\mu = 0$	It measures mean of the distribution *i.e.*, $\mu_1 = \bar{X} = 0$.
2. **Second moment** (μ_2) of frequency distribution about the mean is the variance of the distribution $= \frac{\sum(X-\bar{X})^2}{N}$	$\mu_2 = \sigma^2$	It measures the variance *i.e.*, the spread of the different terms in a distribution.
3. **Third moment** (μ_3) $= \frac{\sum(X-\bar{X})^3}{N}$	It deals with skewness.	It gives an idea about the degree of skewness present.
4. **The fourth** (μ_4): moment $= \frac{\sum(X-\bar{X})^4}{N}$	It highlights on the height of frequency distribution. Whether it is more peaked or flat topped than normal.	It measures kurtosis.

• Problems:

Example 1. Coefficient of Skewness = 3, Mean = 90, Median = 80. Find the value of *S.D.*

Solution:

$$S_k = \frac{3\text{ (Mean} - \text{Median)}}{S.D.} \text{ or } S.D. = \frac{3\text{ (Mean} - \text{Median)}}{S_k}$$

$$S.D. = \frac{3(90-80)}{3} = \frac{3 \times 10}{3} = 10.$$

Example 2. The data on the number of frinch spot on wing of *Anopheles* mosquitoes. 2, 4, 5, 2, 7. Find Skewness & Kurtosis.

Solution:

X	$X - \bar{X} = x$	$(X-\bar{X})^2 = x^2$	$(X-\bar{X})^3 = x^3$	$(X-\bar{X})^4 = x^4$
2	2 – 4 = –2	4	–8	16
4	4 – 4 = 0	0	0	0
5	5 – 4 = 1	1	1	1
2	2 – 4 = –2	4	–8	16
7	7 – 4 = 3	9	27	81
	$\sum x = 0$	$\sum x^2 = 18$	$\sum x^3 = 12$	$\sum x^4 = 114$

$$n = 5, \quad \sum X = 20, \quad \bar{X} = \frac{20}{5} = 4$$

$$\mu_1 = \frac{\sum X}{N} \quad \mu_2 = \frac{\sum X^2}{N} \quad \mu_3 = \frac{\sum X^3}{N} \quad \mu_4 = \frac{\sum X^4}{N}$$

$$\mu_1 = \frac{0}{5} = 0 \quad \mu_2 = \frac{18}{5} = 3.6 \quad \mu_3 = \frac{12}{5} = 2.4 \quad \mu_4 = \frac{114}{5} = 22.8$$

$$\text{Skewness } (\beta_1) = \frac{\mu_3^2}{\mu_2^3} = \frac{5.76}{46.65} = 0.123 = 0.12$$

$$\text{Kurtosis } (\beta_2) = \frac{\mu_4}{\mu_2^2} = \frac{22.8}{12.96} = 1.759 = 1.76$$

8

CHAPTER

SET THEORY & PROBABILITY

A. SET THEORY:

• Set and Element:

• Any well defined collection of all possible distinct objects is called a set.

• Each object of the set is called elements or members.

1. Sets are usually denoted by capital letters (e.g., A, B, C, D.....) and their elements are denoted by small letters (e.g., a, b, c, d........).
2. Description of the elements are enclosed in within curly backets "{.....}".
3. Element of the set are separated by commas.

Example: (*i*) Collection of all consonants of English alphabet.

(*ii*) Collection of all odd numbers less than 50.

> • Every set is a subset of itself.
> • An empty set is a subset of every set.
> • A set containing one element is conceptually distinct from the element itself, but will be represented by the same symbol for the sake of convenience.

• Finite and Infinite Set:

• A set that contain limited (countable) numbers of different elements is called finite.

Example: $A = \{a, b, c, d, e\}$ is finite because it has 5 elements.

• A set that unlimited (*uncountable*) number of different elements is called infinite.

Example: A set of even numbers, $E = \{2, 4, 6, 8......\}$ is infinite because it has no countable elements.

• Null Set:

1. A set which does not contain any elements at all, is called null set.
2. It is also called **void set** or **empty** set.
3. There is only one such set. It is denoted by ϕ or {}.

Example: (*i*) The set {a: a is a person who can jump to a height of 5 miles} is the null set, because none can jump to such height.

(*ii*) {x: x month of a year having less than 20 days} is null set because none month persist.

> • {o} and {ϕ} are not empty set because each of this have one element.
> • Since the empty set has no elements, ϕ is a finite set.

- **Unit Set:**

1. Set having only one element.
2. It is also known as **single tone** set.

- **Equal Set:**

1. Two sets *viz.* A and B are called equal if they have same elements.
2. When A & B are equal, we write $A = B$, if they are not equal we write $A \neq B$.

Example: (*i*) Let $A = \{a, b, c\}$ and $B = \{b, a. c]$, then $A = B$ because the set A has the same element like set B.

(*ii*) Let $A = \{x/x$ is a vowel in word PALMOLIVE$\}$ and $B = \{x/x$ is a vowel in word TEMPTATION$\}$, then $A = B$ because each set can be written as $\{A, O, I, E\}$.

- **Equivalent Sets:**

Two sets viz. A and B are equivalent if the number of elements i.e cardinal numbers are equal.

Example: $A = \{2, 4, 6\}$ and $B = \{a, b, c\}$. Here $n(A) = n(B) = 3$.

All equivalent sets are not always equal, but equal sets are always equivalent.

- **Cardinal number of a Set:**

1. The number of distinct elements in a (finite) set is called its cardinal number.
2. The cardinal number of a finite set A is denoted by $n(A)$.
3. The cardinal number of null set is zero.
4. The cardinal number of infinite set is not defined.

Example: $A = \{2, 3\}$ has 2 elements, so $n(A) = 2$.

$B = \{a, e, i, o, u\}$ has 5 elements, so $n(A) = 5$.

- **Subset and Superset:**

Two sets *viz.* A and B, if each element of a set A is also an element of a set B, then set A is called a subset of B and set B is the superset of A.

This is read as "B contains A" or "A is contained in B" *i.e.*, $A \subseteq B$ or $B \supseteq A$.

Example: The set $A = \{2, 5\}$ is a subset of the set $B = \{2, 5, 7\}$.

Here all the elements of A are also the elements of B.

On the other hand B is a superset of set A.

Properties of Subset:

- Zero set ϕ is the subset of every set.
- Every set (P) is the subset of its own set.
- If $A \subseteq B$ and $B \subseteq C$ then $A \subseteq C$.
- If $A \subseteq B$ and $B \subseteq A$ then $A = B$.

- **Proper and Improper Subset:**

A is the subset of B *i.e.*, $A \subseteq B$ but the sets are not equal ($A \neq B$) then A is said to be a proper subset of B.

We write it as $A \subset B$.

Example: Let $A = \{2, 5\}$ & $B = \{2, 5, 7\}$. A is the proper subset of B because each element of A belongs to B but $A \neq B$.

- **Power Set:**

1. A set 'S' formed by the family of all the subsets of 'S' is called power set of 'S'.
2. It is denoted by $P(S)$.

Example: (*i*) Let 'S' = {2, 3, 4} then the set = {ϕ}, {2}, {3}, {4}, {2, 3}, {2, 4}, {3, 4}, {2, 3, 4}.

(*ii*) If a finite 'S' has an n elements then the power set of 'S' has 2^n elements.

- **Universal Set:**

1. Sometime it happens that all the sets under consideration are subsets of a certain set 'S'. This set 'S' is called the universal set.
2. It is generally denoted by the capital letter 'U'.

Example: (*i*) All the students of the world constitute the universal set of the student of Calcutta University.

- **Toutology:** A statement is called tautology if it is always true.
- Sets are always denoted by capital letters A, B, C etc., the members of the set are denoted by small letters a, b, c etc.

Symbol	*Meaning*
$\in$	belong to or an element of
$x \in A$	x an element of set A
$\notin$	does not belong to
U or S	Universal set
ϕ or {}	Null set or Empty set
$n(A)$ or (A)	A element of finite set
$\subset$ $\subseteq$	subset
$\subset$	proper subset
$A \subset S$	Set A subset of S
$A \not\subset S$	A is not subset of S
$\supset$	Contains
$P(A)$	Power set of A
A^1 or A^c	Complementary set of A
$\cup$	Union/cup
$\cap$	Intersection Cap

- **Representation of Sets:**

The most common method to describe the sets is as follows:

(*a*) Rostar or Tabular form or listing method

(*i*) In this method, elements of a set are separated by commas and are enclosed within braces {}.

(*ii*) Instead of braces even parentheses () or brackets [] may be used.

(*iii*) If a set contains a few elements all elements are written within braces.

Example: $A = \{1, 3, 5, 7\}$.

(*iv*) If a set contains many elements, three dots are used in the middle to include the missing elements.

Example: Set '*A*' of letters of English alphabet can be written as $A = \{a, b, c...., z\}$.

(*v*) When the elements of a set are enumerable then three dots are used at the end within braces.

Example: Set *N* of natural number can be written as $N = \{1, 2, 3, ...\}$.

(*b*) Set builder form or Rule method

(*i*) In this method, a set may be specified by stating properties which the elements of the set must satisfy.

(*ii*) It is written as $S = \{n: P(x)\}$ *i.e.*, the property $P(x)$.

(*iii*) $A = \{x : x$ is a vowel in English alphabet$\}$.

- **Write the following in Roster from:**

1. Set of all letters in "MATHEMATICS".
2. Set of letters in the word "SATELLITE".
3. Set of vowels in the word "INDIA".
4. Set of consonant in the word "AMERICA".

Solution: (*i*) $\{m, a, t, h, e, i, c, s\}$ [It may be noted here that the repeated letter are taken only once each.]

(*ii*) $\{s, a, t, e, l, i\}$.

(*iii*) Vowels in INDIA = $\{i, a\}$.

(*iv*) Consonant in AMERICA *i.e.*, $\{m, r, c\}$.

- **Write the following in set builder form:**

1. {January, March, May, July, August, October, December}.
2. $A = \{101, 103, ...998, 999\}$.
3. Set of planets in our solar system.

Solution: (*i*) $\{x:|\ x$ is a month of a year having 31 days$\}$.

(*ii*) $A\ \{x \mid x$ is a natural member, $100 < n < 1000\}$.

(*iii*) $\{x \mid x$ is a planet in our solar system$\}$.

VENN DIAGRAM:

The pictorial or diagrammatic representation of sets through rectangles and circles are known as Venn Eular diagram, or simply Venn diagrams.

- **Basic Features:**

1. In Venn diagrams, a universal set *U* is represented by a large rectangles and its sub sets are represented by a circular areas drawn within the rectangle.
2. If a set *B* is a sub set of *A*, the circle representing *B* is drawn inside the circle representing *A*.
3. If the sets *A* and *B* are equal, then the same circle represents both *A* and *B*.
4. If the sets *A* and *B* are disjoint, then the circle representing *A* and *B* are drawn in such a way that they have no common area.

5. If the sets A and B are not disjoint, the circles representing end are drawn in such a way that they have some common area to both.
6. When we represent universal set 'U' by a rectangle we write 'U' in one corner of that rectangle. Similarly we write 'A' within the circle representing the set 'A'.

- **Set Operation:**

(a) Union

(i) A set X is the union of two given sets A and B, if and only if any element of X belong to A or B or to both.

(ii) It is written as $X = A \cup B$ and it is read as A union B or A cup B or A join B.

(iii) By Venn diagram the union of two sets A and B is shown with shaded area is given below.

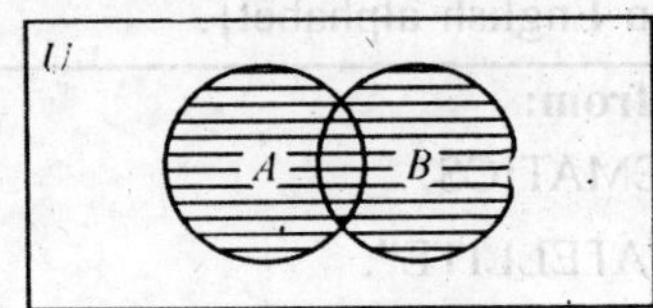

Fig. 8.1 $A \cup B$ is shown with shaded area.

Example: (i) Let $A = \{a, e, i, o, u\}$ and $B = \{a, b, c, d, e\}$.

Then $A \cup B = \{a, e, i, o, u, b, c, d\}$.

(ii) If $A = \{1, 3, 5, 7, 9\}$ and $B = \{2, 4, 6, 8\}$.

Then $A \cup B = \{1, 3, 5, 7, 9, 2, 4, 6, 8\}$.

(iii) If M is the set of all male students of Serampore college and N is the set of all the students in the college. Then $M \cup N = N$.

(b) Intersection

(i) The set X is the intersection or meet of two sets A and B if and only if any element of X is the element of A as well as B.

(ii) It is denoted by $X = A \cap B$ and it is read as A intersection B or A cap B.

(iii) By Venn diagram intersection of sets A and B is represented as below.

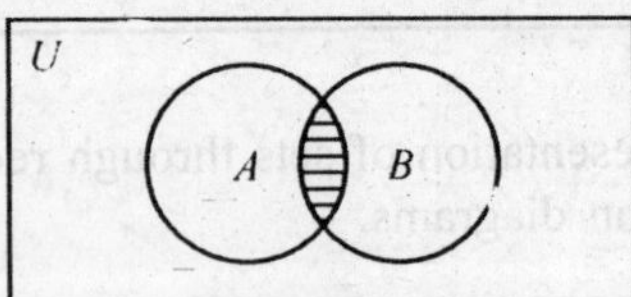

Fig. 8.2 $A \cap B$ is shown with shaded area.

Example: (i) Let $A = \{1, 3, 4, 5, 6\}$ and $B = \{2, 4, 6, 8\}$.

Then $A \cap B = \{2, 4\}$.

(ii) It M is the set of students of R. K. Mission Vidyamandira reading Microbiology and C is the student of the same college reading Chemistry then $M \cap C$ is the set of students reading Microbiology and Chemistry both.

(c) Disjoint

(*i*) Two sets (A & B) which have no common element are said to be disjoint or mutually exclusive.

(*ii*) It A and B are disjoint sets it is written as $A \cap B = \phi$.

(*iii*) Representation of disjoint set is shown in the diagram.

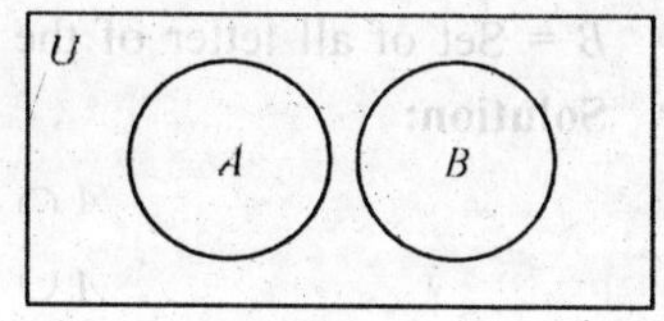

Fig. 8.3 *A* & *B* are disjoint set.

Example: (*i*) Let $A = \{2, 6, 7, 9\}$ and $B = \{1, 3, 5, 8\}$ then A and B are disjoint set, since no element is common to both A and B.

(d) Difference of two sets

(*i*) The difference (X) of the two sets (A & B) is formed by the elements of A which are not common to B, is called difference of the two sets.

(*ii*) We denote the difference of A and B by $A \sim B$ or $A - B$. Which is read as A difference B or A minus B.

(*iii*) By Venn diagram the area corresponding A – B has been shown by shaded line.

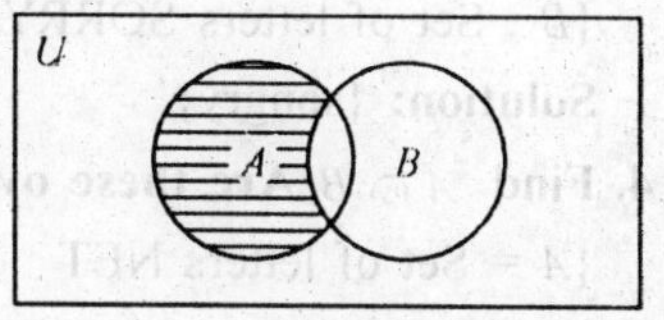

Fig. 8.4 $A \sim B$ shaded area.

Example: (*i*) Let $A = \{1, 2, 5, 6\}$ and $B = \{3, 5, 6, 8\}$ then $A \sim B = \{1, 2\}$ or $B \sim A = \{3, 8\}$.

(e) Complement of a set

(*i*) The complement of set A *i.e.*, A^c with respect to universal set U which have all elements belong to U but are not in A.

(*ii*) It is denoted by A^1(prime) or $\bar{A}$(bar) or A^c (complement).

(*iii*) $A^c = U - A = \{x \in U \text{ and } x \notin A\}$.

(*iv*) By Venn diagram the complement of a set is shown below.

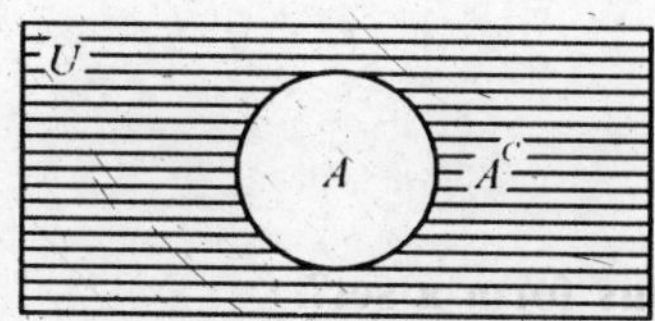

Fig. 8.5 A^c is shown with shaded area.

Example: (*i*) Let universal set $U = \{1, 2, 3, 4, 5, 6, 7, 8\}$ and $A = \{2, 4, 6\}$ $A^c = \{1, 3, 5, 7, 8\}$.

(*ii*) Let $U = \{x : x$ an letter of English alphabet$\}$.

$A = \{x : x$ a consonant letter of English alphabet$\}$.

$A^1 = U - A =$ {All letters of english alphabets} – {All consonant letter of english alphabet}.
= All vowels–letter of english alphabet *i.e.*, $\{a, e, i, o, u\}$.

1. Find $A \cap B$ and $A \cup B$

(*a*) A = Set of all the letters of the word POPULATION.

B = Set of all the letters of the word PEOPLE.

(*b*) A = Set of all letters of the word MANGO.

B = Set of all letter of the word ORANGE.

Solution:

$A \cap B = \{P, O, L\}$ [All the common elements]

$A \cup B = \{P, O, U, L, A, I, N, E\}$ [All the elements of A & B]

2. $A \cap B = \{AGNO\}$

$A \cup B = \{AEGMNOR\}$

3. Find $A \cup B$?

{A : Set of letters SONG

{B : Set of letters SORRY

Solution: {songry}.

4. Find $A \cap B$. Are these overlapping sets?

{A = Set of letters NET

{B = Set of letters NIGHT

Solution: $A \cap B = \{n, t\}$

yes, overlapping.

From the following pairs of sets identify, the disjoint or overlapping sets

1. A = Set of letters the word 'MAD'.

   B = Set of letters the word 'PEN'.
2. R = {Letters in the word 'AMERICA'}.

   S = {Letters in the word INDIA}.
3. E = Set of month having 30 days.

   F = Set of month having 31 days.

Solution:

1. Disjoint.
2. Over lapping.
3. Disjoint.

Which of the following collections form a set?

(*a*) The collection of all tall students of a college?

(*b*) The collection of all immortal men in the world.

(*c*) The collection of all intelligent students of a college.

Ans.

(*a*) The term tall is vague and it is not well defined. So the collection of all tall students in a college do not define set.

(*b*) We know that each man in the world is mortal. So collections of all immortal men, do not define a set.

(*c*) The term intelligent is vague & it is not well defined. So collection of all intelligent students of a college do not define set.

B. PROBABILITY

In the simplest way Probability means the "Chance of occurrence of a certain event when expressed quantitatively"

Probability is a concept which measures the degree of uncertainty. The uncertainty is numerically expressed as probability.

According to "Ya-Lin Chou" Probability is the science of decision making with calculated risks in the face of uncertainty.

- **Important Terms:**
- **Experiment:** A process which results in some well defined outcomes is known as experiment.
- **Random Experiment:**

I. Any natural phenomenon, yielding some results will be termed as random experiment when it is not possible to predict a particular result to turn out.

II. The term random attaches the meaning of unpredictability of an outcome.

- **Trial:** A experiment is called trial.
- **Outcome:** The result of a random experiment will be called an outcome.
- **Event:**

I. The result of an experiment in all possible forms are said to be event of that experiment.

II. Let us toss a coin. The result (event) is either head or tail.

III. It is a sub-collection of a number of sample points under the definite rule.

- **Null Event:**

I. An event having no sample point called a null event.

II. It is denoted by ϕ.

- **Simple Event:**

I. An event consisting of only one sample point of a sample space is called simple event.

II. It corresponds to single possible out come of the experiment.

- **Compound Event:**

I. When an event is decomposable into a number of simple event, then it is called a compound event.

II. It contains more than one sample point.

- **Sample Space:**

I. A set of all possible out comes from an experiment is called a sample space. It's symbol is '*S*'.

II. Each element or component of a sample space is called sample point.

III. Let us toss a coin. The result is either head or tail. Mark the point 0, 1 on a straight line. Thus we get two different points 0 & 1 on a straight line. These points are called sample points or even points.

IV. For a given experiment there are different possible out comes and hence different sample points. The collection of all such points is called sample space.

Example: Toss of one coin the sample space is $[H, T]$.

- **Discrete Sample Space:**

I. A sample space whose elements are finite or infinite but countable is called a *discrete sample space.*

II. If we toss a coin as many as we require for turning up on head then the sequence points $S_1 = (1)$, $S_2 = (0, 1)$, $S_3 = (1, 0, 1)$, $S_4 = (0, 0, 0, 1)$ etc. as a discrete sample space.

- **Continuous Sample Space:**
  I. A sample space whose elements are infinite uncountable or assume all the values on a real line R or on an interval of R is called a continuous sample space.
  II. Here the sample points build up a continuous and the sample space is said to be continuous.
- **Exhaustive Event or Cases:**
  I. It is the total number of all the possible out comes of random experiment.
  II. In an experiment 'E' with a set of events $A_1, A_2, A_3....A_n$, we say events are exhaustive if at least one of the event is sure to occur at every performance of that experiment.
  III. **Example:** When we throw a dice, then any one of the six faces (1, 2, 3, 4, 5, 6) may turn up & therefore there are six possible out comes. Hence there are six exhaustive events in throwing a dice.
- **Mutually Exhaustive Events:**
  I. If in an experiment the occurrence of one event prevents or rules out the happening of all other events in the same experiment then these events are said to be mutually exclusive events. On the other hand, two events A & B are called mutually exclusive if they never occur simultaneously.
  II. **Example:** Tossing a coin, the events head & tail are mutually exclusive because if the out come is head then the possibilities of getting a tail in the same trial is ruled out.
- **Equally Likely Events:**
  The out come of random experiments are said to be equally likely if there is no reason to expect any one in performance to other.
  Example: In a single cast of a fair dice each of the events 1, 2, 3, 4, 5, 6 is equally likely.
- **Favourable Case:**
  The cases which ensure the occurrences of an event are said to be favourable to the events.
- **Independent & Dependent Event:**
  I. When the experiments are conducted in such a way that the occurrence of an event in one trial does not have any effect on the occurrence of the other events at a subsequent experiment, then the events are said to be *independent* event.
  II. When the experiments are conducted in such way that the occurrence of an event in one trial does have some effect on the occurrence of the other events at a subsequent experiment, then the events are said to be *dependent* event.

  Example:
  I. If we draw a card from a pack (well shuffled cards) and again draw a card from the rest of pack of cards (containing 51 cards) then the second draw is *dependent on the first.*
  II. On the other hand, if we draw a second card from the pack by replacing the first card drawn, the second draw is known as *independent of the first.*
- **Statistical (EMPIRICAL) Probability:**

If an experiment is performed repeatedly under essentially homogeneous and identical condition, then limiting value of the ratio of the number of times the event occurs to the total number of trials of the experiment, as the number of trials increases indefinitely, is called the probability of happening of the event.

Symbolically, if N trials an event E happens M times, then the probabilities of the happening of E, denoted by $P(E)$ is given by

$$P(E) = \lim_{N \to \alpha} \frac{M}{N} \quad \text{(Provided the limit is finite and unique)}$$

- **Characteristics:**

  I. Since in the relative frequency approach, the probability is obtained objectively by repetitive empirical observation, it is also known as *"Empirical probabilities"*.

  II. An experiment is unique and non repeating only in the case of subjective probabilities. In other cases there are a large number of experiments to establish the chance of occurrence of an event. This is in the case of empirical probability.

  III. The empirical probability approaches the classical probability as the number of trials becomes indefinitely large.

- **Limitations:**

  I. If an experiment is repeated a large number of times, the experimental conditions may not remain identical homogeneous.

  II. The limit may not attain unique value however large N may be.

Classical Definition of Probability:

If an experiment 'E' has 'n' mutually, exclusive, equally, likely and exhaustive cases, out of which 'm' are favourable to the happening of the event 'A' then the probability of happening of 'A' is denoted by $P(A)$ *i.e.*,

$$P(A) = \frac{m}{n}$$

$$= \frac{\text{Number of cases favourable to } A}{\text{Total number of mutually, exclusive equally likely \& exhaustive cases}}$$

- **Working Procedure:**

  I. Enumerate all the possible out comes of the experiment such that they satisfy three criteria, "mutually exclusive", 'exhaustive' and 'equally likely'. Count the number (n) of such outcomes.

  II. Check how many of these cases are favourable to the event for which the probability is desired. Let this number be 'm'.

  III. Divide 'm' by 'n' and the result gives the probability of the event.

$$P = \frac{m\,(\text{Number of favourable out comes})}{n\,(\text{Total number of possible out comes})}$$

The probability q of failure of the event A is

$$q = P(\bar{A}) = \frac{n-m}{n} = 1 - \frac{m}{n} = 1 - P(A) = 1 - P$$

It is obvious that $P(A) + P(\bar{A}) = P + q = 1$.

- **Characteristics:**

  I. Probability is usually expressed by the symbol 'P'.

  II. It ranges from zero(0) to one (1).

$$0 \le P \le 1$$

  III. When $P(A) = 0$, A is called impossible event. It means there is no chance of happening or its occurrence is impossible.

  IV. If $P(A) = 1$ is called certain event *i.e.*, it mean the chances of an event happening are 100%.

- **Limitation of Classical Probability:**

  I. The classical probability is based on the feasibility of grouping the possible out comes of the experiment into 'mutually exclusive', 'exhaustive' and equally likely cases. Otherwise the classical theory is inapplicable.

II. Classical theory is only applicable when sample space is finitely enumerable.

III. This theory fails when the number of possible out comes is infinitely large.

IV. It may happen that all the out comes of an experiment may become impossible to enumerate. In these cases also classical definition fails.

Example 1. What is the probability of getting tail in a throw of a coin?

Solution: When we toss a coin there are two possible out comes viz. Head & Tail. In this case the number of possible events $n = 2$.

Number of favourable cases $m = 1$.

($\because$ The outcomes of tail is a favourable case).

$\therefore$ Probability $(P) = \frac{m}{n} = \frac{1}{2}$ or 50% of getting tail.

Example 2. What is the probability of getting even number in a single throw with dice?

Solution: The possible cases in the throw of a dice are six-viz. 1, 2, 3, 4, 5, 6. Favourable cases (even) are those 2, 4, 6 and these are these are three number: $m = 3$, $n = 6$.

$\therefore$ Probability (P) of getting even number

$$\therefore \quad \frac{m}{n} = \frac{3}{6} = \frac{1}{2} \text{ i.e., } 50\%.$$

Example 3. A bag contains 6 white balls, 9 block balls. What is the probability of drawing a black ball?

Solution: The total number of equally, likely & exhaustive events $n = 6 + 9 = 15$.

The number of favourable events $m = 9$ ($\because$ the number of black balls = 9).

$$\text{Probability of drawing black ball } (P) = \frac{m}{n} = \frac{9}{15} = \frac{3}{5}$$

• Theorems of Probabiltity:

- The addition theorem or the theorem of total probability.
- The multiplication theorem or theorem on compound probability.
- Binomial law of probability distribution.

(*a*) The Addition Theorem or the Theorem of Total Probability

If the events are mutually exclusive, then the probability of happening of any one of them is equal to the sum of the probabilities of the happening of the separate events.

If two events are A & B are mutually exclusive, then the probability of occurrence of either A or B is given by the sum of their probabilities *i.e.*,

Probability of (A or B) = Probability of A + Probability of B.

i.e., $$P(A + B) = P(A) + P(B).$$

Example 1. If the probability of horse A wining the race is $\frac{1}{5}$ and the probability of horse B wining the same race is $\frac{1}{6}$. What is the probability that one of the horses will win the race?

Solution: Probability of wining of the horse $A = \frac{1}{5}$.

Probability of wining of the horse $B = \frac{1}{6}$.

$$\therefore \quad P(A + B) = P(A) + P(B) = \frac{1}{5} + \frac{1}{6} = \frac{6+5}{30} = \frac{11}{30}$$

The probability of occurence of at-least one of te two events A and B (which may not be) is given by.

$$P(A + B) = P(A) + P(B) - P(AB)$$

In event A & B are independent then

$$P(AB) = P(A).P(B).$$

Example 2. A card is drawn from each of two well shuffled packs of cards. Find the probability that at least one of them is an ace. [*C. U. B. sc (Math), 73*]

Solution: Let us denote by

A = Event that the card from pack I is an ace.

B = Event that the card from pack II is an ace.

It is required to find $P(A + B)$

$$P(A + B) = P(A) + P(B) - P(AB)$$

Since there are 4 aces in a pack of 52 cards.

$$P(A) = \frac{4}{52} = \frac{1}{13}$$

Similarly $P(B) = \frac{4}{52} = \frac{1}{13}$. The events A & B are independent because the drawing of an ace from one pack does not affect the probability of drawing an ace from another pack. So

$$P(AB) = P(A).P(B)$$

$$= \frac{1}{13}.\frac{1}{13} = \frac{1}{169}$$

Then $$P(A + B) = \frac{1}{13} + \frac{1}{13} - \frac{1}{169} = \frac{13+13-1}{169} = \frac{25}{169}$$

(*b*) The Multiplication Theorem or Theorem on Compound Probability

• Simple and Compound Events:

I. A single event is called simple event.

II. When two or more than two simple events occur in connection with each other, then their simultaneous occurrence is called a compound event.

If A and B are two simple events, the simultaneous occurrence of A and B is called compound event and is denoted by AB.

• Conditional Probability:

The probability of happening of an event A, when it is known that B has already happened, is called the conditional probability of A and is denoted by $P(A/B)$ *i.e.*,

$P(A/B)$ = Conditional probability of A given that B has already happened.

$P(B/A)$ = Conditional probability of B given that A has already happened.

• Mutually Independent Event:

An event A is said to be independent of the event B if $P(A/B) = P(A)$ *i.e.*, the probability of the happening of A is independent of the happening of B.

- **Theorem on Compound Probability:**

The probability of simultaneous occurrence of two or more events is called compound probability.

The probability of simultaneous occurrence of two or more events A & B is equal to the probability of one of the events multiplied by the conditional probability of other, given the occurrence of the first *i.e.*,

$$P(AB) = P(A).P(B/A) = P(B).P(A/B)$$

Probability of (A & B) = Probability of A × conditional as probability of B, assuming A.

= Probability of B × conditional probability of A assuming B.

Example: A bag contains 5 white and 3 black balls. Two balls are drawn at random one after the other without replacement. Find the probability that both the balls drawn are black.

Solution: Let A denote the event "first drawing gives 3 black balls" *i.e.*, $P(A)$ and 'B' denotes second attempt *i.e.*, $P(B/A)$.

∴ Probability of drawing a black ball in the first attempt $P(A)$

$$P(A) = \frac{\text{Favourable event}}{\text{Total exhaustive event}} = \frac{m}{n} = \frac{3}{5+3} = \frac{3}{8}$$

Probability of drawing the second ball, given the first ball drawn is black.

$$P(B/A) = \frac{\text{Favourable event}}{\text{Total exhaustive event}} = \frac{m}{n} = \frac{2}{5+2} = \frac{2}{7}$$

Probability that both the balls drawn are black *i.e.*,

$$P(AB) = P(A) \times P(B/A)$$

$$= \frac{3}{8} \times \frac{2}{7} = \frac{6}{56} = \frac{3}{28}.$$

BAYE'S THEORY:

It is associated with the name of British Mathematician **Reverend Thomas Bayes**. This theorem is based on the concept that probability should be revised when new (*additional*) information is supplied by random experiment. It provides a probability law relating a *posterior probability* to *prior probability*.

- **Prior Probabilities (*Priori*):**

Revising a set of old probabilities i.e. probability before revision are called *Prior Probabilities* or simple *Priori*.

- **Posterior Probability:**

Revision of probability with new or added information i.e probability which have been revised in the light of sample information are called *Posterior Probabilities*.

Baye's theorem deals with a statistical method of evaluating new information and revising our prior estimates (*based on limited information*) of the probability.

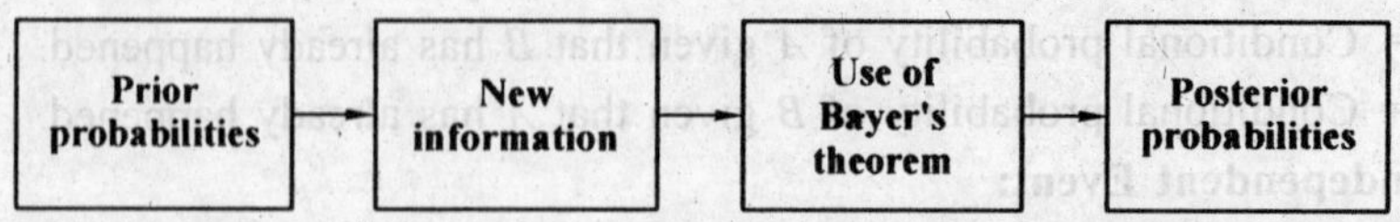

Fig. 8.6

- **Statement:**

1. An event A can occur only if any one of the set of exhaustive and mutually exclusive event $B_1, B_2, B_3....B_n$ occurs.
2. The unconditional probabilities $P(B_1), P(B_2), P(B_3).....P(B_n)$ and the conditional probabilities $P(A/B_i).....\ i = 1, 2, ...n$ for A to occur are known. Then the conditional probability $P\left(\frac{B_i}{A}\right)$ when has actually occurred is given by,

$$P\left(\frac{B_i}{A}\right) = \frac{P(B_i)P\left(\frac{A}{B_i}\right)}{P(B_1).P\left(\frac{A}{B_i}\right) + P(B_2).P\left(\frac{A}{B_2}\right) + ... + P(B_n).P\left(\frac{A}{B_n}\right)}$$

Example 1. Two identical boxes contain respectively 4 white and 3 red balls and 3 white and 7 red balls. A box is chosen random and a ball is drawn from it. If the ball is white what is the probability that it is from the first box?

Solution:

1. Let A be the event that the ball drawn is white and let B_1, B_2 be the events of the choosing the first and the second boxes respectively.
2. Then $P(B_1)$ = Probability of the selecting the first box $= \frac{1}{2}$ and $P(B_2)$ = Probability of the selecting second box $= \frac{1}{2}$.
3. $P(A/B_1)$ = Probability of selecting 1 white ball from the first box $= \frac{4}{7}$

   Similarly,

   $P(A/B_2)$ = Probability of selecting 1 white ball from the second box $= \frac{3}{10}$
4. By Bayes theorem we get

   $P(B_1/A)$ = The required probability that if the ball is white, then it is from the first box.

$$= \frac{P(B_1)P\left(\frac{A}{B_1}\right)}{P(B_1).P\left(\frac{A}{B_1}\right) + P(B_2).P\left(\frac{A}{B_2}\right)}$$

$$= \frac{\frac{1}{2}\times\frac{4}{7}}{\frac{1}{2}\times\frac{4}{7}+\frac{1}{2}\times\frac{3}{10}} = \frac{\frac{2}{7}}{\frac{2}{7}+\frac{3}{20}} = \frac{\frac{2}{7}}{\frac{40+21}{140}} = \frac{\frac{2}{7}}{\frac{61}{140}}$$

$$= \frac{2}{7}\times\frac{140}{61} = \frac{40}{61}.$$

Example 2. There are two identical bottles (A and B) containing respectively 4 red eyed and 2 white eyed *Drosophila* and 5 red eyed and 3 white eyed *Drosophila*. A bottle is chosen randomly and a *Drosophila* fly is taken from it. If the fly is white what is the probability that the red fly is taken from the second box (B)?

Solution:

1. Let E_1 the event that the white eyed fly taken from A.
2. E_2 be the event that the white eyed fly taken from B.

$$\therefore \quad P(E_1) = \frac{1}{2} \text{ and } P(E_2) = \frac{1}{2}$$

Let M denote the event that the fly is white.

$P(M/E_1)$ = Probability of the fly taken from bottle A *i.e.*, $= \frac{2}{6} = \frac{1}{3}$.

Similarly $P(M/E_2) = \frac{3}{8}$

According to Baye's Theories:

$$P(E_2/M) = \frac{P(E_2).P\left(\frac{M}{E_2}\right)}{P(E_1).P\left(\frac{M}{E_1}\right) + P(E_2).P\left(\frac{M}{E_2}\right)}$$

$$= \frac{\frac{1}{2} \times \frac{3}{8}}{\frac{1}{2} \times \frac{2}{6} + \frac{1}{2} \times \frac{3}{8}} = \frac{\frac{3}{16}}{\frac{1}{6} + \frac{3}{16}} = \frac{\frac{3}{16}}{\frac{8+9}{48}}$$

$$= \frac{3}{16} \times \frac{48}{17} = \frac{9}{17}$$

Example 3. Bottle 1 contains 4 curly winged *Drosophila* and 3 Bar eyed *Drosophila*. Box 2 contains 10 curly winged *Drosophila* and 20 Bar eyed *Drosophila*. A fly is drawn at random from bottle 1 and placed in bottle 2. Now take a second fly from bottle 2. Given that second fly is bar eyed, what is the probability that the first fly is also bar eyed?

Solution:

1. In our experiment, taking one fly from bottle 1 placing in an bottle 2 and then taking another fly from bottle 2 (all randomly). Define the events are E_1 and E_2.
2. E_1 = The fly taken first is bar eyed.
   E_2 = The fly taken second is also bar eyed.
3. The number of flies in bottle 2 changes offer the first fly taken is placed in it. Therefore the events E_1 & E_2 are dependent.

$$P(E_1) = \frac{3}{7}.$$

Example 4. (*a*) What is the probability that there are 53 Sundays in a leap year?

(*b*) What is the chance that a non leap year, selected at random, will contain 53 Sundays?

Solution:

1. In a leap year there 52 complete weeks and 2 days over *i.e.*, 52 × 7 + 2 = 366 days.
2. There are seven possible combinations for these two 'over' days.

(*i*) Sunday & Monday, (*ii*) Monday & Tuesday, (*iii*) Tuesday & Wednesday, (*iv*) Wednesday & Thursday, (*v*) Thursday & Friday, (*vi*) Friday & Saturday and (*vii*) Saturday & Sunday.

(*a*) In order that a leap year selected at random should contain 53 Sundays, one of the two over days must be Sunday. Since out of the above 7 possibilities 2 viz. (*i*) & (*vii*) are favourable to this event.

The required probability $= \frac{2}{7}$.

(*b*) A non leap year consist of 365 days. Therefore in a non leap year there are 52 complete weeks & 1 day over. Which can be one of the seven days of a week but there is only one favourable case that it is Sunday. Number of favourable cases (m) = 1. Total number of possible cases (n) = 7.

The required $= \frac{m}{n} = \frac{1}{7}$.

Example 5. (*a*) If $P(A) = 0.4$, $P(B) = 0.3$ and $P(A \cap B) = 0.2$ find $P(B/A)$ & $P(A/B)$

(*b*) If $P(A) = 0.25$, $P(B) = 0.6$ and $P(A \cap B) = 0.2$ are A and B independent events?

Solution:

(*a*)

$$P(B/A) = \frac{P(A \cap B)}{P(A)} = \frac{0.2}{0.4} = \frac{1}{2}$$

$$P(A/B) = \frac{P(A \cap B)}{P(B)} = \frac{0.2}{0.3} = \frac{2}{3}$$

(*b*) A & B are independent if $P(A\&B) = P(A).P(B)$ *i.e.*, $P(A \cap B) = P(A).P(B)$

$$P(A).P(B) = 0.25 \times 0.6 = 0.150 = 0.15$$

$$P(A \cap B) = 0.2 \text{ therefore } P(A \cap B) \neq P(A).P(B)$$

So these are not independent.

Example 6. If $P(A) = \frac{2}{5}$, $P(B) = \frac{1}{3}$ and $P(A \cup B) = \frac{1}{2}$ find $P(A/B)$ and $P(B/A)$.

Solution:

$$P(A \cup B) = P(A) + P(B) - P(A \cap B)$$

$$\frac{1}{2} = \frac{2}{5} + \frac{1}{3} - P(A \cap B)$$

$$\text{or } P(A \cap B) = \frac{2}{5} + \frac{1}{3} - \frac{1}{2} = \frac{12 + 10 - 15}{30} = \frac{7}{30}$$

$$P(A/B) \frac{P(A \cap B)}{P(B)} = \frac{\frac{7}{30}}{\frac{1}{3}} = \frac{7}{30} \times \frac{3}{1} = \frac{7}{10}$$

$$P(B/A) \frac{P(A \cap B)}{P(A)} = \frac{\frac{7}{30}}{\frac{2}{5}} = \frac{7}{30} \times \frac{5}{2} = \frac{7}{12}$$

Example 7. If $P(A) = 0.35$, $P(B) = 0.42$ and $P(A \cap B) = 0.28$ find $P(A/B)$. Are A & B two independent events?

Solution:

$$P(B/A) = \frac{P(A \cap B)}{P(A)} = \frac{0.28}{0.35} = \frac{0.4}{0.5} = \frac{4}{5},\ P(A/B) = \frac{0.28}{0.42} = \frac{0.4}{0.6} = \frac{0.2}{0.3} = \frac{2}{3}$$

A & B are independent if $P(A\&B) = P.A.P(B)$ *i.e.*, $P(A \cap B) = P(A).P(B)$

$$P(A \cap B) = 0.28$$

$$P(A) \times P(B) = 0.35 \times 0.42 = 0.147$$

$P(A \cap B) \neq P(A).P(B)$, so they are not independent.

Example 8. If $P(A) = \frac{1}{2}$, $P(B) = \frac{1}{3}$ and $P(A \cup B) = \frac{2}{3}$ then find $P(B/A)$ and show that A and B are two independent events.

Solution:

$$P(A \cup B) = P(A) + P(B) - P(A \cap B)$$

$$\frac{2}{3} = \frac{1}{2} + \frac{1}{3} - P(A \cap B)$$

$$P(A \cap B) = \frac{1}{2} + \frac{1}{3} - \frac{2}{3} = \frac{3+2-4}{6} = \frac{1}{6} \qquad \therefore P(A \cap B) = \frac{1}{6}$$

$$P(B/A) = \frac{P(A \cap B)}{P(A)} = \frac{\frac{1}{6}}{\frac{1}{2}} = \frac{1}{6} \times \frac{2}{1} = \frac{1}{3}$$

If they are independent *i.e.*, $P(A \cap B) = P(A).P(B)$

$$P(A).P(B) = \frac{1}{2} \times \frac{1}{3} = \frac{1}{6} \quad P(A \cap B) = \frac{1}{6}$$

Therefore $P(A \cap B) = P(A).P(B)$, so they are independent.

Example 9. For two events A and B, $P(A) = 0.5$, $P(B) = 0.6$ and $P(A \cap B) = 0.8$ find $P(A/B)$ and $P(B/A)$.

Solution:

$$P(A \cap B) = P(A) + P(B) - P(A \cap B)$$

$$0.8 = 0.5 + 0.6 - P(A \cap B)$$

$$\text{or } P(A \cap B) = 1.1 - 0.8 = 0.3$$

$$P(A/B) = \frac{P(A \cap B)}{P(\bar{B})} = \frac{0.3}{0.6} = \frac{1}{2} = 0.5$$

$$P(B/A) = \frac{P(A \cap B)}{P(A)} = \frac{0.3}{0.5} = 0.6.$$

Example 10. If $P(E) = 0.40$, $P(F) = 0.35$ & $P(E \cup F) = 0.55$ find $P(F/E)$.

Solution:

$$P(E \cup F) = P(E) + P(F) - P(E \cap F)$$

$$\text{or } P(E \cap F) = 0.40 + 0.35 - 0.55 = 0.75 - 0.55 = 0.20$$

$$P(F/E) = \frac{P(E \cap F)}{P(E)} = \frac{0.2}{0.4} = \frac{1}{2} = 0.5.$$

Example 11. (*a*) A and B are two independent events such that $P(A \cup B) = 0.5$ and $P(A) = 0.2$ find $P(B)$.

(*b*) If A & B are two independent events such that $P(A) = 0.3$ and $P(B) = 0.6$ find $P(A \text{ or } B)$.

Solution:

(*a*) $P(A \cup B) = P(A) + P(B) - P(A \cap B)$

$P(A \text{ or } B)$

$$0.5 = 0.2 + P(B) - P(A).P(B)$$

$$P(B) - 0.2\,P(B) = 0.5 - 0.2$$

$$P(B)\,0.8 = \frac{0.3}{0.8} = \frac{3}{8}.$$

(*b*) $P(A \text{ or } B) = P(A) + P(B) - P(A).P(B)$

$$= 0.3 + 0.6 - 0.3 \times 0.6$$

$$= 0.3 + 0.6 - 0.18 = 0.9 - 0.18$$

$$= 0.72.$$

Example 12. A bag contains 7 red and 5 white flowers. Find probability of taking out of the bag of each type.

Solution:

Total number of flowers = 7 + 5 = 12

Total number of events taking out 1 flower = 12

A = Number of events taking out one red flower = 7

Probability of taking out one red flower *i.e.*, $P(A) = \frac{7}{12}$

B = Number of events taking out one white flower = 5

Probability of taking out one white flower *i.e.*, $P(B) = \frac{5}{12}$

Probability of getting a red and white flower *i.e.*,

$$P(A + B) = P(A) + P(B)$$

$$= \frac{7}{12} + \frac{5}{12} = \frac{12}{12} = 1.$$

Example 13. Of every 100 mosquitoes selected, it is found in average, 25 were in grade A & 10 were in grade B. Use the addition rule of probabilities that the selected mosquitoes will have either grade A or grade B.

Solution: Total number = 100

Probability of grade $A = P(A) = \frac{25}{100} = 0.25$

Probability of grade $B = P(B) = \frac{10}{100} = 0.10$

Since both the events are mutually exclusive, prohabilities one of them happens can be obtained by using the addition rule

$$P(A \text{ or } B) = P(A) + P(B)$$
$$= 0.25 + 0.10 = 0.35.$$

Example 14. In an examination of Serampore College 30% of the students failed in Chemistry and 20% have failed in Botany and 10% have failed in both Chemistry and Botany. A student is selected random.

(*a*) What is the probability that a student has failed in Chemistry if it is known that he has failed in Botany?

(*b*) What is the probability that the student has failed either in Chemistry or in Botany?

Solution:

I. Let the C and B the events that a student selected at random 'fail in Chemistry' and 'fail in Botany' respectively.

II. Them $P(C) = \frac{30}{100} = 0.3$, $P(B) = \frac{20}{100} = 0.2$ and $P(C \cap B) = \frac{10}{100} = 0.1$.

(*a*) The required probability that the student selected has failed in Chemistry given that he has failed in Botany.

$$P(C/B) = \frac{P(C \cap B)}{P(B)} = \frac{0.1}{0.2} = 0.5$$

(*b*) The required probabilities that the student has failed either in Chemistry or in Botany.

$$P(C \cap B) = P(C) + P(B) - P(C \cap B) = 0.3 + 0.2 - 0.1 = 0.4.$$

Example 15. The probability that a student of Bidhan Nager Govt. college passes a Biophysics test is $\left(\frac{2}{3}\right)$ and the probability that he passes in both Biophysics and Statistics test, $\left(\frac{14}{45}\right)$. The probability that he passes at least one test is $\left(\frac{4}{5}\right)$. What is the probability that he passes in Statistics test?

Solution:

I. The student passes in Biophysics test *i.e.*, $P(B)$

The student passes in Statistics test *i.e.*, $P(S)$

II. $P(B) = \frac{2}{3}$, $P(B \cap S) = \frac{14}{45}$

The probability that he passes at least in one test *i.e.*, $\frac{4}{5}$, $P(B \cup S) = \frac{4}{5}$ we have to want $P(S) = ?$

$$P(B \cup S) = P(B) + P(S) - P(B \cap S)$$

$$\frac{4}{5} = \frac{2}{3} + P(S) - \frac{14}{45}$$

$$P(S) = \frac{4}{5} - \frac{2}{3} + \frac{14}{45}$$

$$= \frac{36 - 30 + 14}{45} = \frac{50 - 30}{45}$$

$$= \frac{20}{45} = \frac{4}{9}.$$

Example 16. In a test examination of Charusila Bose girls high school 20% of the students failed in Mathematics, 15% of the students failed in Physical science and 10% of the students failed in both Mathematics and Physical science. A student is selected at random. If she failed in Mathematics, what is the probability that she also failed in Physical science?

Solution:

I. Let M and P_{sc} the events that a student selected random 'failed in Mathematics' & 'failed in Physical science' respectively.

II. Then $P(M) = \frac{20}{100} = 0.2$, $P(P_{sc}) = \frac{15}{100} = 0.15$ & $P(M \cap P_{sc}) = \frac{1}{10} = 0.1$

The required probabilities that student selected has failed in Mathematics & given that she also failed in Physical science.

$$P(M/P_{sc}) = \frac{P(M \cap P_{sc})}{P(P_{sc})} = \frac{0.1}{0.15} = \frac{10}{15} = \frac{2}{3}.$$

Example 17. In a test examination of St. Agnes convent school Howrah, 25% of the students have failed in English, 18% in Mathematics and 10% in both subject. A student is selected random. (*a*) What is the probabilities that the student has failed is English if it is known that she has failed in Mathematics? (*b*)What is the probabilities that the student has failed either in English or in Mathematics?

Solution:

I. Let E and M the events that a student selected random 'fail in English and 'fail in Mathematics' respectively.

II. Then $P(E) = \frac{25}{100} = 0.25$, $P(M) = \frac{18}{100} = 0.18$ and $P(E \cap M) = \frac{1}{10} = 0.1$.

(*a*) The required probabilities that the student selected has failed in English and given that she also failed in Mathematics.

$$P(E/M) = \frac{P(E \cap M)}{P(M)} = \frac{0.1}{0.18} = \frac{10}{18} = \frac{5}{9}.$$

(*b*) The required probability that a student has failed either in English or in Mathematics.

$$P(E \cup M) = P(E) + P(M) - P(E \cap M)$$

$$= 0.25 + 0.18 - 0.1$$

$$= 0.43 - 0.1 = 0.33.$$

- **Using Permutation and Combination:**

Example 1. 10 PG students of Serampore College are to be seated around a circular table. Find the probability that two particular students will always together.

Solution:

I. Total number of ways in which 10 students can sit around circular table = $\underline{|9}$.

II. It two particular students A & B sit together, then the number of ways in which 9 students (considering the pair A & B as unit) can sit around the table $= \lfloor 2 \times \lfloor 8$.

III. P = (The probability that A and B will sit together)

$$= \frac{\text{Favourable case}}{\text{Total number of cases}} = \frac{\lfloor 2 \times \lfloor 8}{\lfloor 9} = \frac{2 \times \lfloor 8}{9 \times \lfloor 8} = \frac{2}{9}.$$

Example 2. If 8 laboratory specimens be distributed among 10 Zoology hons students of R. P. M. College of Uttarpara, find the probability that a particular students receives 5 specimens.

Solution:

I. Every specimen can be given to any one of the 10 student. The first can be distributed in 10 ways, second also in 10 ways and so on.

II. The total number of ways in which 8 specimens can be distributed at random among 10 students.

$$= 10 \times 10 \times 10 \ldots 11 \text{ (8 times)} = 10^8.$$

III. 5 specimens can be given to any particular students in ${}^{10}C_5$. Now $8 - 5 = 3$ specimens are left which have to be distributed among $10 - 1 = 9$ students. This can be done in $9 \times 9 \times 9 = 9^3$ ways.

IV. Required probability $= \dfrac{{}^{10}C_5 \times 9^3}{10^8} = \dfrac{\dfrac{\lfloor 10}{\lfloor 10-5 . \lfloor 5} \times 9^3}{10^8} = \dfrac{\dfrac{10 \times 9 \times 8 \times 7 \times 6 \times \lfloor 5}{\lfloor 5 . 5 \times 4 \times 3 \times 2} \times 9^3}{10^8}$

$$= \frac{9 \times 4 \times 7 \times 9^3}{10^8} = \frac{252 \times 729}{10^8} = \frac{183708}{10^8} = 0.0183708.$$

Example 3. A student of Vidyasagar College, Calcutta is to answer 10 out of 13 questions in Zoology in an examination. Such that he/she must choose at least 4 from first five questions. Find the number of choices available to him/her.

Solution: Two cases are possible.

(*a*) Selecting 4 questions out of first five questions and 6 out of remaining $(13 - 5 = 8)$ 8 questions.

$\therefore$ The number of choice in this case $= {}^5C_4 \times {}^8C_6 = \dfrac{5 \times 8 \times 7}{1 \times 2} = 140.$

(*b*) Selecting 5 out of first five questions and 5 out of remaining 8 questions.

Number of choices $= {}^5C_5 \times {}^8C_5$

$$= \frac{\lfloor 5}{\lfloor 5 \times 1} = \frac{8 \times 7 \times 6 \times \lfloor 5}{\lfloor 5 \times 3 \times 2} = 1 \times 8 \times 7 = 56.$$

$\therefore$ Total number of choices $= 140 + 56 = 196.$

Example 4. In the seminar library of Microbiology department of R. K. Mission Vidyamandira 25 books are placed at random in a shelf. Find the probability that a particular part of books shall be (*a*) Always together and (*b*) Never together.

Solution:

I. Since 25 books can be arranged among themselves in $\lfloor 25$ ways, the exhaustive number of cases $= \lfloor 25$.

II. Let us now regard that the two particular books are tagged together, so that we shall regard them as a single book. Thus we have $(25 - 1) = 24$ books which can be arranged among themselves in $\lfloor 24$ ways. Hence associating these two operations, the number of favourable cases for getting a particular pair of books always together is $\lfloor 24 \times \lfloor 2$.

$\therefore$ Required probability $= \frac{\lfloor 24 \times \lfloor 2}{\lfloor 25} = \frac{\lfloor 24 \times 2 \times 1}{25 \times \lfloor 24} = \frac{2}{25}$

III. Total number of arrangement of 25 books among themselves is $\lfloor 25$. The total number arrangement that particular pair of books will always be together is $\lfloor 24 \times \lfloor 2$. Hence the number of arrangements in which a particular pair of book is never together is

$$\lfloor 25 - 2 \times \lfloor 24 = 25 \times \lfloor 24 - 2 \times \lfloor 24 = \lfloor 24 (25 - 2) = \lfloor 24 \times 23$$

$\therefore$ Required probability $= \frac{23 \times \lfloor 24}{\lfloor 25} = \frac{23 \times \lfloor 24}{25} = \frac{23}{25}$.

Alternative

P = [A particular pair of books shall never be together]

$1 - P$ (A particular pair books is always together)

$$1 - \frac{2}{25} = \frac{25-2}{25} = \frac{23}{25}.$$

Example 5. In how many ways can 9 books of Biostatistics be divided among 3 Zoology honors students?

Solution: 18 books can be dived among 3 students *i.e.*, $9 \div 3 = 3$

Here total number of cases $= \lfloor 9$

Favoural cases $\lfloor 3^3$

The required number of ways $= \frac{\lfloor 9}{\lfloor 3^3}$

$$= \frac{9 \times 8 \times 7 \times 6 \times 5 \times 4 \times \lfloor 3}{\lfloor 3 . \lfloor 3 . \lfloor 3} = \frac{9.8.7.6.5.4 \lfloor 3}{3.2 \times 3.2 \times \lfloor 3} = 1680.$$

Example 6. There are five questions in a question paper. In how many ways can a student solve one or more question?

Solution:

I. The student can dispose of each question in two ways. He may either solve it or leave it. Thus the number of ways of disposing of all questions $= 2^5$.

II. But this includes the cases in which he has left all the question unsolved. Hence total number of ways of solving the paper

$$2^5 - 1 = (2 \times 2 \times 2 \times 2 \times 2) - 1 = 32 - 1 = 31.$$

Example 7. Subir Das has 7 friends. In how many ways he can invite one or more of them to a dinner?

Solution:

I. Subir may invite one or more friends by selecting either 1 friend or 2 friends or 3 friends or 4 friends or 5 friends or 6 friends or 7 friends out of 7 friends.

II. Hence the required number of ways

$${}^7C_1 + {}^7C_2 + {}^7C_3 + {}^7C_4 + {}^7C_5 + {}^7C_6 + {}^7C_7 = 7 + 21 + 35 + 35 + 21 + 7 + 1 = 127.$$

Example 8. A Genetics question paper contains 12 questions, divided into three parts. *A* contains 6 question part *B* & part *C* contain 3 question each. A candidate is required

to attempt 6 questions selecting at least 3 questions from part *A* and at least one from each of the parts of *B* and *C*. In how many ways can the candidate select his 6 questions?

Solution: Satisfying the condition stated in the problem, a candidate can select 6 questions in the following different ways.

(*a*) 3 questions from part *A*, 2 questions from part *B* & 1 question from part *C*.
(*b*) 3 questions from part *A*, 1 question from part *B* & 2 questions from part *C*.
(*c*) 4 questions from part *A*, 1 question from part *B* & part *C*.

Now 3 questions can be selected from part A in ${}^6C_3$ ways and for each such selection 2 questions out of 3 questions of part *B* can be selected in ${}^3C_2$ ways. Hence 3 questions from part *A* and 2 questions from part *B* can be selected in ${}^6C_3 \times {}^3C_2$ ways. Again rest one question can be selected out of 3 questions of part *C* in ${}^3C_1$ ways.

$\therefore$ The number of selection in case (*a*)

$$= {}^6C_3 \times {}^3C_2 \times {}^3C_1 = \frac{6.5.4}{3.2} \times \frac{3.2}{2} \times 3.0 = 20 \times 3 \times 3 = 180$$

Similarly in case of (*b*)

$$= {}^6C_3 \times {}^3C_1 \times {}^3C_2 = \frac{6.5.4}{3.2} \times 3 \times \frac{3.2}{2} = 180$$

Again in case of (*c*)

$$= {}^6C_4 \times {}^3C_1 \times {}^3C_1 = \frac{6.5}{2} \times 3 \times 3 = 135$$

Therefore the candidate can select 6 questions in (180 + 180 + 135) = 495 ways.

Example 9. A Biostatistics question paper contains 10 questions which are divided into two groups each containing 5 questions. A student is asked to answer 6 questions only, add to choose at least 2 questions from each group. In how many different ways can the student make up his or her choice?

Solution: Satisfying the condition stated in the above problem, a student select six questions in the following ways.

Let the groups are *A* and *B* and at least 2 questions should be taken from each group.

I. 4 questions from group *A* and 2 questions from group *B* in

$${}^5C_4 \times {}^5C_2 \text{ ways} = \frac{5 \times \lfloor 4}{\lfloor 4} \times \frac{5 \times 4 \lfloor 3}{2.\lfloor 3} = 50.$$

II. 3 questions from group *A* and 3 questions from group *B* in

$${}^5C_3 \times {}^5C_3 \text{ ways} = \frac{5 \times 4 \times \lfloor 3}{2 \times \lfloor 3} \times \frac{5 \times 4 \times \lfloor 3}{2 \times \lfloor 3} = 100.$$

III. 2 questions from group *A* and 4 questions from group *B* in

$${}^5C_2 \times {}^5C_4 \text{ ways} = \frac{5 \times 4 \lfloor 3}{2.\lfloor 3} \times \frac{5.\lfloor 4}{\lfloor 4.1} = 50.$$

$$= 10 \times 5 = 50$$

$\therefore$ Therefore total number of selection = 50 + 100 + 50 = 200.

Example 10. In how many different ways 6 questions can be selected from a set of 9 questions.

Solution: The required number of ways is ${}^9C_6$ ways

$$= \frac{9.8.7\lfloor 6}{3.2\lfloor 6} = 3 \times 4 \times 7 = 84.$$

Example 11. In how many ways can 18 different Biophysics books can be divided equally among 3 Zoology students?

Solution: The required number of ways $= \frac{\lfloor 18}{\lfloor 6} = \frac{18.17.16.15.14.13.\lfloor 12}{6.5.4.3.2.\lfloor 12} = 18564.$

Example 12. If a man and woman are heterozygous for a gene and if they have three children, what is the probability that all three also be heterozygous?

Solution: Both man and woman are heterozygous they inherit the trait in 1 : 2 : 1 *i.e.*, heterozygous are $\frac{2}{4} = \frac{1}{2}$.

Here the number of children *i.e.*, $n = 3$

$$P = \left(\frac{1}{2}\right)^3 = \frac{1}{8}.$$

Example 13. A DNA segment is 100 base pairs long. (*a*) How many different sequences are possible? (*b*) A polypeptide of four amino acid residues could have different sequences?

Solution:

(*a*) 4 bases are available (*A*, *T*, *G*, *C*)

Therefore sequences are $(4)^{100}$

(*b*) There are 20 amino acids. A polypeptide consist of 4 amino acids residues

Therefore sequence will be $(20)^4$.

Example 14. (In reference to genetic code):

(*a*) How many different triplet arrangements of two four ribonucleotides are possible, if they can be used more than once in a triplet?

(*b*) How many different triplet arrangements of the four ribonucleotides are possible, if they are used only once in each triplet?

(*c*) How many triplet arrangements consist of only one ribonucleotides?

(*d*) How many triplets have two ribonucleotides the same?

Solution:

(*a*) $(4)^3 = 64$.

(*b*) ${}^nP_k = \frac{\lfloor n}{\lfloor n-k}(n = 4, k = 3) \therefore \frac{\lfloor 4}{\lfloor 4-3} = \frac{4.3.2.1}{1} = 24.$

(*c*) 4, *AAA*, *GGG UUU*, *CCC* (as RNA no Thymine).

(*d*) $64 - (24 - 4) = 36$.

Example 15. A cross is made with two *Drosophila* strains in which it is expected that $\frac{1}{4}$ of the progeny will have an ebony coloured body (Chromosome–3) and $\frac{1}{2}$ will have the trait called lobe (Chromosome–2).

(*a*) What is the probability that any progeny will be expected to exhibit both traits?

(*b*) If 4 progeny are sampled at random, what is the probability all four will be lobe progeny?

(*c*) What is the probability that the progeny will not have either of the traits?

(*d*) What is the probability that the progeny will have at least one of the traits or both of them?

Solution:

(*a*) Lobe $= \frac{1}{2}$, ebony $= \frac{1}{4}$

Probabilities of any progeny exhibit both the traits.

$$P = \frac{1}{2} \times \frac{1}{4} = \frac{1}{8}.$$

(*b*) $n = 4$

$$\left(\frac{1}{8}\right)^4 = \frac{1}{4096}.$$

(*c*) Progeny not have either traits

$$\text{Not being lobe} = 1 - \frac{1}{2} = \frac{1}{2}$$

$$\text{Not being ebony} = 1 - \frac{1}{4} = \frac{3}{4}$$

$$\text{Hence the probabilities} = \frac{1}{2} \times \frac{3}{4} = \frac{3}{8}$$

(*d*) $1 - \frac{3}{8} = \frac{8-3}{8} = \frac{5}{8}.$

Example 16. Biochemical evidence shows that tripeptide consists of phenylalanine, methionine & lysine.

(*a*) How many permutations are possible?

(*b*) How many combinations are possible?

Solution: $n = 3$ and $r = 3$

(*a*) $${}^nP_r = \frac{\lfloor n}{\lfloor n-r} = \frac{\lfloor 3}{\lfloor 3-3} = \frac{3.2.1}{\lfloor 0} = \frac{3.2.1}{1} = 6$$

(*b*) $${}^nC_r = \frac{\lfloor n}{\lfloor r \lfloor n-r} = \frac{\lfloor 3}{\lfloor 3 \lfloor 3-3} = \frac{\lfloor 3}{\lfloor 3 \lfloor 0} = \frac{\lfloor 3}{\lfloor 3.1} = 1.$$

Example 17. In a family of six children, what is the probability that at least three are girls?

Solution: $N = 6$

Probability of boy (p) $= \frac{1}{2}$ & girl (q) $= \frac{1}{2}$.

Applying binomial expansion

$$\left(\frac{1}{2} + \frac{1}{2}\right)^6$$

$$= p^6 + 6p^5q + 15p^4q^2 + 20p^3q^3 + 15p^2q^4 + 6pq^5 + q^6$$

$$= \left(\frac{1}{2}\right)^6 + 6\left(\frac{1}{2}\right)^5 \frac{1}{2} + 15\left(\frac{1}{2}\right)^4\left(\frac{1}{2}\right)^2 + 20\left(\frac{1}{2}\right)^3\left(\frac{1}{2}\right)^3 + 15\left(\frac{1}{2}\right)^2\left(\frac{1}{2}\right)^4$$

$$+6\left(\frac{1}{2}\right)\left(\frac{1}{2}\right)^5 + \left(\frac{1}{2}\right)^6$$

$$= \frac{1}{64} + 6\frac{1}{32.2}\frac{1}{2} + 15\frac{1}{16.4}\frac{1}{4} + 20\frac{1}{8.8}\frac{1}{8} + 15\frac{1}{4.16}\frac{1}{16} + 6\frac{1}{2.32}\frac{1}{32} + \frac{1}{64}$$

$$= \frac{1}{64} + \frac{6}{64} + \frac{15}{64} + \frac{20}{64} + \frac{15}{64} + \frac{6}{64} + \frac{1}{64}.$$

Probability of at least 3 girls *i.e.*,

$$= \frac{20}{64} + \frac{15}{64} + \frac{6}{64} + \frac{1}{64} \qquad [20p^3q^3 + 15p^2q^4 + 6pq^5 + q^6]$$

$$= \frac{20 + 15 + 6 + 1}{64} = \frac{42}{64}.$$

Example 18. A normal man (*A*) whose grandfather had galactosemia and normal woman (*B*) whose mother was galactosomic, want to produce a child what is the probability that their first child will be galactosemic?

Solution:

I. The normal man's grandfather had galactosemia. He get this gene from his father $\left(\frac{1}{2}\right)$. So he is heterozygous. The normal woman whose mother was also galactosemic. Therefore, she is heterozygous.

II. So the probability of galactosemic child is $\frac{1}{4}$ & normal is $\frac{3}{4}$.

III. Therefore the probabilities of galactosemic child is $\frac{1}{4} \times \frac{1}{2} = \frac{1}{8}$.

CHAPTER

CHI-SQUARE TEST

CHI-SQUARE TEST:

It is a statistical test involves the calculation of a quantity which is used to compare an "observed" ratio with an "expected" or "theoretical" ratio and to determine how closely the former fits the latter.

The measure of chi-square enables us to find out degree of discrepancy between 'observed' frequencies and 'theoretical' or 'expected' frequencies and thus to determine whether the discrepancy between observed and theoretical frequencies is due to the error of sampling or due to the chance.

The chi-square is computed on the basis of frequencies in a sample and thus the value of chi-square so obtained is a *statistic*. Chi-square is not a parametric test as its value is not derived from the observations in a population. Hence chi-square test is a Non Parametric Test.

Chi-square (X^2) test:

A statistical test to determine if the "observed" numbers deviate from those "expected" or "theoretical" number under a particular hypothesis.

$$X^2 = \frac{\sum_{i=1}^{n}\left[1(O-E)1 - \frac{1}{2}\right]^2}{E}$$

where O = Observed frequency

E = Expected frequency

$\frac{1}{2} = .5$ = Yates correction*

Chi-Square (x^2) test was first used in testing statistical hypothesis by ***Karl Pearson*** in the year 1900.

- **Working Procedure:**

1. Calculate all the expected frequencies *i.e.*, E for all values of $i = 1, 2, 3.....n$.
2. Take the difference between each observed frequency 'O' and the corresponding expected frequency 'E' for each value of i *i.e.*, $(O - E)$.
3. Square the difference for each value of i *i.e.*, $(O - E)^2$ for all values of $i = 1, 2, 3....n$.
4. Divide each square difference by the corresponding expected frequency *i.e.*, Calculate $\frac{(O-E)^2}{E}$ for all values of $i = 1, 2, 3.....n$.

---

* Note: Yates correction: It is $\frac{1}{2}$ or 0.5, which is used to find out the formula with accuracy of probability and used only where degrees of freedom = 1, and expected classes are small.

5. Add all these quotients obtained in the steps '4' *i.e.*,

$$X^2 = \frac{\sum_{i=1}^{n}\left[\,|(O-E)| - \frac{1}{2}\right]^2}{E} \qquad \frac{1}{2} = \text{Yates correction}$$

It is the required value of chi-square.

- **Important Characteristics of Chi-square:**

I. The value of chi-square is always positive as each pair is squared up.

II. X^2 (chi-square) will be zero if each pair is zero and it may assume any value extending to infinity, when the difference between the observed frequency and expected frequency in each pair are unequal. Thus chi-square lies between O and ∞.

III. Chi-square is a statistic not a parameter.

Useful Terms:

- **Statement Hypothesis:**

(*i*) Any statement or assertion about a statistical population or the values of its parameters is called statistical hypothesis.

(*ii*) There are two types of hypothesis-viz. simple and composite.

(*a*) *Simple Hypothesis:* A statistical hypothesis which specifies the population completely (*i.e., the probability distribution and all parameters are known*) is called simple Hypothesis.

(*b*) *Composite Hypothesis:* A statistical hypothesis which does not specify the population completely (*i.e., either the form of probability distribution or some parameters remain unknown*) is called Composite Hypothesis.

- **Test of Hypothesis:**

A test of hypothesis is a procedure which specifies a set of "rule for decision" whether to "accept" or "reject" the hypothesis under consideration (*i.e., null hypothesis*).

- **Null Hypothesis:**

I. A statistical hypothesis which is set up (*i.e. assumed*) and whose validity is tested for possible rejection on the basis of sample observations is called a "Null Hypothesis".

II. It is denoted by H_0 and tested against alternatives.

III. Tests of hypothesis deal with rejection or acceptance of null hypothesis only.

Prof R. A. Fisher remarked "Null hypothesis is the hypothesis which is to be tested for possible rejection under the assumption it is true".

- **Alternative Hypothesis:**

I. The negation of null hypothesis is called the alternative hypothesis.

II. It is denoted by H_1 and H_∞.

III. It is not tested, but its acceptance (rejection) depends on the rejection (acceptance) of the null hypothesis.

IV. Alternative hypothesis contradicts the null hypothesis.

- **Critical Region or Rejection Region:**

I. The set of values of the test statistic which lead to rejection of the null hypothesis is called Critical Region of the test.

II. The probability with which a true null hypothesis is rejected by the test is often referred to as "size of the Critical Region".

III. On the other hand which lead to the acceptance of null hypothesis gives us a region called a "Acceptance Region".

IV. The critical value separates the rejection region from the acceptance region.

- **Test Statistic:**

I. After setting up the null hypothesis and alternative hypothesis, the test statistic is computed.

II. It is a statistic based on appropriate probability distribution.

III. It is used to test whether the null hypothesis set up should be accepted or rejected.

- **Level of Significance:**

I. The maximum probability with which a true null hypothesis is rejected is known as level of significance of the test.

II. It is denoted by α.

III. In framing decision rules, the level of significance is arbitrarily chosen in advance depending on the consequence of statistical decision.

IV. Generally 5% or 1% level of significance is taken although other levels such as 2% or 1/2% are also used.

- **Degrees of Freedom:**

It is the values of a sample which are freely variable without affecting the mean or it is an integer used to determine whether a chi-square value is statistically significant.

The number of data that are given in the form of a series of variables in a row or column or the number of frequencies that are put in cells in a contingency table which can be calculated independently is called the degrees of freedom and is denoted by ***df***.

- **Calculation of Degrees Freedom:**

(*i*) If the data is given in the form of a series of variables in a row or column, then the degrees of freedom (*df*) = (the number of items in the series) –1 *i.e.*, ***df* = *n* – 1**. Where '*n*' is the number of observation.

(*ii*) When the number of frequencies are put in cells in a contingency table, the degrees of freedom will be the product of (*number of rows less one*) and the (*number of columns less one*) *i.e.*, $df = (R - 1)(C - 1)$ where '*R*' is the number of rows '*C*' is the number of columns.

- **Conditions for Using the Chi-square Test:**

I. Each of the observations making up the sample for this test should be independent of each other.

II. The X^2 test applied in a fourfold table, will not give a reliable result with one degree of freedom if the expected value in any cell is less than 5. In such cases, to apply x^2 test yates correction is necessary.

III. The total number of observations used in this test must be large *i.e.*, $n \geq 50$.

IV. This test is used only for drawing inferences by testing hypothesis. It cannot be used for estimation of parameter or any other value.

V. It is wholly dependent on the degrees of freedom.

VI. The frequencies used in x^2 test should be absolute & not relative in terms.

VII. The observation collected for x^2 -test should be on random basis of sampling.

Types:

- **Chi-square:**

(*a*) Goodness of fit.

(*b*) Contingency Chi-square.

(*c*) Homogeneity Chi- square.

1. Test for Goodness of Fit (*Pearsonian* – x^2):

Chi-square test is also applied as a test of "goodness o fit" to determine whether the actual or (*observed*) numbers or frequencies are similar or in "good agreement" with the expected or (*theoretical*) number of frequencies.

Thus the test is called "goodness of fit"

$$x^2 = \frac{\sum\left[|(O-E)| - \frac{1}{2}\right]^2}{E}$$

O = Observed

E = Expected

$\frac{1}{2}$ = Yates correction

- **Working Procedure:**

In order to test "the goodness of fit" of the observed results, it is necessary to find

I. Deviation between the observed and the expected results.

II. If the expected value or frequencies in any observation is less than 5 in one degrees of freedom, Yates correction is necessary *i.e.*, reduction of $\left[(|O-E|) - \frac{1}{2}\right]$. In other words subtract 0.5 from absolute difference between observed and expected frequencies. (*Generally in monohybrid cross*).

III. Here the degrees of freedom (*df*) the number of items in the series – 1 = $n - 1$.

- **Null Hypothesis (*assume*) for Goodness of Fit:**

I. 1 : 1 Monohybrid test cross ratio.

II. 3 : 1 Simple monohybrid Mendelian ratio (F_2).

III. 9 : 3 : 3 : 1 Simple Mendelian dihybrid ratio (F_2).

IV. 1 : 1 : 1 : 1 Dihybrid test cross ratio.

Example: In a cross between black and white coat coloured mice, the F_2 individual segregated into 787 black and 277 white coat coloured individuals. If you have to test that these results agree with the expected ratio 3 : 1. Then apply chi-square P = 5%.

Solution:

(*a*) *Null hypothesis:* 3 : 1

(*b*) *Alternative hypothesis:* 1 : 1

(*c*) *Calculations:* $X^2 = \frac{(|O-E|-.5)^2}{E}$

Observed (O)	*Expected (E)*	*|O – E|*	*{(|O – E|) – .5}*	*{(|O – E|) – .5}²* $= D^2$	$X^2 = \frac{D^2}{E}$
Black = 787	798	787 – 798 = – 11	|11| – .5 = 10.5	110.25	$\frac{110.25}{798} = 0.138$
White = 277	266	277 – 266 = 11	11 – .5 = 10.5	110.25	$\frac{110.25}{266} = 0.414$
Total = 1064					X^2 = .552

(*d*) *Critical Value:* The control value of x^2 at 0.05 for $df = 2 - 1 = 1$ is = 3.84

(*e*) *Decision:* Since calculated value of $X^2 = .55$, so the null hypothesis is accepted. Therefore the data has good fit to the ratio 3 : 1.

Example: Following are tested some of Mendel's reported results with the garden pea. Test each for goodness of fit.

Cross	*Progeny*	*Hypothesis*
(*a*) Green × Yellow pods	(F_2) 428 : 152	3 : 1
(*b*) Violet red × White flower	(F_1) 47 : 40	1 : 1
(*c*) Round yellow × Wrinkled green	(F_1) 31 : 26 : 27 : 26	1 : 1 : 1 : 1

Solution: (*a*)

I. *Null Hypothesis* = 3 : 1

II. *Alternative Hypothesis* = 1 : 1

III. *Calculation:*

Observed (O)	*Expected (E)*	*(O – E)*	*(O – E)²*	$\frac{(O-E)^2}{E}$
428	$\frac{3}{4} \times 580 = 435$	428 – 435 = – 7	49	$\frac{49}{435} = .113$
152	$\frac{1}{4} \times 580 = 145$	152 – 145 = 7	49	$\frac{49}{145} = .338$
Total = 580				$x^2 = .451$

IV. *Critical value:* The control value of X^2 at 0.05 and for 2 – 1 = 1 degrees of freedom is 3.84.

V. *Decision:* Since the calculated value of chi-square (x^2) = .451 < critical value of x^2 for df = 3.84, so the null hypothesis is accepted *i.e.*, there is no significant variation with data. So it is result of F_2-monohybrid cross.

Solution: (*b*)

I. *Null Hypothesis* = 1 : 1

II. *Alternative Hypothesis* = 3 : 1

III. *Calculation:*

Observed (O)	*Expected (E)*	*(O – E)*	*(O – E)²*	$\frac{(O-E)^2}{E}$
47	$\frac{87}{2} = 43.5$	47 – 43.5 = 3.5	12.25	$\frac{12.25}{43.5} = 0.281$
40	$\frac{87}{2} = 43.5$	40 – 43.5 = –3.5	12.25	$\frac{12.25}{43.5} = 0.281$
Total = 87				$x^2 = .562$

IV. *Critical value:* The control value of Chi-Square at 0.05 and for 2 – 1 = 1 degrees of freedom is 3.84.

V. *Decision:* Since the calculated value of chi-square X^2 = .562 < critical value of x^2 for 1 df = 3.84. So the null hypothesis accepted *i.e.*, the variation is non significant.

Solution: (*c*)

I. *Null Hypothesis:* 1 : 1 : 1 : 1

II. *Alternative Hypothesis:* 9 : 3 : 3 : 1

III. *Calculation:*

Observed (O)	*Expected (E)*	*(O – E)*	*(O – E)²*	$\frac{(O-E)^2}{E}$
31	$\frac{1}{4}\times 110 = 27.5$	31 – 27.5 = 3.5	12.25	$\frac{12.25}{27.5} = 0.445$
26	$\frac{1}{4}\times 10 = 27.5$	26 – 27.5 = –1.5	2.25	$\frac{2.25}{27.5} = 0.082$
27	$\frac{1}{4}\times 110 = 27.5$	27 – 27.5 = –0.5	0.25	$\frac{0.25}{27.5} = 0.009$
26	$\frac{1}{4}\times 110 = 27.5$	26 – 27.5 = –1.5	2.25	$\frac{2.25}{27.5} = 0.082$
Total = 110				$x^2 = 0.618$

IV. *Critical value:* The control value of chi-square at 0.05 and 4 – 1= 3 degrees of freedom is 7.82.

V. *Decision:* Since the calculated value of chi-square (x^2) = 0.618 < critical value of X^2 for 3 *df* = 7.82. So the null hypothesis is accepted *i.e.*, the variation is non significant.

2. Test for Independence of Attributes or Contingency Chi-square:

I. It is applied to test the association between the attributes when the sample data is presented in the form of contingency table with any number of rows or columns.

II. It is occasionally desirable to compare one set of observation taken under particular conditions to those of a similar nature taken under different conditions.

III. In this case there are no definite expected values, the question is, whether the results are dependent (*contingent upon*) or independent of conditions under which they are observed. The test is therefore called a test for independence or contingency test.

- **Working Procedure:**

I. Set up the *Null Hypothesis* (H_0): No association exists between the attributes.

Alternative Hypothesis (H_1): An association exists between the attributes.

II. Calculate the expected frequency '*E*' corresponding to each cell by the formula.

$$E_{ij} = \frac{R_i \times C_j}{n}$$

Where R_i = Sum total of the row in which E_{ij} lying.

C_j = Sum total of the column in which E_{ij} is lying.

n = total sample size.

III. Calculate the chi-square by the formula.

$$X^2 = \sum \frac{(O-E)^2}{E}$$

E = Expected value

O = Observed value

Here degrees of freedom

$df = (R - 1)(C - 1)$

R = Number of rows

C = Number of columns in the contingency table.

IV. Find from the table the value of chi-square for the given value of the level of significance (α) and for the degrees of freedom (df).

If no value for α is mentioned, then the table $\alpha = 0.05$.

V. (*i*) Compare the computed values of Chi-square with the tables value of X^2.

(*ii*) If the calculated value of $X^2 <$ tabulated value then accept the hypothesis (H_0).

If $X^2 >$ tabulated value, reject the null hypothesis & accept the alternative hypothesis.

N.B = It is used in (2 × 3) *contingency tests.*

OR

- **Working Procedure:**

I. Set up (*a*) the **Null Hypothesis** (H_0): No associations exist between the attributes.

(*b*) **Alternative Hypothesis** (H_1): An association exists between the attributes.

II. Here contingency table has only 2 rows and 2 columns with four cell frequency viz. *a*, *b*, *c*, *d*. as shown below.

	a	b	R_1
	c	d	R_2
Total =	c_1	c_2	N

III. The frequency '*a*' is placed in the upper left cell and '*d*' is placed in the lower right cell. The frequency '*b*' & '*c*' are placed in other diagonal position.

IV. For simplicity R_1 & R_2 denote row totals and C_1 & C_2 denote the column totals *i.e.,*

$R_1 = a + b,\ R_2 = c + d$

$C_1 = a + c,\ C_2 = b + d$

$N = a + b + c + d = R_1 + R_2 = C_1 + C_2$

V. Therefore

$$X^2 = \frac{N\left\{(|\,ad - bc\,|) - \frac{N}{2}\right\}^2}{R_1 \times R_2 \times C_1 \times C_2} \qquad \frac{N}{2} = \text{Yates correction.}$$

VI. Degrees of freedom = $(R - 1)(C - 1) = 2 - 1 \times 2 - 1 = 1$

[*For* 1 *df, Yates correction is to be done*]

N.B. = It is used in (2 × 2) *contingency test.*

Example: The following table gives the classification of 100 workers according to sex and the nature of work. Justify whether the nature of work is independent of the sex of the worker.

	Skilled	*Unskilled*
Male	40	20
Female	10	30

Solution:

I. *Null Hypothesis:* Nature of work independent of the sex of worker (assume).

II. *Alternative Hypothesis:* Dependent of the sex of worker (assume).

III. *Calculation:*

Sex	Skilled	Unskilled	Total	
Male	40 *a*	20 *b*	60	R_1
Female	10 *c*	30 *d*	40	R_2
	50	50	100	$N = R_1 + R_2$
	C_1	C_2		

$$X^2 = \frac{N\{|ad - bc| - N/2\}^2}{R_1R_2C_1C_2} \qquad ad = 40 \times 30 = 1200$$

$$X^2 = \frac{100\{|1200 - 200| - 100/2\}^2}{60 \times 40 \times 50 \times 50} \qquad bc = 20 \times 10 = 200$$

$$= \frac{100\{1000 - 50\}^2}{60 \times 40 \times 50 \times 50} \qquad R_1 = a + b = 60\ C_1 = 50\ (a + c)$$

$$= \frac{100 \times 950 \times 950}{60 \times 40 \times 50 \times 50} \qquad R_2 = c + d = 40\ C_2 = 50\ (b + d)$$

$$= \frac{95 \times 95}{24 \times 25} = \frac{9025}{600} = 15.04 \qquad N = R_1 + R_2 = C_1 + C_2 = 100$$

- *Degrees of freedom:* $(R - 1)\ (C - 1) = (2 - 1)\ (2 - 1) = 1$
- *Critical value:* The table value of chi-square at 1 *df* at .05 = is 3.84
- *Decision:* Now calculate $X^2 = 15.04 > x^2_{.05,1} > 3.84$. So the null hypothesis is rejected.

Example: Test whether the prevalence of carriers of filaria is associated with sex.

Sex	No. of Carriers	No. of Non Carriers	Total studied
Male	78	412	490
Female	57	553	610

Solution:

- *Null Hypothesis:* Carrier of filaria is not associated with sex (*assume*).
- *Alternative Hypothesis:* Carrier filaria is associated with sex (*assume*).
- *Calculation:*

$$X^2 = \frac{N\{|ad - bc| - N/2\}^2}{R_1R_2C_1C_2}$$

$R_1 = a + b\ a + c = C_1$

$R_2 = c + d\ b + d = C_2$

$N = a + b + c + d = R_1 + R_2 = C_1 + C_2$

$a = 78\ \ R_1 = 490$

$b = 412\ \ R_2 = 610$

$c = 57\ \ C_1 = 78 + 57 = 135$

$d = 553\ \ C_2 = 412 + 553 = 965$

$$X^2 = \frac{1100\{|78 \times 553 - 412 \times 57| - 1100/2\}^2}{490 \times 610 \times 135 \times 965}$$

$$= \frac{1100\{|43134 - 23484| - 550\}^2}{490 \times 610 \times 135 \times 965}$$

$$= \frac{1100\{19650 - 550\}^2}{490 \times 610 \times 135 \times 965}$$

$$= \frac{1100 \times 19100 \times 19100}{490 \times 610 \times 135 \times 965}$$

$$= \frac{11 \times 19100 \times 19100}{2989 \times 130275}$$

$$= 11 \times 6.39 \times .1467$$

$$= 1.6137 \times 6.39 = 10.31$$

- *Degrees of freedom:* $(R - 1)(C - 1) = (2 - 1) \times (2 - 1) = 1 \times 1 = 1.$
- *Critical value:* The table value of X^2 = at .05 for 1 *df* is = 3.84.
- *Decision:* Now calculated x^2 is 10.31, it is greater than tabulated value *.i.e.,* $10.31 > x^2_{0.05} = 3.84$.

Null hypothesis is rejected. Alternative hypothesis is accepted.

Example: A random sample of 500 students was classified according to economic condition of their family and also according to merit.

Merit	Economic condition			Total	
	Rich	Middle Class	Poor		
Meritorious	42	137	61	240	R_1
Not Meritorious	58	113	89	260	R_2
Total	100	250	150	500	N
	C_1	C_2	C_3		

Test whether the two attributes Merit and Economic condition are associated or not. (Given $X^2_{.05} = 5.99$ & $X^2_{.01} = 9.21$ $df = 2$).

Solution: *Null Hypothesis* = Attributes are independent.

Alternative Hypothesis = Attributes are associated (independent).

Calculation:

$$E = \frac{R \times C}{N}$$

	Rich	Middle Class	Poor	Total
Meritorious	$\frac{240 \times 100}{500} = 48$	$\frac{240 \times 250}{500} = 120$	$\frac{240 \times 150}{500} = 72$	240
Not meritorious	$\frac{260 \times 100}{500} = 52$	$\frac{260 \times 250}{500} = 130$	$\frac{260 \times 150}{500} = 78$	260
	100	250	150	500

$$X^2 = \frac{(42-48)^2}{48} + \frac{(137-120)^2}{120} + \frac{(61-72)^2}{72} + \frac{(58-52)^2}{52} + \frac{(113-130)^2}{130} + \frac{(89-78)^2}{78}$$

$$= \frac{36}{48} + \frac{289}{120} + \frac{121}{72} + \frac{36}{52} + \frac{289}{130} + \frac{121}{78}$$

$$= 0.75 + 2.41 + 1.68 + 0.69 + 2.22 + 1.55 = 9.30$$

- Degrees of freedom = 3 – 1 = 2
- *Critical Value:* The table value of X^2 at .05 for 2 df = 5.99
- *Decision:* Now calculated X^2 is 9.30, it is greater than tabulated value. *i.e.*, 9.30 > $x^2_{.05,2}$ = 5.99, therefore null hypothesis is rejected. Alternative hypothesis is accepted. Therefore merit and economic conditions are associated.

3. Homogeneity Chi-square:

A test of "homogeneity" must be performed to decide whether the separate samples are sufficiently uniform to be added together.

- **Working Procedure:**

I. The chi-square of each individual sample should be calculated based on expected ratio. Since these chi-squares are to be added, they Yates correction factor should not be used, even though only 1 degree of freedom may be involved in each calculation.

II. The individual chi-square should be summed to give a total chi-square. In this process the total chi-square accumulated a number of degrees of freedom equal to the sum of the degrees of freedom in the individual chi-square.

The total chi-square value has two components.

(*a*) The chi-square contributed by the departure of the pooled data from the expected ratio.

(*b*) The chi-square contributed by the differences between individual samples.

To calculate '*b*', & '*a*' and subtract the value from total chi-square.

III. Calculate chi-square for the summed data of all samples.

IV. Substract the chi-square of for the summed data from the summed chi-square to obtain the homogeneity chi-square.

V. Accompany this, by subtracting the number of degrees of freedom between these respective values to obtain degrees of freedom for homogeneity chi-square.

Example: Four of the self fertilized F_1 plants that Mendel observed for segregating of yellow and green seeds colour showed the following results among their seeds.

Plants	*1*	*2*	*3*	*4*
Yellow Seeds	25	32	14	70
Green Seeds	11	7	5	27

Test the homogeneity of the four plants for the 3:1 ratio, and determine whether the data can be summed to calculate chi-square.

Solution:

I. *Null Hypothesis:* 3 : 1 (*assume*)

II. *Calculation:*

	Observed (O)	Expected (E)	(O – E)	$(O-E)^2$	$\frac{(O-E)^2}{E}$	X^2	df
1	25	27	–2	4	$\frac{4}{27} = .148$		2 – 1 = 1
	11	9	.2	4	$\frac{4}{9} = .44$	.588	
2	32	29.25	2.75	7.56	$\frac{7.56}{29.25} = .258$		2 – 1 = 1
	07	9.75	–2.75	7.56	$\frac{7.56}{9.75} = .775$	1 03	
3	14	14.25	–.25	.0625	$\frac{.0625}{14.25} = .004$		2 – 1 = 1
	5	4.75	.25	.0625	$\frac{.0625}{4.75} = .013$	.017	
4	70	72.75	–2.75	7.56	$\frac{7.56}{72.75} = .103$		2 1 = 1
	27	24.25	2.75	7.56	$\frac{7.56}{24.25} = .312$	.415	
						$\sum x^2 = 2.06$	= 4

Summed:

141	143.25	2.25	5.06	$\frac{5.06}{143.25} = 0.035$		
					.141	2–1 = 1
50	47.75	2.25	5.06	$\frac{5.06}{47.75} = .106$		

	Chi-square	Degrees of Freedom	Probabilities
Total	2.06	4	
Summed Data	.14	1	
Homogeneity	1.92	3	.70 – .50

III. *Critical Value:* The control value at 0.05 for 3 df is 7.82.

IV. *Decision:* Since calculated x^2 value at 3df is 1.92 < critical value. So the null hypothesis is accepted, *i.e.,* homogeneity persist among the samples.

Example: The following results were obtained by four experimenters in a genetics laboratory class who back crossed a stock of *Drosophila melanogaster* heterogeneous for the recessive genes black body colour and sepia eye colour (*b*/ +; *se*/+) to a stock homogeneous for the recessives (*b*/*b*; *se*/*se*)

	Phenotypes			
Experimenters	Wild type	Black	Sepia	Blacksepia
1	42	38	38	36
2	28	30	25	25
3	102	94	100	93
4	75	80	84	72

Test these data to see whether they homogeneous and can be pooled.

Solution: *Hypothesis:* Since this is a back cross to a homozygous recessive stock. The expected ratio will be 1 wild type: 1 black: 1 sepia: 1black sepia.

i.e., 1 : 1 : 1 : 1 (*assume*)

Calculation:

Experimenter	*Observed (O)*	*Expected (E)*	*(O – E)*	$(O-E)^2$	$\frac{(O-E)^2}{E}$	X^2	*df*
1	42	38.5	3.5	12.25	$\frac{12.25}{38.5} = 0.318$		4 – 1 = 3
	38	38.5	–0.5	.25	$\frac{.25}{38.5} = .006$	0.492	
	38	38.5	–0.5	.25	$\frac{.25}{38.5} = .006$	0.49	
	36	38.5	–2.5	6.25	$\frac{6.25}{38.5} = .162$		
2	28	27	1	1	$\frac{1}{27} = 0.37$		
	30	27	3	9	$\frac{9}{27} = 0.33$	0.667	
	25	27	–2	4	$\frac{4}{27} = 0.15$	= 0.67	4 – 1 = 3
	25	27	–2	4	$\frac{4}{27} = 0.15$		
3	102	97.25	4.75	22.56	$\frac{22.56}{97.25} = .23$	0.608	4 – 1 = 3
	94	97.25	–3.25	10.56	$\frac{10.56}{97.25} = .108$		
	100	97.25	2.75	7.56	$\frac{7.56}{97.25} = 0.08$	= .61	
	93	97.25	–4.25	18.06	$\frac{18.06}{97.25} = .19$		
4	75	77.75	–2.75	7.56	$\frac{7.56}{77.75} = .097$		
	80	77.75	2.25	5.06	$\frac{5.06}{77.75} = .06$	1.077	4 – 1 = 3
	84	77.75	6.25	39.06	$\frac{6.25}{77.75} = .50$	= 1.08	
	72	77.75	–5.75	33.06	$\frac{33.06}{77.75} = .42$		
						$\sum x^2 = 2.85$	$\sum df = 12$

Summed

Observed	Expected	O – E	$(O-E)^2$	$\frac{(O-E)^2}{E}=x^2$		df
247	240.5	6.5	42.25	$\frac{42.25}{240.5}$ =	.175	
242	240.5	1.5	2.25	$\frac{2.25}{240.5}$ =	0.009	4 – 1 = 3
247	240.5	6.5	42.25	$\frac{42.5}{240.5}$ =	.175	
226	240.5	14.5	210.25	$\frac{210.25}{240.5}$ =	.874 $\Sigma x^2 = 1.233$	

Critical value: The control value at 0.05 for 9 *df* is = 16.92.

Decision: Since calculated X^2 value at 9 *df* is 1.62 < critical value. So the null hypothesis is accepted.

• Uses of Chi-square:

The Chi-square test is very powerful test for testing the hypothesis of a number of statistical problems.

(*a*) Test of Goodness of Fit:

With the help of this test, probabilities of association between two attributes are measured.

(*b*) Test of Independence of attributes:

I. In this test attributes are classified into a two way table or contingency table.

II. This test discloses whether there is any association or relationship between two or more variables.

(c) Test of Homogeneity:

I. This test may be used to test the homogeneity of the attributes in respect of particular characteristics.

II. It may also be used to test the population variance.

SOLVED PROBLEMS

Q1. In *Drosophila* wild type allele (+) is responsible for red coloration of the eye, is dominant over its mutant allele white (w). From a cross the following result was obtained. Red eyed female 126 white eyed male 112. Test the result & comment on segregation.

- It is a result from a cross between homozygous white eyed female & homozygous red eyed male.

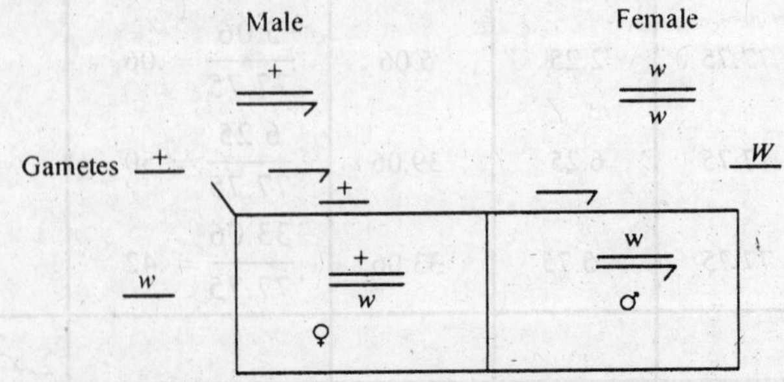

- *Result:* Red eyed female $\frac{\pm}{w}$ & white eyed male $\frac{w}{\rightarrow}$
- *Null Hypothesis:* It is a result of cross between homozygous white eyed female & homozygous red eyed male. Such cross naturally produces offspring in 1 : 1 ratios.
- *Alternative Hypothesis:* 3 : 1

  Calculation:

Observed (O)	Expected (E)	$\left(\lvert O-E \rvert - \frac{1}{2}\right)$	$\left(\lvert O-E \rvert - \frac{1}{2}\right)^2$	$\frac{\left(\lvert O-E \rvert - \frac{1}{2}\right)^2}{E}$	x^2
Red eyed Female = 126	119	$\lvert 126-119 \rvert - \frac{1}{2}$ = 7 − 0.5 = 6.5	42.25	$\frac{42.25}{119} = 0.36$	0.72
White eyed Male = 112	119	$\lvert 112-119 \rvert - \frac{1}{2}$ = 7 − 0.5 = 6.5	42.25	$\frac{42.25}{119} = 0.36$	
Total = 238					

- Degrees of freedom = 2 − 1 = 1.
- Critical value = The table value of chi-square for 2 − 1 = 1*df* is 3.84.
- Decision: Since the calculated value of chi-square (x^2) is 0.72 < critical value of x^2 for 1 *df* 3.84. So the null hypothesis is accepted *i.e.,* there is no significant variation with the result.

Q2. In the evening primrose pure red flowered plants are crossed with white coloured flowered plants, F_1 are all pink coloured. Inbreed F_1 plants produced Red 113 pink 242 white 129. This phenotypic ratio also seems to be genotypic ratio of the F_2 of a Mendelian monohybrid cross involving a gene responsible for flower pigmentation.

Analyse the result with suitable statistical test.

Solution:

- *Null Hypothesis:* 1 : 2 : 1
- *Alternative Hypothesis:* 3 : 1
- *Calculation:*

Observed (O)	Expected (E)	$\left(\lvert 0-E \rvert = -\frac{1}{2}\right) = D$	D^2	$\frac{D^2}{E}$	X^2
Red = 113	121	$\lvert 113-121 \rvert - \frac{1}{2}$ = $\lvert -8 \rvert$ − 0.5 = 7.5	56.25	$\frac{56.25}{121} = 0.46$	
Pink = 242	242	$\lvert 242-242 \rvert - \frac{1}{2}$ = 0 − 0.5 = 0.5	0.25	$\frac{0.25}{242} = 0.001$	0.921
White = 129	121	$129-121 \rvert - \frac{1}{2}$ = 8 − 0.5 = 7.5	56.25	$\frac{56.25}{121} = 0.46$	
Total = 484					

- Degrees of freedom = 3 – 1 = 2
- *Critical value:* The control value of chi-square (X^2) at 0.05 for 2 *df* is 5.99.
- *Decision:* Since the calculated value of chi-square (X^2) is 0.921 < critical value of X^2 for 2 *df* is 5.99. So the null hypothesis is accepted *i.e.,* the variation is non significant.

Q3. Crossing of purple eyed straight winged *Drosophila* with red eyed curved wing one produced dihybrid red eyed, straight winged female in F_1. On crossing such F_1 females with double recessive purple eyed, curved winged males gives following phenotypes. Red eyed straight wing 339, purpled eyed straight wing 612, red eyed curved wing 725, purple eyed curved wing 348. Find out whether or not F_2 generation obey the test cross ratio?

Solution:

- *Null Hypothesis:* 1 : 1 : 1 : 1
- *Alternative Hypothesis:* 9 : 3 : 3 : 1
- *Calculation:*

Observed (O)	*Expected (E)*	*O – E*	*(O – E)²*	$\frac{(O-E)^2}{E}$	x^2
Red eyed & straight wing = 339	506	339 – 506 = – 167	27889	$\frac{27889}{506} = 55.116$	
Purpled eyed straight wing = 612	506	612 – 506 = + 106	11236	$\frac{11236}{506} = 22.205$	221.44
Red eyed & curved wing = 725	506	725 – 506 = + 219	47961	$\frac{47961}{506} = 94.784$	
Purpled eyed curved wing = 348	506	348 – 506 = – 158	24964	$\frac{24964}{506} = 49.335$	
					2024

- Degrees of freedom = 4 – 1 = 3
- *Critical value:* The control value of x^2 for 3 *df* at 0.05 is 7.82.
- *Decision:* Since the calculated value of chi-square (x^2) for 3 *df* is 221.4 > critical value 7.82. So the hypothesis is rejected *i.e.,* there is significant difference between the observed & expected result.

Q4. Mendel self fertilized pea plants with round & yellow peas. In the next generation he recovered the following number of peas.

315 round & yellow peas.

108 round & green peas.

101 wrinkled & yellow peas.

32 wrinkled & green peas.

What is your hypothesis about the genetic control of the phenotype? Does the data support this hypothesis?

Solution:

I. Hypothesis about the genetic control *i.e.,* Null hypothesis = 9 : 3 : 3 : 1

II. *Calculation:*

Observed (O)	Expected (E)	(O − E)	$(O-E)^2$	$\frac{(O-E)^2}{E} = x^2$	X^2	df
315	$\frac{9}{16} \times 556$ = 312.75	315 − 312.75 = 2.25	5.06	$\frac{5.06}{312.75} = 0.016$		
108	$\frac{3}{16} \times 556$ = 104.25	108 − 104.25 = 3.75	14.06	$\frac{14.06}{104.25} = 0.135$		
101	$\frac{3}{16} \times 556$ = 104.25	101 − 104.25 = − 3.25	10.56	$\frac{10.56}{104.25} = 0.101$	0.470	3
32	$\frac{1}{16} \times 556$ = 34.75	32 − 34.75 = 2.75	7.56	$\frac{7.56}{34.75} = 0.218$		
Total = 556						

III. *Critical value:* The control value of chi-square at 0.05 for 3 *df* is 7.82

IV. *Decision:* Since calculated x^2 value at 3 *df* is 0.470, it is less than the critical chi-square value *i.e.,* $0.470 < x^2_{0.05}$, 3 = 7.82 so the null hypothesis is accepted.

Q5. Two curly winged flies when mated, produce 61 curly & 35 straight wing progeny. Use a chi-square test to determine whether these numbers fit a 3 : 1 ratio.

Solution:

I. *Null Hypothesis:* 3 : 1

II. *Alternative Hypothesis:* 2 : 1

III. *Calculation:*

Observed (O)	Expected (E)	| O − E | − 0.5	(| O − E | − 0.5)^2	$\frac{(O-E)^2}{E} = x^2$	x^2	df
Curly winged = 61	$\frac{3}{4} \times 96 = 72$	| 61 − 72 | − 0.5 = | − 11.0 | − 0.5 = 11.0 − 0.5 = 10.5	110.25	$\frac{110.25}{72} = 1.531$		1
Straight wing = 35	$\frac{1}{4} \times 96 = 24$	| 35 − 24 | − 0.5 = 11 − 0.5 = 10.5	110.25	$\frac{110.25}{24} = 4.593$	6.124	
Total = 96						

IV. *Critical value:* The control value at 0.05 for 1*df* is 3.84

V. Decision: Since the calculated x^2 value at 1*df* is 6.723, it is greater than the critical value *i.e.,* $6.124 > x^2_{,0.05} = 3.84$, so the null hypothesis is rejected.

Q6. Find whether or not the following phenotype distribution in a sample of 96 flies from a particular fruit fly population has a goodness of fit with the Mendelian 9 : 3 : 3 : 1 distribution.

Phenotypes:	Grey body red eye	Black body red eye	Grey body scarlet eye	Black body scarlet eye
Frequencies:	60	16	15	5

Solution:

I. *Null Hypothesis:* 9 : 3 : 3 : 1

II. *Alternative Hypothesis:* 1 : 1 : 1 : 1

III. *Calculation:*

Observed (O)	*Expected (E)*	*(O – E)*	*$(O-E)^2$*	*$\frac{(O-E)^2}{E}$*	*X^2*
60	54	+ 6	36	$\frac{36}{54} = 0.67$	
16	18	– 2	04	$\frac{4}{18} = 0.22$	
15	18	– 3	09	$\frac{9}{18} = 0.50$	1.56
5	6	– 1	01	$\frac{1}{6} = 0.17$	
Total = 96					

IV. *Critical value:* The control value of chi-square at 0.05 for 4 – 1 = 3 degrees of freedom is 7.82.

Decision: Since the calculated value of chi-square (x^2) 1.56 < critical value of x^2 for 3 *df* is 7.82, so the null hypothesis is accepted *i.e.,* the result is good fit for 9 : 3 : 3 : 1 hypothesis.

Q7. Random testing of *ABO* blood group in the offspring of only *AB* couples in an European population obtained the following distribution of blood groups.
***A* 312 , *AB* 575 & *B* 313.**
Test whether the data is consistent with the normal segregation of alleles in the population.

Solution:

- *Null Hypothesis:* These is no significant difference between the observed & expected number of progeny blood group as *AB* × *AB* produces *AA* : *AB* : *BB i.e.,* 1 : 2 : 1 ratio.
- *Calculation:*

	AA	*AB*	*BB*	*Total*
Observed (O)	312	575	313	1200
Expected (E)	300	600	300	1200
O – E = D	312 – 300 = 12	575 – 600 = – 25	313 – 300 = 13	
D^2	144	625	169	
$\frac{D^2}{E}$	$\frac{144}{300} = 0.48$	$\frac{625}{600} = 1.04$	$\frac{169}{300} = 0.56$	

$$X^2 = 0.48 + 1.04 + 0.56 = 2.08$$

- *Critical value:* The control value of x^2 at 0.05 for 3 – 1 = 2*df* is 5.99.

- *Decision:* Since the calculated x^2 value at $2df$ is 2.08 < critical value, so the null hypothesis is accepted *i.e.,* the variation between observed & expected data is non significant.

Q8. A sampling of south Indian population has found 40 individuals have homozygous curly hair, 40 individuals have heterozygous curly hair and 20 individuals have straight hair. Test statistically whether this sample is in equilibrium or not.

Solution: To test the sample for genetic equilibrium we have to follow the Hardy-Weinberg law of population genetics for the calculation of gene frequency or allelic frequency for this generation. Then we can calculate the genotypic frequency for the next generation.

- *Calculation of gene or allele frequency:*

(*i*) A gene '*A*' for curly hair, and gene '*a*' for straight hair.

(*ii*) *P* the gene frequency for allele *A* and *q* the gene frequency for allele '*a*'.

(*iii*) p^2 represent the frequency of homozygous dominant genotype *i.e., AA*. $2pq$ represent the frequency heterozygous genotype *Aa*. q^2 represent the frequency of homozygous *aa*.

(*iv*) In the present problem gene frequency of allele *A i.e.,* $P_A = p^2 + \frac{1}{2}pq$. & gene frequency of allele a *i.e.,* $q_a = q^2 + \frac{1}{2}.pq$.

	AA	*Aa*	*aa*	Total
	40	40	20	100
Genotypic frequency	0.4	0.4	0.2	

Gene frequency of *A i.e.,* $p_A = 0.4 + \frac{1}{2} \times 0.4 = 0.4 + 0.2 = 0.6$

Gene frequency of *a i.e.,* $q_a = 0.2 + \frac{1}{2} \times 0.4 = 0.2 + 0.2 = 0.4$

(*v*) Therefore gene frequency in the present generation is $pA = 0.6$ & $q_a = 0.4$

(*vi*) Calculation of genotypic frequency for the next generation

$$(p_A + q_a)^2 = p^2_A + 2p_Aq_a + q^2_{aa}$$

$$(0.6 + 0.4)^2 = 0.36 + 0.48 + 0.16$$

- *Null Hypothesis:* (*i*) If the genotypic frequency of present generation is exactly or more or less close to the genotypic frequency of next generation then the population is called in equilibrium.

(*ii*) Here in present generation genotypic frequency is 0.40, 0.40 & 0.20 & the next generation will be 0.36 , 0.48 & 0.16.

Therefore there is no significant deviation between observed & expected genotypic frequency of this population.

- *Calculation:*

	AA	*Aa*	*aa*
Observed (*O*)	0.4	0.4	0.2
Expected (*E*)	0.36	0.48	0.16
$O - E = D$	0.4 – 0.36 = 0.04	0.4 – 0.48 = 0.08	0.2 – 0.16 = 0.04
D^2	0.0016	0.0064	0.0016
$\frac{D^2}{E}$	$\frac{0.0016}{0.36} = 0.0044$	$\frac{0.0064}{0.48} = 0.0133$	$\frac{0.0016}{0.16} = 0.01$

$$X^2 = 0.0044 + 0.0133 + 0.012 = 0.0277$$

- *Critical value:* The control value of x^2 at 0.05 for $df = 2 - 1 = 1$ is 384.
- *Decision:* Since the calculated value of chi-square $X^2 = 0.0277 <$ critical value of X^2 for 1*df* = 3.84. So the null hypothesis is accepted *i.e.,* deviation is non significant.

Q9. Find whether or not the following observed distribution of phenotypes in a sample of 384 *Drosophila* flies have a significant goodness of fit with proposed Mendelian 9 : 3 : 3 : 1 distribution. (α = 0.05).

Phenotypes:	AB	Ab	aB	ab	Total
Number of animals:	232	76	58	18	384

Solution:

- *Null Hypothesis:* 9 : 3 : 3 : 1
- *Alternative Hypothesis:* 1 : 1 : 1 : 1
- *Calculation:*

Observed (O)	*Expected (E)*	*O – E = D*	*D²*	$\frac{D^2}{E}$
232	$\frac{384}{16} \times 9 = 216$	232 – 216 = 16	256	$\frac{256}{216} = 1.185$
76	$\frac{384}{16} \times 3 = 72$	76 – 72 = 04	16	$\frac{16}{72} = 0.223$
58	$\frac{384}{16} \times 3 = 72$	58 – 72 = – 14	196	$\frac{916}{72} = 2.723$
18	$\frac{384}{16} \times 1 = 24$	18 – 24 = – 6	36	$\frac{36}{24} = 1.5$
Total = 384				5.625

- *Critical value:* The control value of chi-square at 0.05 levels & for 4 – 1 = 3 degrees of freedom is 7.82.
- *Decision:* Since the calculated value of chi-square (X^2) = 5.625, < critical value of chi-square for 3*df* is 7.82, so the null hypothesis is accepted *i.e.,* the variation is non significant.

Q10. In *Drosophila*, the wild type allele (+) responsible for normal bristle is dominant over mutant allele forked (*f*). From a cross the following progeny are obtained. Normal bristle female 238 Normal bristle male 108 forked male 122. Test the result.

Solution: In the present cross, the uneven distribution of dominant character between the two sexes attempted as to speculate that the gene under consideration resides on the *x* chromosome.

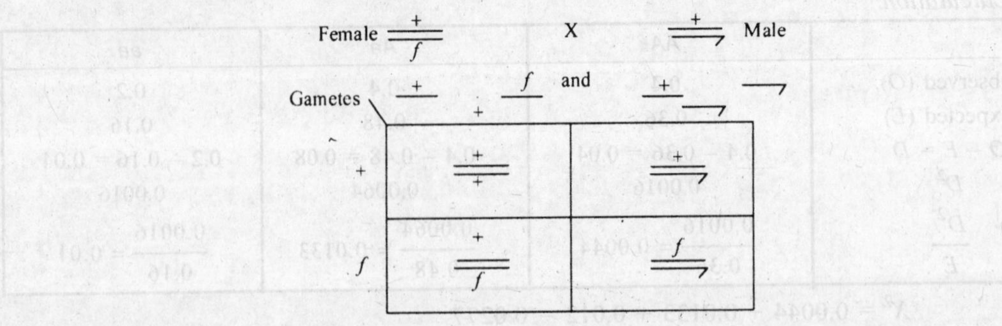

$\frac{+}{+}$ Normal female }
$\frac{+}{f}$ Normal female } 2] 3
$\frac{+}{\rightarrow}$ Normal male } 1]
$\frac{f}{\rightarrow}$ forked male = } 1] 1

As a result of cross between heterozygote female $\left(\frac{+}{f}\right)$ & wild type male $\frac{+}{\rightarrow}$ three types of attributes appeared (3 : 1) or [1 : 2 : 1].

- *Null Hypothesis:* 3 : 1 *i.e.,* 1 : 2 : 1 (male : female: male)
- *Alternative Hypothesis:* 1 : 1

Observed (O)	*Expected (E)*	*O – E*	$(O-E)^2$	$\frac{(O-E)^2}{E}$	x^2
wild type male = 108	117	108 – 117 = – 9	81	$\frac{81}{117} = 0.69$	
wild type female = 238	234	238 234 = + 4	16	$\frac{16}{234} = 0.06$	0.96
forked male = 122	117	122 – 117 = + 5	25	$\frac{25}{117} = 0.21$	

- *Degrees of freedom* = 3 – 1 = 2
- *Critical value:* The table value of X^2 at 0.05 level for 2*df* is 5.99
- *Decision:* Since the calculated Chi-square (X^2) value 0.96 < critical value of X^2 for 2*df* = 5.99, so the null Hypothesis is accepted *i.e.,* deviation is non significant.

Q11. Evaluate the association between the abilities to test PTC & sex from the following scores. Tester male 80, Tester female 60, Non tester male 40 & Non tester female 20.

Solution:

- *Null Hypothesis:* No close relationship between PTC tester non tester with sex (male & female).
- *Alternative Hypothesis:* Relationship between taster non tester & sex (male & female).
- *Calculation:*

P.T.C.	*Male*	*Female*	*Total*	
Tester	80(*a*)	60(*b*)	140	$(a + b)\ R_1$
Non-tester	40(*c*) $a + c = 120$	20(*d*) $b + d = 80$	60 $a + b + c + d = 200$	$(c + d)\ R_2$
	C_1	C_2		

$$X^2 = N\frac{\{|\,ad - bc\,| - N/2\}^2}{R_1.R_2.C_1.C_2}$$

$ad = 80 \times 20 = 1600$

$bc = 60 \times 40 = 2400$

$$= \frac{200 \times \left\{ |1600 - 2400| - \frac{200}{2} \right\}^2}{140 \times 60 \times 120 \times 80}$$

$R_1 = a + b = 140$, $C_1 = a + c = 120$

$R_2 = c + d = 60$, $C_2 = b + d = 80$

$N = R_1 + R_2 = C_1 + C_2 = 200$

$$= \frac{200 \times \{|-800| - 100\}^2}{140 \times 60 \times 120 \times 80} = \frac{200 \times 700 \times 700}{140 \times 60 \times 120 \times 80} = 1.215$$

- *Degrees of freedom:* $(R - 1)(C - 1) = (2 - 1)(2 - 1) = 1$
- *Critical value:* The table value of Chi-square (X^2) at $1 df$ at 0.05 is 3.84.
- *Decision:* Since calculated values of Chi-square at $1 df$ is 1.215 < critical value = 3.84, so the null hypothesis is accepted *i.e.,* deviation is non significant.

Q12. In a laboratory among 60 *Drosophila*, 30 are male, & 30 are female found. Further segregated by the presence of mutation 21 mutated male, 25 mutated female, 9 normal male & 5 normal are found. Find out the relationship between mutation & sex.

Solution: In this case there is no definite expected value, the question is whether the result is dependent or independent of the condition under which they are observed.

- *Null Hypothesis:* No significant relationship between the sex & mutation.
- *Alternative Hypothesis:* Presence of significant relationship between mutation & sex.
- *Calculation:*

phenotype

Sex	Mutated	Normal	Total
Male	21(*a*)	9(*b*)	30
Female	25(*c*)	5(*d*)	30
	$a + c = 46$	$b + d = 14$	60
	C_1	C_2	

$$X^2 = \frac{N\left\{ |ad - bc| - \frac{N}{2} \right\}^2}{R_1 R_2 C_1 C_2}$$

$ad = 21 \times 5 = 105$

$bc = 9 \times 25 = 225$

$$= \frac{60\left\{ |105 - 225| - \frac{60}{2} \right\}^2}{30 \times 30 \times 46 \times 14}$$

$R_1 = a + b = 30$, $C_1 = a + c = 46$

$R_2 = c + d = 30$, $C_2 = b + d = 14$

$N = R_1 + R_2 = C_1 + C_2 = 60$

$$= \frac{60\{|-120| - 30\}^2}{30 \times 30 \times 46 \times 14}$$

$$= \frac{\cancel{60}^2 \times \cancel{90}^3 \times 90}{\cancel{30} \times \cancel{30} \times 46 \times 14} = \frac{540}{644} = 0.838$$

$= 0.84$

- *Degrees of freedom:* $(R - 1)(C - 1) = (2 - 1)(2 - 1) = 1$

- *Critical value:* The table value of Chi-square (x^2) at 0.05 for 1*df* is 3.84.
- *Decision:* Since the calculated values of Chi-square (x^2) is 0.84 < critical value 0.384, so the null hypothesis is accepted *i.e.,* variation is non significant.

Q13. In a sample of owls, it is found that red male 35, red female 70, grey male 50, grey female 45. Colouration is due to the plumage. Is the colouration independent of sex of the sample?

Solution:

- *Null Hypothesis:* The experimental result may show that colouration of the individuals are independent or not contingent of the sex.
- *Alternative Hypothesis:* The result may show that the colouration of the individuals are not independent of sex. *i.e.,* dependent or contingent of sex.
- *Calculation:*

Sex	*Red*	*Grey*	*Total*
Male	35(*a*)	50(*b*)	85
Female	70(*c*)	45(*d*)	115
	($a + c$) = 105 C_1	($b + d$) = 95 C_2	200

$$X^2 = N\frac{\{|ad - bc| - N/2\}^2}{R_1R_2C_1C_2}$$

$ad = 35 \times 45 = 1575$

$bc = 50 \times 70 = 3500$

$$= \frac{200\left\{|1575 - 3500| - \frac{200}{2}\right\}^2}{85 \times 115 \times 105 \times 95}$$

$R_1 = a + b = 85,\ C_1 = a + c = 105$

$R_2 = c + d = 115,\ C_2 = b + d = 95$

$N = R_1 + R_2 + C_1 + C_2 = 200$

$$= \frac{200\{|-1925| - 100\}^2}{85 \times 115 \times 105 \times 95}$$

$$= \frac{200 \times 1825 \times 1825}{85 \times 115 \times 105 \times 95} = \frac{200 \times 3330625}{97505625}$$

$$= 200 \times 0.0341582 = 6.831$$

- *Degrees of freedom* = (R – 1) (C – 1) = (2 – 1) (2 – 1) = 1
- *Critical value:* The table of Chi-square at 1*df* at 0.05 is 3.84
- *Decision:* Now calculated $X^2 = 6.831 > X^2_{0.05,1} > 3.84$

So the null hypothesis is rejected.

Q14. In *Drosophila sp* wild type gene (+) responsible for normal wing shape dominant over mutant allele vestigial (vg) responsible for abnormal wing shape. From a cross between two phenotypically wild type males and females, following progeny were observed—(*i*) Wild type male–155, (*ii*) Vestigeal male–52, (*iii*) Wild type female–154, (*iv*)Vestigeal female–56.

From the above observation determine—(*a*) whether the gene under consideration is *X*-linked, *Y*-linked or autosomal, (*b*) Genotype of the offspring, (*c*) Genotype of parents and (*d*) Substantiate your claim statistically.

Solution:

(*a*) (*i*) In the given problem only wing type character is under consideration. The sex ratio is roughly 1 : 1 (207 : 210).

It indicates absence of any sex linked recessive or lethal or sex specific lethal genes are involved in this cross.

(*ii*) Here dominate recessive ratio is roughly 3 : 1 (309 : 210). Hence both the characters are expressed equally.

Therefore the data indicates the gene under consideration is autosomal.

(*b*) Genotypes of offsprings:

Probable genotypes: (*i*) Wild type male or female: $\frac{+}{vg}$ or $\frac{+}{+}$

(*ii*) Vestigial male or female: $\frac{vg}{vg}$

(*c*) Since vestigial male and females are recessive homozygous, they have received each vestigial gene from their parents. Therefore parents have at least one vestigial allele (gene) in their genetic configuration. Since dominant to recessive ratio among individual offspring is 3 : 1; the cross is between two heterozygous *i.e.*, $\frac{+}{vg} \times \frac{+}{vg}$ (parents)

off springs: $\frac{+}{+}, \frac{+}{vg}, \frac{+}{vg}, \frac{vg}{vg}$

Therefore genotype of each parent is $\frac{+}{vg}$

(d) Statistical Analysis:

- *Null hypothesis:* The result indicates 3:1 ratio of Mendelian monohybrid cross.
- *Alternative hypothesis:* 1 : 1
- *Calculation:*

Observed (O)	Expected (E)	$(\{\|O-E\|\} - 0.5\})$ = D	$\{(\|O-E\|) - 0.5\}^2 = D^2$	$\frac{D^2}{E}$	x^2	df
Wild type = 309	$\frac{3}{4} \times 417$ = 312.75	| 309 – 312.75 | – 0.5 = {|3.75| – 0.5}	$(3.25)^2$	$\frac{10.56}{312.75}$	0.3376	
Vestigeal = 108	$\frac{1}{4} \times 417$ = 104.25	|.108 – 140.25| – 0.5 = {|3.75| – 0.5}	$(3.25)^2$	$\frac{10.56}{104.25}$	0.1012	
Total = 417					$X^2 = 0.1349$	

- *Critical value:* The control value of x^2 at 0.05 for 2 – 1 = 1*df* is = 3.84
- *Decision:* Since the calculated value of $x^2 = 0.13$

  So the null hypothesis is accepted.

  Therefore the data is good fit to the ratio 3 : 1

Q15. In *Drosophila* sp two eye colour mutation brown (*bw*) and cinnabar (*cn*) are recessive to the respective wild type alleles (+) and responsible for brown and cinnabar colouration respectively. Interestingly, when both mutants are present in homozygous conditions, colouration of eye will be white. The mutant allele responsible for upward bending of the wing curly (*cy*) dominant over wild type allele (+) and is homozygous lethal. From a cross following progenies are recovered—(*i*) curly male–108, (*ii*) curly female –112, (*iii*) white male –106 and (*iv*) white female–118.

From the above observation, determine (*a*) Whether genes under consideration are *X*-linked, *Y*-linked or autosomal, (*b*) Genotype of parents, (*c*) Genotype of off-spring and (*d*) Substantiate your claim statistically.

Solution:

(*a*) (*i*) In this problem bending of wing and colour of eye are under consideration. So it is a dihybrid cross.

(*ii*) Here sex ratio is roughly 1 : 1 (114 : 230). It indicates absence of any sex linked recessive or lethal, sex specific lethal or sex transforming mutant involved in this cross.

(*iii*) Here phenotypic ratio (curly : white) 1 : 1 : 1 : 1. Hence both characters are expressed equally in both sexes.

These observations indicate that genes under consideration are autosomal. Brown and cinnabar are alleles of eye colour. Hence they occupy separate loci on the chromosome.

(*b*) White female are of genotype $\frac{bw\,cn+}{bw\,cx+}$, it indicates absence of genes for curly (*cy*) *i.e.*, upward bending or wing. Therefore, they have homozygous recessive gene for wing. Parents have at least one strand of white eye colour gene in their genetic configuration. Now if parents both have curly gene, then offspring will have $\frac{cy}{cy}$ genotype which is lethal. Then sex ratio will be 2 : 1 but the ratio is 1 : 1 : 1 : 1. Therefore it is cross between homozygous and heterozygous recessive parents.

Therefore genotype of parents is

Heterozygous: $\frac{++cy}{bw\,cn+} \times \frac{bw\,cn+}{bw\,cn+}$ (Homozygous)

(*c*) Genotypes of offspring: (*i*) $\frac{++cy}{bw\,cn+}$ (*ii*) $\frac{++cy}{bw\,cn+}$ (*iii*) $\frac{bw\,cn+}{bw\,cn+}$ (*iv*) $\frac{bw\,cn+}{bw\,cn+}$

(*i*) Curly male (*ii*) Curly female (*iii*) White male (*iv*) White female

(d) Statistical Analysis:

• *Null Hypothesis:* The result indicates 1 : 1 : 1 : 1 test cross of dihybrid cross.

• *Calculations:*

Observed (O)	Expected (E)	$\lvert O - E \rvert = D$	D^2	$\frac{D^2}{E}$	x^2
Curly male = 108	111	108 – 111 = – 3	9	$\frac{9}{11}$	0.08
Curly female = 112	111	112 – 111 = + 1	1	$\frac{9}{111}$	0.009
White male = 106	111	106 – 111 = – 5	25	$\frac{25}{111}$	0.225
White female = 118	111	118 – 111 = + 7	49	$\frac{49}{111}$	0.44
Total = 444					0.754

- *Degrees of freedom:* 4 – 1 = 3
- *Critical value:* The control value for 3*df* at 0.05 is 7.82.
- *Decision:* Since the calculated value of chi-square (x^2) for 3*df* is 0.754 < critical value of x^2 for 3*df* is 7.82. So the null hypothesis is accepted *i.e.*, variation is non significant.

Q16. In *Drosophila sp* two eye colour mutations, brown (*bw*) and cinnabar (*cn*) are recessive to the respective wild type alleles (+) and responsible for brown and cinnabar colour respectively. When both mutants are present together in homozygous condition, eye colour is white. The mutant allele responsible for inward bending of the wing curly (*cy*) is dominant over its wild type allele(+) and is homozygous lethal, from a cross the following progenies are recovered (*i*) curly male–208, (*ii*) white male–106, (*iii*) curly female–212 and (*iv*) white female–118.

Determine (*a*) whether gene under consideration is *X*-linked, *Y*-linked or autosomal, (*b*) Genotype of parents, (*c*) Genotype of offspring and (*d*) Substantiate your claim statistically.

Solution:

(*a*) (*i*) In this problem both eye colour and wing shape are under consideration. Hence it is a dihybrid cross.

(*ii*) Sex ratio is roughly 1 : 1 (330 : 314). It indicates absence of any *X*-linked recessive or lethal gene or sex transforming lethal genes are involved in this cross.

Here dominant and recessive ratio is roughly 2 : 1 (420 : 224).

Hence both dominant and recessive characters are expressed equally in both sexes.

The above mentioned observation indicate, that gene under considerations is ***autosomal***.

(*b*) (*i*) Now brown and cinnabar colour genes occupy separate loci on the same strand of the chromosome curly males and females are heterozygous $\left(\frac{+}{cy}\right)$ in genetic configuration (as curly homozygous lethal *i.e.*, $\frac{cy}{cy}$). White males and females are homozygous for brown and cinnabar and recessive forewings shape.

(*ii*) Since curly : white ratio is 2 : 1 both parents are curly. Then the offspring with $\frac{cy}{cy}$ genotypes becomes absent and phenotypic ratio becomes 2 : 1.

$$\frac{+\ cy\ +}{bw\ +\ cn} \times \frac{+\ cy\ +}{bw\ +\ cn}$$

Genotype of parents: $\frac{+\ cy\ +}{bw\ +\ cn}$ (curly);

(*c*) Genotype of offsprings: $\frac{+\ cy\ +}{bw\ +\ cn}$ (curly); $\frac{bw\ +\ cn}{bw\ +\ cn}$ white.

(*d*) Statistical Analysis:

- ***Null hypothesis:*** Result indicates 1 : 2 : 2 : 1 (dihybrid, autosomal dominant homozygous lethal).

• *Calculation:*

Observed (O)	Expected (E)	(|O − E|) = D	|O − E| = D^2	(|O − E|)^2 = D^2	$\frac{D^2}{E}$	X^2	df
Curly male = 208	$\frac{2}{6} \times 644$ = 214.66	208 − 214.66 = 6.66	44.35	$\frac{44.35}{214.66} = 0.2$	0.2		
Curly female = 212	$\frac{2}{6} \times 644$ = 214.66	212 − 214.66 = 2.66	7.07	$\frac{7.07}{214.66} = 0.03$	0.03	1.30	4 − 1 = 3
White male = 106	$\frac{1}{6} \times 644$ = 107.33	106 − 107.33 = 1.33	1.76	$\frac{1.76}{107.33} = 0.016$	0.016		
White female = 118	$\frac{1}{6} \times 644$ = 107.33	118 − 107.33 = 10.67	113.84	$\frac{113.84}{107.33} = 1.06$	1.06		
Total = 644							

• *Critical value:* The control value of X^2 for 3*df* at 0.05 levels is 7.82.

• *Decision:* Since the calculated value of chi-square (x^2) for 3*df* is 1.30 < critical value 7.82. So the hypothesis is accepted i.e. deviation is non-significant.

Q17. In *Drosophila sp* a wild type gene (+) responsible for normal wing shape is recessive to mutant allele curly (*cy*). This dominant mutation like any other dominant mutation of *Drosophila* is homozygous lethal, From a cross following results are recovered (*i*) Wild type male–56, (*ii*) Wild type female–52, (*iii*) Curly male –108 and (*iv*) Curly female–112. From above observation, determine (*a*) whether gene under consideration is *X*-linked, *Y*-linked or autosomal, (*b*) Genotype of parents, (*c*) Genotype of offspring and (*d*) Substantiate your claim statistically.

Solution:

(*a*) (*i*)In this problem both eye colour and wing shape are under consideration. Hence it is a **dihybrid** cross.

(*ii*) Here sex ratio is 1 : 1 (164 : 164). It indicates absence of any sex linked recessive, lethal gene or sex specific lethal or sex transforming mutants are involved in this cross.

(*iii*) Here dominant and recessive are present in 2 : 1 ratio, therefore both characters are equally expressed in both sexes.

The above mentioned observation indicates that gene under consideration is ***autosomal.*** Since gene for wild type is recessive, then individuals with wild type phenotype will be homozygous recessive.

Parents have at least one strand of wild type allele in their genetic configuration (Since dominant mutation is homozygous lethal)

Curly: Wild type 2 : 1 indicates that both parents have gene for curly shape wing in their genetic configuration. Hence parents are heterozygous for this particular gene.

(*b*) *Genotype of parents:* $\frac{cy}{+}$

(*c*) *Genotype of offsprings:* $\frac{cy}{+}$ $\frac{cy}{+}$ $\frac{+}{+}$ $\frac{cy}{cy}$ (Lethal)

phenotypicration: (curly) : (wildtype) : 2 : 1

(*d*) ***Statistical Analysis:***

- *Null Hypothesis:* The result shows dominant : recessive phenotype = 2 : 1
- *Calculation:*

Observed (O)	Expected (E)	(|O – E|) = D	D^2	$\frac{D^2}{E}$	x^2	Df
Curly male = 108	$328 \times \frac{2}{6}$ = 109.33	108 – 109.33 = 1.33	1.7689	$\frac{1.7689}{109.33}$	0.016	4 – 1 = 3
Curly female = 112	$328 \times \frac{2}{6}$ = 109.33	112 – 109.33 = 2.67	7.1289	$\frac{7.1289}{109.33}$	0.065	
Wildtype male = 56	$328 \times \frac{1}{6}$ = 54.66	56 – 54.66 = 1.34	1.795	$\frac{1.795}{54.66}$	0.033	
Wildtype female = 52	$328 \times \frac{1}{6}$ = 54.66	52 – 54.66 = 2.66	7.075	$\frac{7.075}{54.66}$	0.129	
Total = 328					Total = 0.243	

- *Critical value:* The control value of chi square (x^2) for 3df at 0.05 is 7.82.
- *Decision:* Since the calculated value of chi square for 3*df* is 0.243 < critical value 7.82. So the null hypothesis is accepted *i.e.,* there is no significant variation between observed and expected result.

Q18. In *Drosophila sp* wild type gene responsible for grey colouration of body is dominant over mutant allele yellow. From a cross following progenies are recovered (*i*) Yellow females 88 and (*ii*) Wild females 78.

Determine (*a*) Whether gene under consideration is autosomal or sex-linked, (*b*) Genotype of parents, (*c*) Genotype of off-spring and (*d*) Substantiate your claim statistically.

Solution:

(*a*) (*i*) Here only colour of the body is under consideration. Hence it is a **monohybrid** cross.

(*ii*) Since among progenies males are absent *i.e.,* sex ratio ≠ 1 : 1. Moreover distribution of dominant and recessive characters among sexes is not equal. Therefore the gene under consideration is **sex linked**.

(*b*) (*i*) Since both yellow and wild type females are present we can conclude that none of the allele are lethal.

(*ii*) Therefore the genetic constitution of the offspring and that of the parents must have two different loci for lethal genes which complement each other and produce healthy offspring in **trans condition** (females) but become lethal when present in **cis condition**.

(*iii*) Now, female parents are healthy, they have genes in trans condition $\frac{l_1 +}{+ l_2}$ and in males no lethal genes are present $\frac{+\ +}{\longrightarrow}$

(*iv*) Since both recessive (yellow) and dominant (grey) phenotypes are present among off springs, the parent female is heterozygous for colouration of the body as well as trans heterozygous for lethal genes. The male parent must carry recessive gene for colour of the body and lethal genes are absent.

Genotype of parents: $\frac{y\, l_1 +}{+ + l_2}$ ♀ and $\frac{y + +}{\longrightarrow}$ ♂

(*c*) *Genotype of offspring:* $\frac{y\, l_1 +}{\longrightarrow}$, $\frac{y + l_2}{\longrightarrow}$, $\frac{y\, l_1 +}{y + +}$, $\frac{y + l_2}{y + +}$

♂ \ ♀	$\frac{y\, l_1 +}{}$	$\frac{+ + l_2}{}$
$\frac{y + +}{}$	$\frac{y\, l_1 +}{y + +}$ yellow ♀	$\frac{+ + l_2}{y + +}$ wild ♀
$\longrightarrow$	$\frac{y\, l_1 +}{\longrightarrow}$ ♂ lethal	$\frac{y + l_2}{\longrightarrow}$ ♂ lethal

• ***Statistical Analysis:***

• *Null Hypothesis:* Ratio of wild and yellow females are 1 : 1.

• *Calculation:*

Observed (O)	Expected (E)	(|O − E| − 0.5) = D	D^2	$\frac{D^2}{E}$	x^2	df
Yellow female = 88	83	(83 − 88) − 0.5 = 4.5	20.25	$\frac{20.25}{83}$	0.243	2 − 1
Wild female = 78	83	(78 − 83) − 0.5 = 4.5	20.25	$\frac{20.25}{83}$	0.243	1
Total = 166					Total = 0.486	

• *Critical value:* The control value of the chi-square for 1*df* at 0.05 is 3.84.

• *Decision:* Since the calculated value of chi square (X^2) for 1*df* is 0.486 < critical value of x^2 for 1*df* is 3.84. So the null hypothesis is accepted *i.e.*, variation is non significant.

Q19. In *Drosophila sp* a wild type allele responsible for normal shape of bristles (+) is dominant to its mutant allele forked (*f*). From a cross the following results are obtained—(*i*) wild male 72, (*ii*) wild female 148 and (*iii*) forked male–66.

Determine—(*i*) whether gene is *X*-linked, *Y*-linked or autosomal, (*b*) Genotype of parents, (*c*) Genotype of offsprings and (*d*) Substantiate your answer statistically.

Solution:

(*a*) (*i*) Only shape of bristle is considered, so it is **monohybrid cross** (*ii*) Sex ratio is roughly 1 : 1 (138 : 148)

Dominant recessive ratio is 3 : 1 (220 : 66)

(*iii*) But characters show uneven distribution, *i.e.*, forked male present and forked females are absent This indicates that gene under consideration is **sex linked**.

(*b*) (*i*) Since forked males have $\frac{f}{\rightharpoondown}$ genetic configuration. They receive *Y* chromosome from their male parent and forked character is inherited from their mother (female parent).

(*ii*) Since forked females are absent, the female parent is heterozygous for that gene. Cross between heterozygous female and wild type male parent gives such data.

$$\frac{f}{+} \times \frac{+}{\rightharpoondown} \rightarrow \frac{f}{\rightharpoondown} \quad \frac{+}{\rightharpoondown} \quad \frac{f}{+} \quad \frac{+}{+}$$

Forked male Normal male Normal female Normal female

Genotype of parents: 1) female heterozygous $\frac{f}{+}$

2) male wild type $\frac{+}{\rightharpoondown}$

(*c*) *Genotype of offspring:*

1. forked male $\frac{f}{\rightharpoondown}$
2. wild male $\frac{+}{\rightharpoondown}$
3. wild female $\frac{+}{+} / \frac{+}{f}$

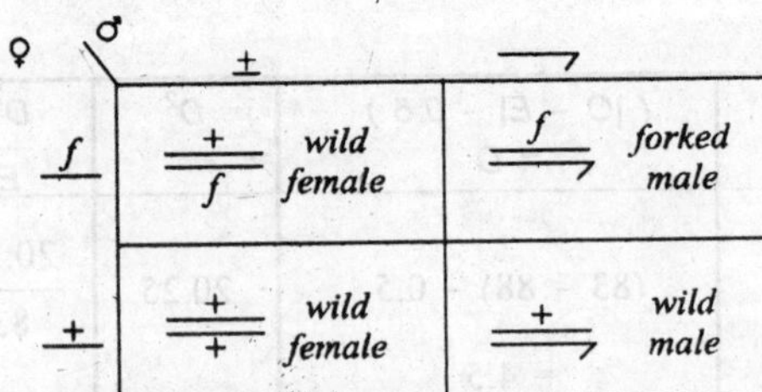

(*d*) ***Statistical Analysis:***

- *Null Hypothesis:* The result shows 1 : 2 : 1 ratio of sex linked recessive character.
- *Calculation:*

Observed (O)	Expected (E)	|O – E| = D	D^2	$\frac{D^2}{E}$	X^2
Wildmale = 72	$286 \times \frac{1}{4} = 71.5$	(72 – 71.5) = 0.5	0.25	$\frac{.25}{71.5} = 0.003$	0.003
Wild female = 148	$286 \times \frac{2}{4} = 143$	(148 – 143) = 5	25	$\frac{25}{143} = 0.174$	0.174
Forked male = 66	$286 \times \frac{1}{4} = 71.5$	(66 – 71.5) = 5.5	30.25	$\frac{30.25}{17.5} = 0.423$	0.423
Total = 286		*df* = 3 – 1 = 2			= 0.6

- *Critical value:* The control value of chisquare (x^2) for 2*df* at 0.05 is 7.82
- *Decision:* Since the calculated value of chisquare (x^2) for 3*df* is 0.6 < critical value 7.82. So the null hypothesis is accepted.

Q20. In *Drosophila sp* genetic distance between the red (*r*) responsible for red colouration of *malpighian* tubule and ebony (*e*) responsible for black colouration of the body, is stated to be 3.5 cross over unit apart. In order to determine the validity of this claim, the student of P.G. of Serampore college crossed females heterozygous for the two mutants with male homozygous for this two mutants, such a cross gives following offsprings—(*i*) Red 1790, (*ii*) Ebony 1812, (*iii*) Wild type 62 and (*iv*) Red ebony 68.

Substantiate the claim statistically.

Solution:

I. Male homozygous for the mutants have the genotype $\frac{r\,e}{r\,e}$ r = red, e = ebony genes for the corresponding characters.

II. Female heterozygous for mutants have the genotype $\frac{r\,+}{+\,e}$ (trans heterozygous).

III. It is well known that crossing over takes place in female *Drosophila* flies.

Here assumed genetic distance between red and ebony loci is 3.5

IV. Number of crossing over progeny (between red & ebony)

$$3.5 = \frac{x \times 100}{\text{Total}} = \frac{x \times 100}{3732}$$

$$\text{or} \quad x = \frac{3.5 \times 3732}{100} = 130.62$$

$\frac{r\,+}{+\,e}$ crossing over leads to red ebony $\left(\frac{r\,e}{r\,e}\right)$ & wild type $\left(\frac{+\,+}{r\,e}\right)$ = 130.62

$$\text{Number of expected red ebony} = \frac{130.62}{2} = 65.31$$

$$\text{Number of expected wild type} = \frac{130.62}{2} = 65.31$$

Remaining number of offsprings 3732 – 130.62 = 3601.38

It is the expected red and ebony

$$\text{red} = \frac{3601.38}{2} = 1800.69$$

$$\text{ebony} = \frac{3601.38}{2} = 1800.69$$

- ***Statistical Analysis:***
- *Null Hypothesis:* **Genetic distance between red and ebony loci is 3.5 cu (cross-over unit) (mu) (Morgan unit).**

- *Calculation:*

Observed (O)	Expected (E)	(|O − E|) = D	D^2	$\frac{D^2}{E}$	x^2	df
Red ebony = 68	65.31	68 − 65.31 = 2.69	7.236	$\frac{7.236}{65.31}$	0.110	4 − 1
Wild type = 62	65.31	62 − 65.31 = 3.31	10.956	$\frac{10.956}{65.31}$	0.167	= 3
Red = 1790	1800.69	1790 − 1800.69 = 10.69	114.276	$\frac{114.276}{1800.69}$	0.063	
ebony = 1812	1800.69	1812 − 1800.69 = 11.31	136.656	$\frac{136.656}{1800.69}$	0.0758	
Total = 3732					0.4158	

- *Critical value:* The control value of chi square (x^2) for 3*df* at 0.05 is 7.82.
- *Decision:* Since the calculated value of chi square for 3*df* is 0.415 < 7.82. So, the null hypothesis is accepted *i.e.,* deviation is non significant.

Q21. Map distance between A – B is 20mu. The following progenies are obtained *AB* = 85, *ab* = 75, *Ab* = 25, *aB* = 15. Find out whether the actual map distance represent *C.O* distance.

Solution:

20 mu

A ———————————— *B*

We know 1% crossing over represent, 1 map unit. Map distance between *A* – *B* 20mu crossing over classes are *Ab* = 25, *aB* = 15.

- *Null Hypothesis:* The actual map distance represent the crossing over distance.
- *Alternative Hypothesis:* Actual map distance does not represent the *C.O* distance.
- *Calculation:*

Observed (O)	Expected (E)	(|O − E|) = D	D^2	$\frac{D^2}{E}$	x^2	df
AB = 68	80	85 − 80 = 5	25	$\frac{25}{80} = 0.3125$		
ab = 75	80	75 − 80 = − 5	25	$\frac{25}{80} = 0.3125$	3.125	4 − 1 = 3
Ab = 25	20	25 − 20 = 5	25	$\frac{25}{20} = 1.25$		
aB = 15	20	15 − 20 = − 5	25	$\frac{25}{20} = 1.25$		
Total = 200						

- *Critical value:* The control value of x^2 for 3*df* at 0.05 is 7.82.
- *Decision:* Since the calculated chi square (x^2) for 3*df* is 3.125 < critical value of x^2 = 7.82 so the null hypothesis is accepted, Therefore actual map distance represent crossing over distance.

Q22. Map distance between *b–vg* is 17.5mu the following progenies are observed. ++ = 415, *bvg* = 400, *b*+ = 95, +*vg* = 90.

Find out whether actual map distance represent *C.O* (crossing over) distance.

Solution:

b ——— 17.5 *mu* ——— *vg*

We know 1% crossing over represent, map unit. 17.5% *C.O* present between b and *vg*. Crossing over classes are *b*+ = 95 and +*vg* = 90.

- *Null Hypothesis:* The actual map distance represent the crossing over distance.
- *Alternative Hypothesis:* Actual map distance does not represent the *C.O* distance.
- *Calculation:*

Observed (O)	Expected (E)	(|O − E|) = D	D^2	$\frac{D^2}{E}$	x^2	df
++ = 415	412.5	415 – 412.5 = 2.5	6.25	$\frac{6.25}{412.5} = 0.015$		4 – 1 = 3
b vg = 400	412.5	400 – 412.5 = 12.5	156.25	$\frac{156.25}{12.5} = 0.378$	1.106	
b + = 95	87.5	95 – 87.5 = 7.5	56.25	$\frac{56.25}{87.5} = 0.642$		
+ *vg* = 90	87.5	90 – 87.5 = 2.5	6.25	$\frac{6.25}{87.5} = 0.071$		
Total = 1000						

- *Critical value:* The control value of chisquare (x^2) for 3*df* at 0.05 is 7.82
- *Decision:* Since the calculated chisquare for 3*df* is 1.106 < critical value of chi square 7.82. So the null hypothesis is accepted, we can conclude that actual map distance represent *C.O* distance.

Q23. In pea plant, the wild type gene responsible for normal length is dominant over recessive allele short, responsible for shortness of plant. From a cross between two tall plants the following results are obtained.

Experiment	*Tall*	*Short*
1	102	33
2	129	42
3	148	51
4	70	24
5	108	38

Recheck the results of the data whether they can be pulled together or not.

Solution:

- *Null Hypothesis:* Since this is a monohybrid cross (Tall × Tall). The expected ratio will be 3 : 1.

• *Calculation:*

	Observed (O)	Expected (E)	(|O − E|) = D	D^2	$\frac{D^2}{E}$	x^2	df
1.	Tall = 102	101.25	102 − 101.25 = 0.75	0.5625	$\frac{0.5625}{101.25} = 0.005$		1
	Short = 33	33.75	33 − 33.75 = 0.75	0.5625	$\frac{0.5625}{33.25} = 0.016$	0.021	
2.	Tall = 129	128.25	129 − 128.25 = 0.75	0.5625	$\frac{0.5625}{128.25} = 0.004$	0.017	1
	Short = 42	42.75	42 − 42.75 = 0.75	0.5625	$\frac{0.5625}{42.25} = 0.013$		
3	Tall = 148	149.25	148 − 149.25 = 1.25	1.56	$\frac{1.56}{149.25} = 0.01$	0.041	
	Short = 51	49.75	51 − 49.75 = 1.25	1.56	$\frac{1.56}{49.75} = 0.031$		1
4.	Tall = 70	70.5	70 − 70.5 = 0.5	0.25	$\frac{0.25}{70.5} = 0.003$	0.013	
	Short = 24	23.5	24 − 23.5 = 0.5	0.25	$\frac{0.25}{23.5} = 0.070$		1
5.	Tall = 108	109.5	108 − 109.5 = 1.5	2.25	$\frac{2.25}{109.5} = 0.02$	0.081	
	Short = 38	36.5	38 − 36.5 = 1.5	2.25	$\frac{2.25}{36.5} = 0.061$		1
						$x^2 = 0.173$	$df = 5$

• *Summed:*

Observed (O)	Expected (E)	(|O − E|) = D	D^2	$\frac{D^2}{E}$	x^2	df
Tall = 557	558.75	557 − 558.75 = 1.75	3.06	$\frac{3.06}{558.75} = 0.005$	0.021	2 − 1
Short = 188	186.25	188 − 186.25 = 1.75	3.06	$\frac{3.06}{186.25} = 0.016$		1

	Chi square	*Degrees of freedom*
Total:	0.173	5
Summed:	0.021	1
Homogenity:	0.152	4

• *Critical value:* **The control value of chi square (x^2) for 4*df* at 0.05 is 9.49.**

• *Decision:* **Since the calculated chi square (x^2) value is 0.152 < critical value = 9.49. So the null hypothesis is accepted *i.e.*, deviation is non significant. We can conclude that homogeneity persist among the sample.**

Q24. According to latest census report (2010) the number of women to every thousand men in the following five states/union territories is as follows: in Andaman 846, in Delhi 821, in Dadra and Nagar Haveli 812, in Chandigarh 777 and in Daman and Diu 710.

By appropriate statistical method determine whether 1 : 1 sex ratio among human population could be attributed to above five regions.

Solution:

- *Null Hypothesis:* Sex ratio is 1 : 1 among human population of five regions.
- *Calculation:*

	Observed (O)	Expected (E)	(|O – E|) = D	D^2	$\frac{D^2}{E}$	x^2	df
A	Men = 1000	923	1000 – 923 = 77	5929	$\frac{5929}{77} = 6.42$	12.84	1
	Women = 846	923	846 – 923 = – 77	5929	$\frac{5929}{77} = 6.42$		
D	Men = 1000	910.5	1000 – 910.5 = 89.5	8010.25	$\frac{8010.25}{910.5} = 8.79$	17.58	1
	Women = 821	910.5	821 – 910.5 = –89.5	8010.25	$\frac{8010.25}{910.5} = 8.79$		
DN	Men = 1000	906	1000 – 906 = 94	8836	$\frac{8836}{906} = 9.75$		1
	Women = 812	906	812 – 906 = – 94	8836	$\frac{8836}{906} = 9.75$	19.5	1
C	Men = 1000	888.5	1000 – 888.5 = 111.5	12432.25	$\frac{12432.25}{888.5} = 13.99$	27.98	1
	Women = 777	888.5	777 – 888.5 = –111.5	12432.25	$\frac{12432.25}{888.5} = 13.99$		
DD	Men = 1000	855	1000 – 855 = 145	21025	$\frac{21025}{855} = 24.59$		
	Women = 710	855	710 – 855 = –145	21025	$\frac{21025}{855} = 24.59$	49.18	1
	Male = 5000 Female = 3966					x^2 = 127.08	df = 5

- *Summed:*

Observed (O)	Expected (E)	(|O – E|) = D	D^2	$\frac{D^2}{E}$	x^2	df
Male = 5000	4483	5000 – 4483 = 517	267289	$\frac{267289}{4483}$	59.62	2 – 1
Female = 3966	4483	3966 – 4483 = 517	267289	$\frac{267289}{4483}$	59.62	1
					119.24	

	Chi square	Degress of freedom
Total:	127.08	5
Summed:	119.2	1
Homogenity:	7.88	4

- *Critical value:* The control value of chi square (x^2) for 4*df* at 0.05 is 9.49.
- *Decision:* Since the calculated value of x^2 for 4*df* at 0.05 is 7.88 < critical value of x^2 for 4*df* 9.49. So the null hypothesis is accepted.

Q25. In *Drosophila sp* mutant vestigial (*vg*) is recessive to its wild type allele (+) responsible for normal wing shape. From a cross conducted during midsummer 123 *vg* flies, 138 wild type females appeared. The same experiment in winter produce 374 *vg* and 262 wild type progenies. Using appropriate statistical method, determine whether environmental factors are responsible for such deviation.

Solution:

According to given data, the flies obtained after crossing.

	Wild	Vestigial
Summer	138	123
Winter	262	374

- *Null Hypothesis:* The data is independent of environmental condition.
- *Alternative Hypothesis:* The data is dependent on environmental condition

	Wild	Vestigial	Total	
Summer	138 (*a*)	123 (*b*)	261 *a* + *b*	R_1
Winter	262 (*c*)	374 (*d*)	636 *c* + *d*	R_2
	(*a* + *c*) 400	497 (*b* + *d*)		

$$X^2 = \frac{N\{|ad - bc| - N/2\}^2}{R_1.R_2.C_1.C_2}$$

$ad = 138 \times 374 = 51612$

$bc = 123 \times 262 = 32226$

$$X^2 = \frac{\left[(138 \times 374) - \left(123 \times 262) - \frac{N}{2}\right)^2\right] \times 897}{(138 + 123)(262 + 374)(138 + 262)(123 + 374)}$$

$$X^2 = \frac{[(|51612 - 32226|) - 897/2]^2 \times 897}{261 \times 636 \times 400 \times 497}$$

$$= \frac{(19386 - 448.5)^2 \times 897}{165996 \times 198800}$$

$$= \frac{18937.5 \times 18937.5 \times 897}{165996 \times 198800}$$

$$= 0.095259 \times 0.114084 \times 897$$

$$= 9.748 = 9.75$$

$R_1 = a + b = 261,$

$R_2 = c + d = 636$

$C_1 = a + c = 400$

$C_2 = b + d = 497$

$N = R_1 + R_2 = C_1 + C_2 = 897$

- *Degrees of freedom* = (*R* – 1) (*C* – 1) = (2 – 1) (2 – 1) = 1

- *Critical value:* The table value of chi square (x^2) for 1*df* at 0.05 is 3.84
- *Decision:* Since the calculated value of chi square (x^2) at 1*df* is 9.75 < critical value 3.84. So the null hypothesis is rejected *i.e.*, deviation is significant.

  Therefore we can conclude that environment has some impact on that result.

Q26. In garden pea plant heterozygote for mutant for shortness was self crossed. Results obtained from two different localities give the following results—locality 1 Tall–218, Short 70, Locality 2–Tall 346, Short 116.

Using appropriate statistical method determine whether any environmental factors responsible for such deviation.

Solution:

According to given data the following plants are obtained after crossing.

	Tall	*Short*
Locality–1	218	70
Locality–2	346	116

- *Null Hypothesis:* The result is independent of environmental factor.
- *Alterative Hypothesis:* Environment has some impact on the result.

	Tall	*Short*	*Total*	
Locality–1	218 (*a*)	70 (*b*)	288 (*a* + *b*)	R_1
Locality–2	346 (*c*)	116 (*d*)	462 (*c* + *d*)	R_2
	(*a* + *c*) 564	186 (*b* + *d*)	750	

$$X^2 = \frac{N\{(ad - bc) - N/2\}^2}{R_1 \times R_2 \times C_1 \times C_2}$$

$ad = 218 \times 116 = 25288$

$bc = 70 \times 346 = 24220$

$$= \frac{750\{(25288 - 24220) - 750/2\}^2}{288 \times 462 \times 564 \times 186}$$

$R_1 = a + b = 288,\ C_1 = a + c = 564$

$R_2 = c + d = 462,\ C_2 = b + d = 186$

$$= \frac{(1068 - 375)^2 \times 750}{133056 \times 104904}$$

$N = R_1 + R_2 = C_1 + C_2 = 750$

$$= \frac{693 \times 693 \times 750}{133056 \times 104904}$$

$= 0.0052 \times 0.006606 \times 750$

$= 0.0000343 \times 750$

$= 0.0257 = 0.026$

- *Degrees of freedom:* $(R - 1)(C - 1) = (2 - 1)(2 - 1) = 1$
- *Critical value:* The table of chi square for 1*df* at 0.05 is 3.84
- *Decision:* Since the calculated value of chi square for 1*df* at 0.05 is 0.026 < critical value = 3.84, so the null hypothesis is accepted *i.e.*, deviation is non significant. We can conclude that environment has no impact on this result.

Q27. The gene producing sickle shaped red blood cells in humans is called Hb^s and normal allele Hb^A. Various investigations have tested incidence of heavy infections of malarial parasite *Plasmodium falciparum* in African children who are heterozygous for sickle cell gene (Hb^s/Hb^A) and in children who are homozygous normal (Hb^A/Hb^A). In an investigation by *Allison and Clyde* following observations are procured:

	Heavy infections	*Noninfected lightly infected*
Hb^s/Hb^A	36	100
Hb^s/Hb^A	152	255

Test whether heterozygotes in the sample are better protected against heavy malarial infections.

Solution:

According to the given data, infection incidence is as follows:

Genotype	*Heavy infections*	*Non/lightly infected*
Hb^s/Hb^A	36	100
Hb^s/Hb^A	152	255

- *Null Hypothesis:* Results of the investigation are independent of genetic configuration.
- *Alternative Hypothesis:* Results of investigation are dependent on genetic configuration:

	Heavy infection	*Non/light infection*	*Total*	
Hb^s/Hb^A	36 (*a*)	100 (*b*)	136 (*a* + *b*)	R_1
Hb^A/Hb^A	152 (*c*)	255 (*d*)	407 (*c* + *d*)	R_2
	C_1 = 188 (*a* + *c*)	C_2 = 355 (*b* + *d*)		

$$X^2 = \frac{N\{(|ad - bc|) - N/2\}^2}{R_1 \times R_2 \times C_1 \times C_2}$$

$ad = 36 \times 255 = 9180$

$bc = 100 \times 152 = 15200$

$$X^2 = \frac{\{(9180 - 15200) - 543/2\}^2 \times 543}{136 \times 407 \times 188 \times 355}$$

$R_1 = a + b = 136$

$R_2 = c + d = 407$

$$= \frac{\{6020 - 271.5\}^2 \times 543}{55352 \times 66740}$$

$C_1 = a + c = 188$

$C_2 = b + d = 355$

$N = R_1 + R_2 = C_1 + C_2 = 136 + 407 = 543$

$$= \frac{5748.5 \times 5748.5 \times 543}{55352 \times 66740} = \frac{33045252 \times 543}{55352 \times 66740}$$

$= 597.0 \times 0.0081 = 4.83$

$= 4.8$

- *Degrees of freedom* = (*R* – 1) (*C* – 1) = (2 – 1) × (2 – 1) = 1
- *Critical alue:* The control value of chi square for 1*df* at 0.05 is 3.84.
- *Decision* Since the calculated value of chi square (x^2) at 1*df* is 4.8 > critical value of chi square 3.84. So the null hypnosis is rejected *i.e.*, deviation significant. So the null hypothesis is rejected *i.e.*, deviation significant. So we conclude that heterozygous are better protected.

Q28. Arambagh sub divisional hospital reported that the birth of 100 female and 106 male babies for the year 2007. Is this statistically significant deviation from the expected 1 : 1 sex ratio? At the same time, exactly the same ratio was reported for a larger population that is about 10,000 female babies and 10,600 male babies. Is this latter case a statistically significant deviation from the expected 1 : 1 ratio? Comment.

Solution:

Report I

I. *Null Hypothesis:* 1 : 1

II. *Calculation:*

Observed (O)	Expected (E)	(|O – E|)	(|O – E|)^2	$\frac{(O-E)^2}{E}$	x^2
Female = 100	103	(|100 – 103|) = 3	9	$\frac{9}{103}$ = 0.0873	0.174
Male = 106	103	(|106 – 103|) = 3	9	$\frac{9}{103}$ = 0.0873	
Total = 206					

III. *Critical value:* The control value of x^2 at 0.05 for *df* 2 – 1 = 1 is 3.84.

IV. *Decision:* Since the calculated value of chi square (x^2) is 0.174 < critical value of chi square at 1*df* is 3.84, So the null hypothesis is accepted *i.e.,* 0.17 is not a significant deviation from 1 : 1 ratio the corresponding probabilities is 0.68 *i.e.,* 68%.

Report II

I. *Null Hypothesis:* 1 : 1

II. *Calculation:*

Observed (O)	Expected (E)	(O – E)	(O – E)^2	$\frac{(O-E)^2}{E}$	x^2
Female = 10,000	10,300	10,000 – 10,300 = – 300	9000	$\frac{9000}{10300}$ = 8.737	17.474 = 17.5
Male = 10,600	10,300	10,600 – 10,300 = 300	9000	$\frac{9000}{10300}$ = 8.737	
Total = 20,600					

III. *Critical value:* The control value of chi square at 0.05 for 2 – 1 = 1*df* is 3.84

IV. *Decision:* Since the calculated chi square is 17.5 > critical value of chi square at 1*df* is 3.84. So the null hypothesis is rejected. Here the deviation is significant from 1 : 1 ratio.

The two ratios are identical but the x^2 values are different. Therefore the conclusions are very different. It indicates that chi square test is sensitive to sample size.

Q29. In the F_2 progeny from a cross of a white flowered plant with a yellow flowered plant, 570 yellow to 155 white flowered progeny resulted. What is the probability of obtaining a deviation from 3 : 1 ratio greater than these results due to only random chance?

Solution:

- *Hypothesis:* 3 : 1

• *Calculation:*

Observed (O)	Expected (E)	(|O − E|)	(|O − E|)^2	$\frac{(O\ E)^2}{E}$	x^2
Yellow = 570	543.75	(570 − 543.75) = 26.25	689.06	$\frac{689.06}{543.75}$ = 1.267	5.0687
White = 155	181 25	155 − 181.25 = 26.25	689.06	$\frac{689.06}{181.25}$ = 3.801	= 5.07
Total = 725					

• *Critical value:* The control value of chi square at 0.05 for 2 − 1 = 1*df* is 3.84.

• *Decision:* Since the calculated x^2 value at 2*df* is 5.07 > critical value, so the null hypothesis is rejected *i.e.,* variation between observed and expected data is significant it indicates that the data is deviated from 3 : 1 ratio.

The x^2 value of 5.07 corresponds to probabilities of 0.024. This means that the probabilities of getting a deviation from 3 : 1 ratio as great or greater than these observed data by random chance is only 0.024 × 100 = 2.4 %.

Q30. In laboratory researchers had repeated some of Mendel's experiments. For example, the following F_2 results were shown with seed shape in peas. Wrinkled 884 and Round 288 calculate the goodness of fit for these data.

Solution:

• *Null Hypothesis:* 3 : 1

• *Alternative Hypothesis:* 1 : 1

• *Calculation:*

Observed (O)	Expected (E)	(O − E)	(O − E)^2	$\frac{(O - E)^2}{E}$	x^2
Wrinkled = 884	$1172 \times \frac{3}{4}$ = 879	884 − 879 = 5	25	$\frac{25}{879}$ = 0.0284	0.1137
Round = 288	$1172 \times \frac{1}{4}$ = 293	288 − 293 = − 5	25	$\frac{25}{293}$ = 0.0853	= 0.114
Total = 1172					

• *Critical value:* The control value of chi square at 0.05 for *df* (2 − 1) = 1 is 3.84

• *Decision:* Since the calculated value of chi square (x^2) is 0.114 < critical value of x^2 for 1*df* is 3.84, So the null hypothesis is accepted.

x^2 of 0.114 corresponds to a probabilities of 0.73 *i.e.,* good fit.

Q31. In a Botanical laboratory researchers repeated two of Mendel's experiments and reported the following result. (*a*) 705 plants with violet red flowers and 224 plants with white flowers and (*b*) 428 plants with green pods and 152 plants with yellow pods. Hypothesize the 3 : 1 ratio in each case & test the goodness of fit.

Solution:

(*a*) • *Null Hypothesis* = 3 : 1

• *Calculation:*

Observed (O)	*Expected (E)*	*(O – E)*	*$(O - E)^2$*	*$\frac{(O-E)^2}{E}$*	*x^2*
Violet red = 705	$705 \times \frac{3}{4} = 696.75$	70.5 – 696.75 = 8.25	68.062	$\frac{68.062}{696.75} = 0.09768$	0.39073
White = 224	$929 \times \frac{1}{4} = 232.25$	224 – 232.25 = – 8.25	68.062	$\frac{68.062}{232.25} = 0.29305$	= 0.391
Total = 929					

• *Critical value:* The control value of chi square at 0.05 for *df* (2 – 1) = 1 is 3.84

• *Decision:* Since the calculated value of chi square is 0.391 < critical value of x^2 for 1*df* is 3.84. so the null hypothesis is accepted. This is a good fit.

(*b*) • *Null Hypothesis* = 3 : 1

• *Calculation:*

Observed (O)	*Expected (E)*	*(O – E)*	*$(O - E)^2$*	*$\frac{(O-E)^2}{E}$*	*x^2*
Greenpods = 428	$580 \times \frac{3}{4} = 435$	428 – 435 = 7	49	$\frac{49}{435} = 0.1126$	0.4505
Yellowpods = 152	$580 \times \frac{1}{4} = 145$	152 – 145 = 7	49	$\frac{49}{145} = 0.3379$	= 0.451
Total = 580					

• *Critical value:* The control value of chi square at 0.05 for *df* (2 – 1) = 1 is 3.84

• *Decision:* Since the calculated value of chi square is 0.451 < critical value of x^2 for 1*df* is 3.84 so the null hypothesis is accepted. This is a good fit.

Q32. In Genetics Research unit laboratory Calcutta, we get a 103 normal female flies and 53 normal male flies from a cross. Explain the result with suitable cross & statistical test.

Solution: Cross result female 103 and male 53 *i.e.*, 2 : 1 ratio.

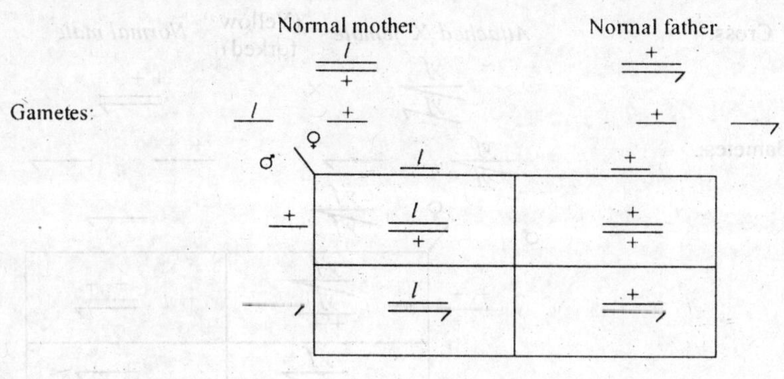

Result:

	Genotype	Phenotype	Phenotypic ratio
1.	$\frac{l}{+}$	Normal female	2
2.	$\frac{+}{+}$	Normal female	
3.	$\frac{+}{\rightharpoondown}$	Normal male	1
4.	$\frac{l}{\rightharpoondown}$	Dead male	0

- *Null Hypothesis:* 2 : 1
- *Calculation:*

Observed (O)	Expected (E)	(O – E)	$(O - E)^2$	$\frac{(O-E)^2}{E}$	x^2
Normal female = 103	104	103 – 104	1	$\frac{1}{104} = 0.0096$	0.0285
Normal male = 53	52	52 – 53	1	$\frac{1}{53} = 0.01886 = 0.0189$	= 0.029
Total = 156					

- *Degrees of freedom* = 2 – 1 = 1
- *Critical value:* The table value of chi square at 1*df* at 0.0*f* is 3.84
- *Decision:* Since the calculated value of chi square for 1*df* is 0.029 < critical value, 3.84. So the null hypothesis is accepted, so we conclude that normal female parent has a lethal gene on its one chromosome and wild type gene on other homologous.

Q33. In a cross of *Drosophila* flies from a Genetics laboratory, we get 106 yellow forked female and 104 normal male. Explain the result with suitable cross & statistical test.

Solution:

- In this problem both body colour and shape of the bristles are under consideration. Hence it is a dihybrid cross.
- Sex ratio is roughly 1 : 1 (106 : 104)

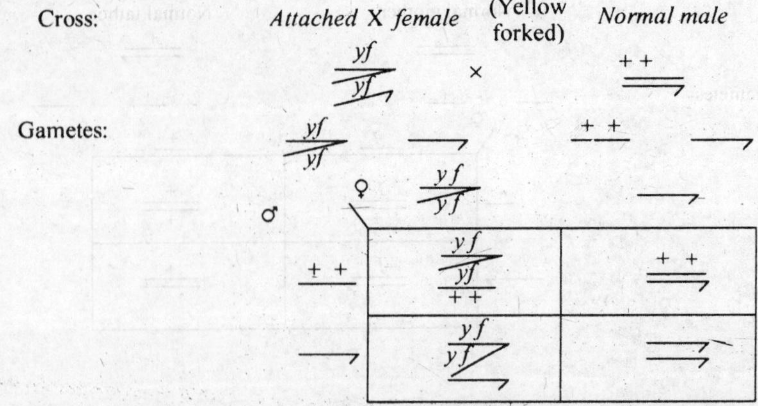

Results:

	Genotype	Phenotype	Phenotypic frequency
1.	$\frac{++}{\frac{yf}{yf}}$	Meta female (Dies)	0
2.	$\frac{++}{\rightharpoondown}$	Normal male	1
3.	$\frac{yf}{yf}$ / $\rightharpoondown$	Yellow forked female	1
4.	$\rightharpoondown$	Dies	0

- *Hypothesis:* 1 : 1
- *Calculation:*

Observed (O)	Expected (E)	(O – E)	$(O - E)^2$	$\frac{(O-E)^2}{E}$	x^2
Yellow forked female = 106	105	106 – 105 = 1	$(1)^2$	$\frac{1}{105} = 0.009$	0.018
Normal male = 104	105	104 – 105 = –1	$(-1)^2$	$\frac{1}{105} = 0.009$	
Total = 210					

- *Degrees of freedom:* 2 – 1 = 1
- *Critical value:* The table value of chi square for 1*df* at 0.05 level is 3.84
- *Decision:* Since the calculated value of chi square for 1*df* is 0.018 < critical value = 384 so the null hypothesis accepted.

  So we can conclude that it is a cross of attached *X* female with normal male flies.

Q34. In Drosophila Bar eye (*B*) is dominant over normal (+) eye. From a cross the following flies are obtained.

104 Bar female, 102 normal male and 103 normal female. explain the result with a suitable cross & statistical test.

Solution:

I. Here only eye shape is under consideration.

II. Here the ratio Bar eye female, normal eye male & normal eye females are roughly 1 : 1 : 1 (104 : 103 : 102)

	Female	Male
	$\frac{ClB}{+++}$	$\frac{+++}{\rightharpoondown}$
Gametes:	$\underline{ClB}$ $\underline{+++}$	$\underline{+++}$ $\rightharpoondown$

♂ \ ♀	$\underline{ClB}$	$\underline{+++}$
$\underline{+++}$	$\frac{ClB}{+++}$	$\frac{+++}{+++}$
$\rightharpoondown$	$\frac{ClB}{\rightharpoondown}$	$\frac{+++}{\rightharpoondown}$

Genotypes	*Phenotypes*	*Phenotypic frequency*
$\frac{ClB}{+++}$	Bar female	1
$\frac{+++}{+++}$	Normal female	1
$\frac{ClB}{\rightharpoondown}$	Male (dies)	0
$\frac{+++}{\rightharpoondown}$	Normal male	1

- *Null Hypothesis:* 1 : 1 : 1
- *Calculation:*

Observed (O)	*Expected (E)*	*(O – E)*	*(O – E)²*	$\frac{(O-E)^2}{E}$	x^2
Bar female = 104	103	104 – 103	$(1)^2$	$\frac{1}{103} = 0.009$	
Normale female = 103	103	103 – 103	$(0)^2$	$\frac{0}{103} = 0.0$	0.018
Normal male = 102	103	102 – 103	$(-1)^2$	$\frac{1}{103} = 0.009$	
Total = 309					

- *Degrees of freedom:* 3 – 1 = 2
- *Critical value:* The table value of chi square for 3*df* at 0.05 level is.
- *Decision:* Since the calçulated value of x^2 for 2*df* is 0.018 < the critical value of x^2 2*df*, 5.99 so the null hypothesis is accepted.

Therefore we can conclude that the above mentioned result comes from a cross of $\frac{ClB}{+++}$ female $\frac{+++}{\rightharpoondown}$ with normal male.

Q35. A geneticist crossed wild grey coloured mice with albino (white) mice. All the F_1 progeny were grey. These progeny were intercrossed to produce an F_2. The F_2 progenies were 198 gray and 72 albino mice. Propose an hypothesis to explain those results, Design the crosses and compare the results with the predictions of the hypothesis.

Solution: The data suggest that the coat colour is controlled by a single gene with two alleles *C* (grey) and *c* (albino) and that *C* is dominant over *c*.

Grey coloured mice $\frac{C}{C}$ *albino mice* $\frac{c}{c}$

Gametes $\underline{C}$ $\underline{c}$

F_1 → $\frac{C}{c}$ (Grey)

F_2 = Grey × Grey = $\frac{C}{c} \times \frac{C}{c}$

Gametes: $\underline{C}$ $\underline{c}$ $\underline{C}$ $\underline{c}$

	C	c
C	$\frac{C}{C}$	$\frac{c}{C}$
c	$\frac{C}{c}$	$\frac{c}{c}$

Results:

Genotypes	Phenotypes	Phenotypic frequency
$\frac{C}{C}$	Grey	
$\frac{C}{c}$	Grey	$3 = \frac{3}{4}$
$\frac{C}{c}$	Grey	
$\frac{c}{c}$	albino	$1 = \frac{1}{4}$

The data (F_2) showed that 198 grey mice and 72 albino mice.
The ratio is roughing 3 : 1 (198 : 72)

- *Null Hypothesis:* 3 : 1
- *Calculation:*

Observed (O)	Expected (E)	(O – E)	$(O - E)^2$	$\frac{(O-E)^2}{E}$	x^2
Grey = 198	$270 \times \frac{3}{4} = 202.5$	198 – 202.5 = – 4.5	$(4.5 - 0.5)^2$	$\frac{16}{202.5} = 0.079$	0.316
albino = 72	$270 \times \frac{1}{4} = 67.5$	72 – 67.5 = 4.5	$(4.5 - 0.5)^2$	$\frac{16}{67.5} = 0.237$	
Total = 270					

- *Degrees of freedom:* 2 – 1 = 1
- *Critical value:* The control value of chi square for 1*df* 0.05 level is 3.84
- *Decision:* Since the calculated chi square value is 0.316 < critical value at 1*df* 3.84. So the null hypothesis is accepted.

Q36. In tomatoes cut leaf is dominant over potato leaf, purple stem is dominant over green. A true breeding cut, green tomato plants is crossed with true breeding potato purple plant and the F_1 plants were allowed to interbreed. The 320 F_2 plants were phenotypically 189 cut purple, 67 cut green, 50 potato purple and 14 potato green. Test the result with the help of chi square test.

Solution:

I. In this problem two characters (cut/potato & purple/green) are under consideration. So it is a dihybrid cross.

II. The results showed that cut purple 189, cut green 67, potato purple 50 and potato green 14 *i.e.,* ratio is roughly 9 : 3 : 3 : 1.

- *Null Hypothesis:* 9 : 3 : 3 : 1.
- *Calculation:*

Observed (O)	Expected (E)	(O – E)	$(O - E)^2$	$\frac{(O-E)^2}{E}$	x^2
Cut purple = 189	$320 \times \frac{9}{16} = 180$	189 – 180 = 9	81	$\frac{81}{180} = 0.45$	
Cut green = 67	$320 \times \frac{3}{16} = 60$	67 – 60 = 7	49	$\frac{49}{60} = 0.816$ =0.82	4.74
Potato purple = 50	$320 \times \frac{3}{16} = 60$	50 – 60 = 10	100	$\frac{100}{60} = 1.67$	
Potato green = 14	$320 \times \frac{1}{16} = 20$	14 – 20 = – 6	36	$\frac{36}{20} = 1.8$	
Total = 320					

- *Degrees of freedom:* 4 – 1 = 3
- *Critical value:* The table value of chi square for 3*df* at 0.05 level is 7.82
- *Decision:* Since the calculated chi square value for 3*df* is 4.74 < critical value for 1*df* 7.82 so the null hypothesis is accepted.

  Therefore we can conclude the problem is a result of dihybrid cross & followed independent assortment law.

Q37. In *Drosophila* '*b*' is recessive autosomal mutation that results in black body colour and *vg* is a recessive autosomal mutation results in vestigial wing. Wild type flies have grey bodies and long wings. True breeding black normal ($b/b\ vg^+/vg$) flies were crossed with true breeding grey vestigial ($b^+/b^+\ vg/vg$) flies. F_1 grey normal female ($b^+/b\ vg^+/vg$) flies were test cross to black vestigial ($b/b\ vg/vg$) male flies. (The female is the heterozygote in this test cross because in *Drosophila* no crossing over occurs between homologous pair of chromosome in males. The test cross progeny data were as follows) 283 grey, normal, 1, 294 grey vestigial, 1,418 black normal, 241 black vestigial total 3, 236 flies. Use chi square test to test the hypothesis that the two genes are unlinked.

Solution: If the two genes are unlinked, then a test cross should result in a 1 : 1 : 1 : 1

- *Null Hypothesis:* 1: 1 : 1 : 1
- *Calculation:*

Observed (O)	Expected (E)	(O – E)	$(O - E)^2$	$\frac{(O-E)^2}{E}$	x^2
Grey normal = 283	$3236 \times \frac{1}{4} = 809$	283 – 809 = 526	276676	$\frac{276676}{809} = 341.99$	
Grey vestigial = 1294	$3236 \times \frac{1}{4} = 809$	1294 – 809 = 485	235225	$\frac{235225}{809} = 290.76$	1489.98
Black normal = 1418	$3236 \times \frac{1}{4} = 809$	1418 – 809 = 609	370881	$\frac{370881}{809} = 458.44$	
Black vestigial = 241	$3236 \times \frac{1}{4} = 809$	241 – 809 = – 568	322624	$\frac{322624}{809} = 398.79$	
Total = 3236					

• *Degrees of freedom:* 4 – 1 = 3

• *Critical value:* The table value for $3df$ at 0.05 level is 7.82

• *Decision:* Since the calculated chi square value is 1489.98 > critical value 7.82 for $3df$ at 0.05 so the null hypothesis is rejected *i.e.,* deviation is significant.

Therefore we would reject the independent assortment hypothesis (1 : 1 : 1 : 1) and genetically, the only alternative hypothesis that could logically apply is that genes are linked.

Q38. From several crosses of the general type *A/A B/B* × *a/a b/b* the F_1 individuals of the type *A/a B/b* were test crossed to *a/a b/b* The results are shown below.

	Test cross progeny			
Test cross of F_1 from cross	*A/a B/b*	*a/a b/b*	*A/a b/b*	*a/a B/b*
1	310	315	287	288
2	36	38	23	23
3	360	380	230	230
4	74	72	50	44

For each set of progeny use the x^2 test to decide if there is evidence of linkage.

Solution:

1. • *Null Hypothesis:* 1 : 1 : 1 : 1 *i.e.,* no linkage.

• *Alternative Hypothesis* = loci are linked

• *Calculation:*

Observed Number (O)	Expected Number (E)	(O – E)	$(O - E)^2$	$\frac{(O-E)^2}{E}$	x^2
A/a B/b = 310	$1200 \times \frac{1}{4} = 300$	310 – 300 = 10	100	$\frac{100}{300} = 0.333$	
a/a b/b = 315	$1200 \times \frac{1}{4} = 300$	315 – 300 = 15	225	$\frac{225}{300} = 0.750$	2.126 = 2.13
A/a b/b = 287	$1200 \times \frac{1}{4} = 300$	287 – 300 = – 13	169	$\frac{169}{300} = 0.563$	
a/a B/b = 288	$1200 \times \frac{1}{4} = 300$	288 – 300 = – 12	144	$\frac{144}{300} = 0.48$	
Total = 1200					

• *Degrees of freedom:* 4 – 1 = 3

• *Critical value:* The control value of x^2 for $3df$ at 0.05 is 7.82.

• *Decision:* Since the calculated value of chi square for $3df$ is 2.13 < critical value of x^2 7.82. So the null hypothesis is accepted *i.e.,* genes are not linked.

2. • *Null Hypothesis:* 1 : 1 : 1 : 1 *i.e.,* no linkage.

• *Alternative Hypothesis:* Loci are linked.

• *Calculation:*

Observed Number (O)	Expected Number (E)	(O − E)	$(O-E)^2$	$\frac{(O-E)^2}{E}$	x^2
A/a B/b = 36	$120 \times \frac{1}{4} = 30$	36 − 30 = 6	36	$\frac{36}{30} = 1.2$	
a/a b/b = 38	$120 \times \frac{1}{4} = 30$	38 − 30 = 8	64	$\frac{64}{30} = 2.13$	6.59 = 6.6
A/a b/b = 23	$120 \times \frac{1}{4} = 30$	23 − 30 = − 7	49	$\frac{49}{30} = 1.63$	
a/a B/b = 23	$120 \times \frac{1}{4} = 30$	23 − 30 = − 7	49	$\frac{49}{30} = 1.63$	
Total = 120					

• *Degrees of freedom:* 4 − 1 = 3

• *Critical value:* The control value of chi square (x^2) for 3*df* at 0.05 is 7.82

• *Decision:* Since the calculated value of chi square is 6.6 < critical value of x^2 7.82. So the null hypothesis is accepted *i.e.,* loci are not linked

3. • *Null Hypothesis:* 1 : 1 : 1 : 1 loci are not linked.

• *Alternate Hypothesis:* Loci are linked.

• *Calculation:*

Observed Number (O)	Expected Number (E)	(O − E)	$(O-E)^2$	$\frac{(O-E)^2}{E}$	x^2
A/a B/b = 360	$1200 \times \frac{1}{4} = 300$	360 − 300 = 60	3600	$\frac{3600}{300} = 12.0$	
a/a b/b = 380	$1200 \times \frac{1}{4} = 300$	380 − 300 = 80	6400	$\frac{6400}{300} = 21.33$	50.90 = 50.6
A/a b/b = 230	$1200 \times \frac{1}{4} = 300$	230 − 300 = − 70	4900	$\frac{4900}{300} = 16.33$	
a/a B/b = 230	$1200 \times \frac{1}{4} = 300$	230 − 300 = − 70	4900	$\frac{4900}{300} = 16.33$	
Total = 1200					

• *Degrees of freedom:* 4 − 1 = 3

• *Critical value:* The table value of chi square for 3*df* at 0.5 is 7.82

• *Decision:* Since the calculated value of chi square for 3*df* at 0.05 is 50.6 > critical value of x^2 7.82 *i.e.,* significant.

Therefore the null hypothesis is rejected i.e. loci are not linked.

4. • *Null Hypothesis:* 1 : 1 : 1 : 1 loci are not linked (No evidence of linkage).

• *Alternative Hypothesis:* Loci are linked.

• *Calculation:*

Observed Number (O)	Expected Number (E)	(O − E)	$(O-E)^2$	$\frac{(O-E)^2}{E}$	x^2
$A/a\ B/b = 74$	$240 \times \frac{1}{4} = 60$	74 − 60 = 14	196	$\frac{196}{60} = 3.266$	
$a/a\ b/b = 72$	$240 \times \frac{1}{4} = 60$	72 − 60 = 12	144	$\frac{144}{60} = 2.4$	11.598 = 11.60
$A/a\ b/b = 50$	$240 \times \frac{1}{4} = 60$	50 − 60 = − 10	100	$\frac{100}{60} = 1.606$	
$a/a\ B/b = 44$	$240 \times \frac{1}{4} = 60$	44 − 60 = − 16	256	$\frac{256}{60} = 4.266$	
Total = 240					

• *Degrees of freedom:* 4 − 1 = 3

• *Critical value:* The control value of x^2 for 3*df* at 0.05 is 7.82

• *Decision:* Since the calculated value of chi square for 3*df* is 11.60 > critical value of chi square 7.82, so the null hypothesis is rejected *i.e.*, deviation is significant. Therefore no evidence of linkage is rejected, *i.e.*, deviation is significant. Therefore no evidence of linkage is rejected *i.e.*, loci are linked.

Q39. In the Genetics Research Unit laboratory Calcutta, a research student crossed flies with normal long wings to flies with mutant dumpy wings which he believed was a recessive trait. In the F_1 all flies had long wings. In the F_2 the following results were obtained. 792 long wings and 208 dumpy winged flies. The student tested the hypothesis that dumpy wing is inherited as a recessive trait by performing x^2 analysis of the F_2 data.

(*a*) What ratio did the student hypothesized?

(*b*) Did the x^2 analysis support the hypothesis?

(*c*) What do the data suggest about dumpy mutation?

Solution:

(*a*) The student hypothesized that the F_2 data (792 : 208) fit Mendel's 3 : 1 monohybrid ratio for recessive genes.

(*b*) Chi square analysis to support the hypothesis 3 : 1

• *Null Hypothesis:* 3 : 1

• *Calculation:*

Observed Number (O)	Expected Number (E)	(O − E)	$(O-E)^2$	$\frac{(O-E)^2}{E}$	x^2
Longwing = 792	$1000 \times \frac{3}{4} = 750$	792 − 750 = 42	1764	$\frac{1764}{750} = 2.352$	9.408 = 9.41
Dumpywing = 208	$1000 \times \frac{1}{4} = 250$	208 − 250 = − 42	1764	$\frac{1764}{250} = 7.056$	
Total = 1000					

- *Degrees of freedom:* 2 – 1 = 1
- *Critical value* The table value of chi square for 1*df* at 0.05 levels is 3.84.
- *Decision:* Since the calculated chi square value is 9.41 > critical value of chi square at 1*df* 3.84. So the null hypothesis is rejected.

Therefore the data is not statistically fit for 3:1 ratio.

(*c*) I. At the time of data collection we must assume that all genotypes are equally viable.

II. The genotypes yielding long wings are equally likely to survive from fertilization through adulthood as the genotype yielding *dumpy* wings.

III. Further study would reveal that *dumpy* flies are some what less viable than normal flies. As a result we expect less than $\frac{1}{4}$ of the total offspring to express *dumpy*.

This observation was highlighted in this data although we have not proved it.

Q40. The following F_2 results of two of Mendel's monohybrid crosses. (*a*) Full pods = 882 and constricted pods = 299, (*b*)Violet flowers = 705 and white flower = 224. State Null hypothesis to be tested using chi square analysis. Calculate x^2 value and determine the *p* value for both. Interpret *p* value. Can the deviation in each case be attributed to chance or not? Which of the two crosses shows a greater amount of deviation?

Solution:

- *Null Hypothesis:* 3 : 1
- *Calculation:*

(*a*)

Observed (O)	Expected (E)	(|O – E|) = D	D^2	$\frac{D^2}{E}$
Full pods = 882	885.75	882 – 885.75 = 3.75	14.0625	$\frac{14.6025}{885.75} = 0.01587$
Constricted pods = 299	295.25	299 – 295.25 = 3.75	14.0625	$\frac{14.6025}{295.25} = 0.0476$
Total = 1181				

$$X^2 = 0.01587 + 0.04762 = 0.06349$$

(*b*)

Observed (O)	Expected (E)	(|O – E|) = D	D^2	$x^2 = \frac{D^2}{E}$	x^2
Violet flower = 705	696.75	705 – 696.75 = 8.25	68.0625	$\frac{68.0625}{696.75} = 0.0976$	0.3906
White flower = 224	232.25	224 – 232.25 = 8.25	68.0625	$\frac{68.0625}{232.25} = 0.2930$	
Total = 929					

- *Critical value:* The control value of x_2 at 0.05 for *df* 3 – 1 = 2 is 3.84

- *Decision:*

I. Since the calculated value of chi square (x^2) is 0.06349 & 0.3906 respectively for (*a*) & (*b*).

II. In case of (*a*) 0.0634 < critical value, so the null hypothesis is accepted, (*b*) 0.3906 < critical value so the null hypothesis also accepted.

Here in both cases deviation is **nonsignificant** although a slight deviation persists. In case of (*b*) deviation is greater than (*a*).

Q41. On crossing F_1 flies, the F_2 generation gave the following phenotypes.

Red straight = 339 Purple straight = 632

Red curved = 725 Purple curved = 384

Test whether the genetic theory is compatible with Mendelian 9 : 3 : 3 : 1 distribution.

[Given x^2 .05(3) = 7.82] *[M. Sc (Zool) C.U. = 2007]*

Solution:

- *Null Hypothesis* = 9 : 3 : 3 : 1
- *Calculation:*

Observed (O)	Expected (E)	(O – E)	$\frac{(O-E)^2}{E}$	x^2
Red curved = 725	$\frac{9}{10} \times 2080 = 1170$	725 – 1170 = – 445	$\frac{1980.25}{1170} = 169.252$	169.252
Purple curved = 384	$\frac{3}{10} \times 2080 = 390$	384 – 390 = – 6	$\frac{36}{390} = 0.092$	0.092
Red straight = 339	$\frac{3}{10} \times 2080 = 390$	339 – 390 = 51	$\frac{2601}{390} = 6.669$	6.669
Purple straight = 632	$\frac{1}{10} \times 2080 = 130$	632 – 130 = 502	$\frac{252004}{130} = 1938.492$	1938.492
Total = 2080				

- *Degrees of freedom.* 4 – 1 = 3
- *Critical value:* The tabulated value of chi square for (4 – 1) = 3 of is 7.82.
- *Decision:* Since the calculated x^2 value is 2114.5 which is more than the tabulated chi square value 2114.5 > critical value *i.e.,* 7.8.

Therefore in this problem the genetic theory is incompatible with the Mendelian 9 : 3 : 3 : 1 distribution.

Q42. A pure breeding cross of genotype *A/A B/B* and *a/a b/b* and result a dihybrid *A/a B/b*. Which is again test crossed to *a/a b/b*. A total of 500 progeny are classified as follows.

***AB* (parental) = 140, *ab* (parental)= 135**

***Ab* (recombinant) = 110, *aB* (recombinant) = 115**

Use chi square test under the hypothesis of independent assortment.

Solution:

- *Null Hypothesis:* The cross is under independent assortment *i.e.,* the ratio is 1 : 1 : 1 : 1 (attributes are independent).

- *Calculation:* Here the proportion of parental phenotype and recombinant phenotype is not exactly 1 : 1 : 1 : 1. We now place the phenotype in contingency table.

$$\text{Expected Phenotype } (E) = \frac{R \times C}{N}$$

	B	*b*	
A	*AB* = 140 (*a*)	*Ab* = 110 (*b*)	R_1
a	*aB* = 115 (*c*)	*ab* = 135 (*d*)	R_2
	C_1	C_2	

$R_1 = a + b = 140 + 110 = 250$

$R_2 = c + d = 115 + 135 = 250$

$C_1 = a + c = 140 + 115 = 255$

$C_2 = b + d = 110 + 35 = 245$

$N = R_1 + R_2 = 250 + 250 = 500$

Observed (O)	Expected (E)	(O − E)	$\frac{(O-E)^2}{E}$	x^2
AB = 140	$\frac{250 \times 255}{500} = 127.5$	140 − 127.5 = 12.5	$\frac{156.25}{127.5}$	1.225 = 1.23
ab = 135	$\frac{245 \times 250}{500} = 122.5$	135 − 122.5 = 12.5	$\frac{156.25}{122.5}$	1.275 = 1.28
Ab = 110	$\frac{245 \times 250}{500} = 122.5$	110 − 122.5 = − 12.5	$\frac{156.25}{122.5}$	1.275 = 1.28
aB = 115	$\frac{250 \times 250}{500} = 127.5$	115 − 127.5 = − 12.5	$\frac{156.25}{127.5}$	1.225 = 1.23
				Total = 5.02

- *Degrees of freedom:* (2 − 1) (2 − 1) = 1
- *Critical value:* The table value of chi square at 0.05 for 1*df* is 3.84.
- *Decision:* Now calculated chi square is 5.02 > 3.84 ($x^2_{.05,3}$) so the hypothesis (1 : 1 : 1 : 1) is rejected. We can conclude that the loci are **linked**.

Q43. In Uttarpara General hospital, out of 55 hypercholesterolemia cases, 25 suffer from high blood pressure while out of 45 cases with normal serum cholesterol, 15 are of high blood pressure. Use chi square test to find whether or not there is a significant association between hypercholesterolemia & high blood pressure. (α = 0.05).

Solution:

- *Null Hypothesis:* No association between cholesteromia & high blood pressure *i.e.*, independent.

• *Contingency table:*

	High blood pressure	*Normal blood pressure*
Hyper cholesteromic cases (55)	25	30 = 55 R_1
Normal seran cholesterol (45)	15	30 = 45 R_2
	C_1 = 40	C_2 = 60

• *Calculation:*

Observed (O)	Expected (E)	(O – E)	$(O - E)^2$	$\frac{(O-E)^2}{E}$	x^2
• Hyper cholesteromic cases high B.P. = 25	$\frac{55 \times 40}{100} = 22$	25 – 22 = 3	9	$\frac{9}{22} = 0.41$	
• Hyper Cholesteromic with low B.P. = 30	$\frac{55 \times 60}{100} = 33$	30 – 33 = 3	9	$\frac{9}{33} = 0.27$	1.52
• Normal serum with high B.P. = 15	$\frac{45 \times 40}{100} = 18$	15 – 18 = 3	9	$\frac{9}{18} = 0.50$	
• Normal serum with low B.P. = 30	$\frac{45 \times 60}{100} = 27$	30 – 27 = 3	9	$\frac{9}{27} = 0.34$	

• *Degrees of freedom* = $(r - 1)(r - 1) = (2 - 1)(2 - 1) = 1$

• *Critical value:* The table value of chi square at 0.05 for 1*df* is 3.84.

• *Decision:* Since the calculated chi square value is 1.54 < critical value of chi square for 1*df* is 3.84. So the null hypothesis can not be rejected. Therefore, data indicates that there is no significant association between the two.

Q44. In Arambagh sub divisional hospital 100 patients are admitted out of 40 diabetic patients. 20 were hypertensive while out of 60 non diabetic patients 15 were hypertensive. Find is there any significant association between diabetes and hypertension.

Solution:

• *Null Hypothesis:* No association between diabetes and hypertension.

• *Contingency table:*

	Nonhyper tensive	*Hyper tensive*	*Total*
Diabetic	20	20	40 = R_1
Non diabetic	45	15	60 = R_2
	C_1 = 65	C_2 = 35	

Calculation:

Observed (O)	Expected (E)	(O – E)	$(O-E)^2$	$\frac{(O-E)^2}{E}$	x^2
Diabetic & nonhyper tensive = 20	$\frac{40 \times 65}{100} = 26$	20 – 26 = – 6	36	$\frac{36}{26} = 1.38$	
Diabetic & hyper tensive = 20	$\frac{40 \times 35}{100} = 14$	20 – 14 = 6	36	$\frac{36}{14} = 2.57$	
Non diabetic & non hypertensive = 45	$\frac{60 \times 65}{100} = 39$	45 – 39 = 6	36	$\frac{36}{39} = 0.92$	
Non diabetic & hypertensive = 15	$\frac{60 \times 35}{100} = 21$	15 – 21 = – 6	36	$\frac{36}{21} = 1.71$	
				6.58	

- *Degrees of freedom:* (R – 1) (C – 1) = (2 – 1) (2 – 1) = 1
- *Critical value:* The table value of chi square for 1*df* at 0.05 level is 3.84
- *Decision:* Since the calculated value of chi square for 1*df* 0.05 level is 6.58 > critical value. Therefore the null hypothesis is rejected, so we can conclude that there is a significant association between diabetes and hypertension.

Q45. A study of 160 families with 4 children in Serampore subdivision showed the following:

No. of girls & boys:	4 girls 0 boy	3 girls 1 boy	2 girls 2 boys	1 girl 3 boys	0 girl 4 boys
No. of families:	7	50	55	32	16

Using a x^2 (chi square) test at 0.05 level of significance test the hypothesis that female and male births occur with same probability.

(Given the critical value of x^2 for 4 degrees of freedom at the 0.05 level of significance are 9.49).

Solution: Here the boys and girls represent the two set of terms and the problems can be solved by expanding the binomial $(p + q)^n$.

Expanding of $(p + q)^n = p^n + p^{n-1}q + p^{n-2}q^2 + p^{n-3}q^3 + \ldots + q^n$.

Each term of the expansion has an appropriate coefficient and it can be calculated from the general formula.

In the present problem $(p + q)^n$ can be written as $(p + q)^4$ where p = girl and q = boy. Hypothesis to be assumed is girls and boys are equally (*randomly/consistent*) distributed in the family *i.e.*, the probability of girls is $\frac{1}{2}$ and that of boy is also $\frac{1}{2}$

- *Null Hypothesis* = 1 : 1

$$(p + q)^4 = p^4 + 4p^3q + 6p^2q^2 + 4pq^3 + q^4$$

Total families = 160

$\begin{bmatrix}4\text{ girls}\\0\text{ boy}\end{bmatrix}$	$\begin{bmatrix}3\text{ girls}\\1\text{ boy}\end{bmatrix}$	$\begin{bmatrix}2\text{ girls}\\2\text{ boys}\end{bmatrix}$	$\begin{bmatrix}1\text{ girl}\\3\text{ boys}\end{bmatrix}$	$\begin{bmatrix}0\text{ girl}\\4\text{ boys}\end{bmatrix}$
$= p^4$	$= 4p^3q$	$= 6p^2q^2$	$= 4pq^3$	$= q^4$
$= \left(\frac{1}{2}\right)^4$	$= 4 \times \left(\frac{1}{2}\right)^3 . \left(\frac{1}{2}\right)$	$= 6 \times \left(\frac{1}{2}\right)^2 \times \left(\frac{1}{2}\right)^2$	$= 4 \times \left(\frac{1}{2}\right)\left(\frac{1}{2}\right)^3$	$= \left(\frac{1}{2}\right)^4$
$= \frac{1}{16}$	$= 4 \times \frac{1}{8} \times \frac{1}{2}$	$= 6 \times \frac{1}{4} \times \frac{1}{4}$	$= 4 \times \frac{1}{2} \times \frac{1}{8}$	$= \frac{1}{16}$
$= \frac{1}{16} \times 160$	$= \frac{1}{4} \times 160$	$= \frac{6}{16} \times 160$	$= \frac{1}{4} \times 160$	$= \frac{1}{16} \times 160$
$= 10$	$= 40$	$= 60$	$= 40$	$= 10$

	Families					Total
Observed (O)	7	50	55	32	16	160
Expected (E)	10	40	60	40	10	160

$$X^2 = \Sigma\frac{(O-E)^2}{E}$$

$$X^2 = \frac{(7-10)^2}{10} + \frac{(50-40)^2}{40} + \frac{(55-60)^2}{60} + \frac{(32-40)^2}{40} + \frac{(16-10)^2}{10}$$

$$X^2 = \frac{9}{10} + \frac{100}{40} + \frac{25}{60} + \frac{64}{40} + \frac{36}{10}$$

$$= 0.9 + 2.5 + 0.4166 + 1.6 + 3.6$$

$$= 9.0166$$

Here degrees of freedom $(n - 1)$ *i.e.,* $5 - 1 = 4$.

- *Critical value:* The control value of chi square (x^2) for $4df$ at 0.05 is 9.49.
- *Decision:* Since the calculated chi square for $4df$ is 9.0166 < critical value of chi square 9.49. So the null hypothesis is accepted. Thus the observed and expected frequencies for families are good fit and the distribution of boys and girls with in families are **equal** i.e. **random** in nature.

Q46. A study of 640 families with 3 children in Bankura district showed the following.

No. of girls & boys:	3 girls 0 boy	2 girls 1 boy	1 girl 2 boys	0 girl 3 boys
No. of families:	72	222	264	82

Using a x^2 (chi square) test at the 0.05 level of significance, test the hypothesis that female and male births occur with the same probabilities.

(Given that the critical value of x^2 for 3 degrees of freedom at 0.05 level of significance is 7.81).

Solution: Here the boys and girls represents the two set terms and the problems can be solved by expanding the binomial $(p + q)^n$.

Expansion of $(p + q)^n = p^n + p^{n-1}q + p^{n-2}q^2 + p^{n-3}q^3 + \ldots\ldots + q^n$

Each of the expansion has an appropriate coefficient and it can be calculated from the general formula.

In the present problem $(p + q)^n$ can be written as $(p + q)^3$ where p = girl and q = boy.

- Hypothesis to be assumed is girls and boys are equally (*randomly/consistent*) distributed in the family *i.e.*, the probability of girl is $\frac{1}{2}$ and that of boy is also $\frac{1}{2}$.
- *Null Hypothesis:* Girl : Boy *i.e.*, $\frac{1}{2} = \frac{1}{2} = 1:1$.

$$(p + q)^3 = p^3 + 3p^2q + 3pq^2 + q^3$$

Total families = 640.

3 girls 0 boy	2 girls 1 boy	1 girl 2 boys	0 girl 3 boys
$= p^3$	$= 3p^2q$	$= 3pq^2$	$= q^3$
$= \left(\frac{1}{2}\right)^3$	$= 3\left(\frac{1}{2}\right)^2 \cdot \frac{1}{2}$	$= 3 \times \frac{1}{2} \times \left(\frac{1}{2}\right)^2$	$= \left(\frac{1}{2}\right)^3$
$= \frac{1}{8}$	$= 3 \times \frac{1}{4} \times \frac{1}{2}$	$= 3 \times \frac{1}{2} \times \frac{1}{4}$	$= \frac{1}{8}$
$= \frac{1}{8} \times 640$	$= \frac{3}{8} \times 640$	$= \frac{3}{8} \times 640$	$= \frac{1}{8} \times 640$
$= 80$	$= 240$	$= 240$	$= 80$

	Families				Total
Observed (O)	72	222	264	82	640
Expected (E)	80	240	240	80	640

$$X^2 = \sum \frac{(O-E)^2}{E} = \frac{(72-80)^2}{80} + \frac{(222-240)^2}{240} + \frac{(264-240)^2}{240} + \frac{(82-80)^2}{80}$$

$$= \frac{64}{80} + \frac{18 \times 18}{240} + \frac{24 \times 24}{240} + \frac{2 \times 2}{80}$$

$$= 0.8 + 1.35 + 2.4 + 0.05$$

$$= 4.6$$

Degrees of freedom $(n - 1)$ *i.e.*, $4 - 1 = 3$.

- *Critical value:* The control value of chi square (x^2) for $3df$ at 0.05 is 7.8%.
- *Decision:* Since the calculated chi square $3df$ is 4.6 < critical value of chi square 7.81. So the null hypothesis is accepted. Thus the observed and expected frequencies for families are good fit and the distribution of boys and girls with in the families are **equal** *i.e.*, **random** in nature.

EXERCISE

Q1. Bridges & Morgan gave the following results for the F_2 generation of the cross between wild type (+ +) with vestigial ebony (*vg .e*).

++ = 268; *vg* + = 94 +*e* = 79 *vg e* = 24 = 465

Apply chi square (X^2) test & comment on the result.

Q2. Crossing gray bodied scarlet eyed drosophila with black bodies red eyed one produces all grey bodies fly in F_1. The F_1 flies are inbreeded & produce the following progeny: Grey body red eye–362, black body red eye–128, grey body scarlet eye–122 & black body scarlet eye–44. Analyses & test the data for goodness of fit.

Q3. In Genetics Research Unit laboratory we get black male & grey female flies. The offspring obtained – 224 black & 705 grey coloured. Calculate chi-square & give your inference on ratio of black & grey offspring.

Q4. In an experiment the effect of concentration of pesticide dimecron was studied in relation to the mortalities of fishes in two different aquarium. The results were noted after 24 hours. It was observed that some fishes died in both the aquarium & the percentage of death was 27% in lower concentration & 39% in higher concentration. It is to test whether the death in higher concentration is significantly different from that of lower concentration or both are independent from the following data.

Categories of Observation

Subject	*Lower concentration*	*Higher concentration*	*Total*
Dead	23 (*a*)	35 (*b*)	58 (*a* + *b*)
Alive	62 (*c*)	55 (*d*)	117 (*c* + *d*)

Ans. 1. $x^2 = 2.329$

2. $x^2 = 0.562$

3. $x^2 = 0.39$

4. $x^2 = 2.252$

Test of Significance

Test done to measure significance:

1. x^2 – **test** (Chi-square test)
2. ***t*-test** (upaired & paired)
3. ***Z*-test**
4. ***F*-test** (or Fisher's test or analysis of variance ANOVA
5. ***r*-test** (correlation coefficient test)

CHAPTER

STUDENT 'T' DISTRIBUTION

The '*t*' test as used with small sample ($n < 30$) and was worked out by **W.S. Gossett** whose pen name was "student". Hence this test is also called "*Student's 't' test*".

The 't' may be defined as quantity representing the difference between the sample mean and true mean or population mean expressed in terms of the standard error.

$$t = \frac{\text{Difference between sample mean}}{\text{Standard error of the difference between means}}$$

$$t = \frac{|\bar{X}_1 - \bar{X}_2|}{SE} \quad \bar{X}_1 \text{ and } \bar{X}_2 = \text{Mean}$$

$SE =$ Standard error

- **Type:**
  (*a*) Unpaired *t* test.
  (*b*) Paired *t* test.
- **Criteria for applying *t* test:**
  I. Random samples are drawn from normal population.
  II. For testing the equality of two population means, the population variances are regarded as equal.
  III. The samples are less than 30.
  IV. In case of two samples some adjustments in degrees of freedom for '*t*' are made.
- **Properties of '*t*' Distribution:**
  I. The shape of the't' distribution curve varies with the degrees of freedom. (*The degrees of freedom is defined as* size of sample minus one).
  II. '*t*'-distribution is symmetrical distribution with mean zero.
  III. '*t*'-distribution is asymptotic to *X*-axis *i.e.*, it extends to infinity on either side.
  IV. Its graph is similar to that of normal distribution. '*t*'-distribution has a greater spread than normal distribution.
  V. The larger the number of degrees of freedom the more closely '*t*'-distribution resembles standard normal distribution.
- **Application of '*t*' Distribution:**
  1. To test the significance of a single mean when the population variance is unknown.
  2. To test the significance of difference between two samples means the population variance being equal and unknown.
  3. To test the significance of an observed sample correlation coefficient or difference between means of two sample (*dependent sample or paired observation*).
- **Illustration:**
  (*a*) **Unpaired '*t*' test:** This test is applied to unpaired data of independent observations made on individuals of two different groups or samples drawn from two populations.

Working Procedure:

I. Calculate means of the two samples and find the observed difference between means of two samples $(\bar{X}_1 - \bar{X}_2)$.

II. Calculate the standard error of difference between the two means.

III. Calculate the '*t*' value *i.e.*, the ratio between the observed difference and its standard error by substituting the above values in the formula.

$$t = \frac{|\bar{X}_1 - \bar{X}_2|}{SE} \qquad \bar{X}_1 \text{ and } \bar{X}_2 = \text{mean}$$

S_{x_1} and S_{x_2} = SD

n_1 and n_2 = size of sample

$$SE \text{ of } (\bar{X}_1 - \bar{X}_2) = SD\sqrt{\frac{1}{n_1} + \frac{1}{n_2}}$$

$$SD = \sqrt{\frac{\sum(X_1 - \bar{X}_1)^2 + \sum(X_2 - \bar{X}_2)^2}{n_1 + n_2 - 2}}$$

IV. Determine the pooled degrees of freedom from the formula.

$$df = (n-1) + (n_2 - 1) = n_1 + n_2 - 2$$

V. Compare the calculated value with the table value at particular degrees of freedom.

Example: In a nutritional study, 13 children were given a usual diet plus vitamins *A* & *D* tables. While the second comparable group of 12 children was taking the usual diet. After 12 months, the gain in weight in pounds was noted as given in the table. Can you say that vitamins A & D were responsible for this difference?

A =	5	3	4	3	2	6	3	2	3	6	7	5	3
B =	1	3	2	4	2	1	3	4	3	2	2	3	—

Solution: Null Hypothesis: Vitamins (A & D) are responsible for the gain weight difference.

Alternative Hypothesis: Vitamins are not responsible for the gain weight difference.

Calculation:

Sl. No.	Gr A (x)	$X - \bar{X} = D$	$(X - \bar{X})^2 = D_1^2$	Gr B (Y)	$Y - \bar{Y} = D_2$	$(Y - \bar{Y})^2 = D_2^2$
1.	5	5 – 4 = 1	1	1	1 – 2.5 = –1.5	2.25
2.	3	3 – 4 = –1	1	3	3 – 2.5 = 0.5	.25
3.	4	4 – 4 = 0	0	2	2 – 2.5 = –0.5	.25
4.	3	3 – 4 = –1	1	4	4 – 2.5 = 1.5	2.25
5.	2	2 – 4 = –2	4	2	2 – 2.5 = 0.5	.25
6.	6	6 – 4 = 2	4	1	1 – 2.5 = –1.5	2.25
7.	3	3 – 4 = –1	1	3	3 – 2.5 = 0.5	.25
8.	2	2 – 4 = –2	4	4	4 – 2.5 = 1.5	2.25
9.	3	3 – 4 = –1	1	3	3 – 2.5 = 0.5	.25
10.	6	6 – 4 = 2	4	2	2 – 2.5 = –0.5	.25
11.	7	7 – 4 = 3	9	2	2 – 2.5 = –0.5	.25
12.	5	5 – 4 = 1	1	3	3 – 2.5 = +0.5	.25
13.	3	3 – 4 = –1	1	—		
	$\sum 52$		$\sum D_1^2 = 32$	$\sum 30$		$\sum D_2^2 = 11$

Gr A	*Gr B*
$n = 13$	$n = 12$
$\overline{X} = 4$	$\overline{Y} = 2.5$
$D_1^2 = 32$	$D_2^2 = 11$

$$SD\ (A - B) = \sqrt{\frac{\sum (X - \overline{X})^2 + \sum (Y - \overline{Y})^2}{n_1 + n_2 - 2}}$$

$$= \sqrt{\frac{D_1^2 + D_2^2}{13 + 12 - 2}} = \sqrt{\frac{32 + 11}{23}} = \sqrt{\frac{43}{23}}$$

$$SE = SD\sqrt{\frac{1}{n_1} + \frac{1}{n_2}} = \sqrt{\frac{43}{23}} \times \sqrt{\frac{1}{13} + \frac{1}{12}} = \sqrt{\frac{43}{23}} \times \sqrt{\frac{12 + 13}{156}}$$

$$= \sqrt{\frac{43}{23}} \times \sqrt{\frac{25}{156}} = \sqrt{1.87} \times \sqrt{.160} = 1.37 \times .4$$

$$= .548 = .55$$

$$t = \frac{\overline{X} - \overline{Y}}{SE} = \frac{4 - 2.5}{.55} = \frac{1.5}{.55} = 2.72$$

- Level of significance: 5% Level *i.e.*, 0.05.
- Critical value: Tabulated value at 0.05 for *df* 23 is 2.07.
- Decision: The calculated value of $(t) = 2.72$.

$$|t| = 2.72 > t_{0.05,\ 23} = 2.07$$

So the Null Hypothesis is rejected.

(*b*) Paired '*t*' test: It is applied to paired data of independent observation from one sample only when each individual gives a pair observations.

I. To study the role of a factor or cause when the observations are made before and after its play.

II. To compare the effect of two drugs, given to the same person.

III. To compare the results of two different laboratory technique.

IV. To compare observations made at two different sites in the same body.

Working Procedure:

I. Find the difference in each set of paired observations before and after $(X_1 - X_2 = D)$.

II. Calculate the mean of the difference $(\overline{D})$.

III. Work out the *SD* of differences and then the S.E of mean from the same.

$$\frac{SD}{\sqrt{n}}$$

IV. Calculate '*t*' value by substituting the above values in the formula.

$$t = \frac{\overline{D} - 0}{SD/\sqrt{n}} = \frac{\overline{D}}{\frac{SD}{\sqrt{n}}}$$

$$SD = \sqrt{\frac{\sum D^2 - \frac{\left(\sum D\right)^2}{n}}{n - 1}}$$

V. Determine the degrees of freedom. Being one and the same sample df should be $n - 1$.

VI. Compare the calculated value with the table value at a particular degrees of freedom.

Example: Ten students were given intensive coaching in Statistics. The scores obtained in 1st & 5th test are given below:

Sl. No.	1	2	3	4	5	6	7	8	9	10
Marks in 1st	50	52	53	60	65	67	48	69	72	80
Marks in 5th	65	55	65	65	60	67	49	82	74	86

Does the score from test 1st to test 5th show an improvement? Test at 5% level of significance.

Solution: Null Hypothesis: No improvement has been occurred (*assume*).

Alternative Hypothesis: Improvement (*assume*).

Calculation:

Sl. No.	*Marks in 1st Test*	*Marks in 5th Test*	*Difference* $X_1 - X_2 = D$	D^2
1.	50	65	−15	225
2.	52	55	−3	09
3.	53	65	−12	144
4.	60	65	−5	25
5.	65	60	5	25
6.	67	67	0	0
7.	48	49	−1	1
8.	69	82	−13	109
9.	72	74	−2	04
10.	80	86	−6	36
			$\sum D = -52$	$\sum D^2 = 638$

$$\bar{D} = \frac{\sum D}{n} = \frac{-52}{10} = -5.2$$

$$SD = \sqrt{\frac{\sum D^2 - \frac{\left(\sum D\right)^2}{n}}{n-1}}$$

$$= \sqrt{\frac{638 - \frac{(-52 \times -52)}{10}}{10-1}}$$

$$= \sqrt{\frac{638 - 270.4}{9}}$$

$$= \sqrt{\frac{367.6}{9}} = \frac{19.17}{3} = 6.391$$

$$SE \text{ of difference} = \frac{S}{\sqrt{n}} = \frac{6.391}{\sqrt{10}} = \frac{6.391}{3.16} = 2.022$$

$$t = \frac{\bar{D}}{SE} = \frac{-5.2}{2.022} = 2.57$$

Level of significance = 0.05

Critical value = The critical value of 't' at 0.05 for 10 − 1 = 9df = *i.e.*, $t_{0.05,\,9}$ = 1.833.

Decision: Since calculated 't' = 2.57 > tabulated $t_{0.05,\,9}$ = 1.833. So the null hypothesis is rejected *i.e.*, alternative hypothesis is accepted.

Example 1: An *I-Q* Test was administered to 5 persons before and after they are trained the results are given below:

Candidates	*I*	*II*	*III*	*IV*	*V*
IQ before training	110	120	123	132	125
IQ after training	120	118	125	136	121

Test whether there is any change in *IQ* after training programme. It is given that $t_{0.01}$ = 4.6 for *df*. 4.

Solution: (*a*) Null Hypothesis: No change in *IQ* after the training (assume).

(*b*) Alternative Hypothesis: Change in IQ after the training (assume).

(*c*) Calculation:

Candidates	*IQ before training*	*IQ after training*	*Difference*	D^2
I	110	120	110 − 120 = −10	100
II	120	118	120 − 118 = +2	04
III	123	125	123 − 125 = −2	04
IV	132	136	132 − 136 = −4	16
V	125	121	125 − 121 = +4	16
			$\sum D = -10$	$\sum D^2 = 140$

$$\text{Mean difference } (\bar{D}) = \frac{-10}{5} = -2$$

$$\text{Estimated } SD\ (S) = \sqrt{\frac{D^2 - \frac{\left(\sum D\right)^2}{n}}{n-1}}$$

$$= \sqrt{\frac{140 - \frac{(-10 \times -10)}{5}}{5-1}}$$

$$= \sqrt{\frac{140 - \frac{100}{5}}{4}}$$

$$= \sqrt{\frac{120}{4}} = \sqrt{30}$$

SE i.e., Standard Error of difference

$$= \frac{S}{\sqrt{n}} = \frac{\sqrt{30}}{\sqrt{5}} = \sqrt{6}$$

$$t = \frac{\bar{D}}{SE} = \frac{-2}{2.45} = -.816 \quad i.e.,\ |t| = 0.816$$

(*d*) Level of significance = 0.01.

(*e*) Critical value: Critical value of '*t*' at 0.05 for *df* 4 *i.e.*, 5 – 1 is 4.6.

(*f*) Decision: The calculated value of $|t| = 0.816 < t_{0.01} = 4.6$.

So the null hypothesis is accepted *i.e.*, no significant difference after training programmer.

Example 1: A research minded student started fish breeding & wanted to see whether there is any growth difference at 100th day of hybrid fishes at reciprocal parental crosses. The data (wt.) of fishes are given below:

Male × Female	1	2	3	4	5	6	7	8	9	10
X : Rohu × Mrigal:	15	12	8	14	16	16	9	5	11	14
Y : Mrigal × Rohu:	12	9	13	10	8	12	13	14	9	10

Find out whether sex of parents of given crosses have any effect on hybrid fishes growth or not by applying suitable statistical method.

Solution:

- *Null Hypothesis:* Sex of the parents have no effect on the growth of the hybrid fishes.
- *Alternative Hypothesis:* Sex of the parents have some effect on the growth of the hybrid fishes.
- Calculation:

Sl. No.	*R × M: (X)*	$X - \bar{X} = D_1$	$(X - \bar{X})^2 = D_1^2$	*M × R: (Y)*	$Y - \bar{Y} = D_2$	$(Y - \bar{Y})^2 = D_2^2$
1.	15	15 – 12 = +3	09	12	12 – 11 = +1	01
2.	12	12 – 12 = 0	00	9	9 – 11 = –2	04
3.	8	8 – 12 = –4	16	13	13 – 11 = +2	04
4.	14	14 – 12 = +2	04	10	10 – 11 = –1	01
5.	16	16 – 12 = +4	16	8	8 – 11 = –3	09
6.	16	16 – 12 = +4	16	12	12 – 11 = +1	01
7.	9	9 – 12 = –3	09	13	13 – 11 = –2	04
8.	5	5 – 12 = –7	49	14	14 – 11 = +3	09
9.	11	11 – 12 = –1	01	9	9 – 11 = –2	04
10.	14	14 – 12 = +2	04	10	10 – 11 = –1	01

$\Sigma X = 120$ $\quad \Sigma D_1^2 = 124$ $\quad \Sigma Y = 110$ $\quad \Sigma D_2^2 = 38$

$N = 10$ $\quad N = 10$

$\bar{X} = \frac{120}{10} = 12$ $\quad \bar{Y} = \frac{110}{10} = 11$

$\Sigma (X - \bar{X})^2 = D_1^2 = 124$ $\quad \Sigma (Y - \bar{Y})^2 = D_2^2 = 38$

$$SD = \sqrt{\frac{D_1^2 + D_2^2}{N_1 + N_2 - 2}} = \sqrt{\frac{124 + 38}{10 + 10 - 2}} = \sqrt{\frac{162}{18}} = \sqrt{9} = 3$$

$$SE\ (X - Y) = S.D. \sqrt{\frac{1}{N_1} + \frac{1}{N_2}} = 3\sqrt{\frac{1}{10} + \frac{1}{10}} = 3\sqrt{\frac{2}{10}} = 3\sqrt{0.2}$$

$$= 3 \times 0.447 = 1.341$$

$$|t| = \frac{\bar{X} - \bar{Y}}{SE} = \frac{12 - 11}{1.34} = \frac{1}{1.34} = 0.75$$

- Level of significance = 5% level *i.e.*, 0.05.
- **Critical value:** Tabulated value of $|t|$ at 0.05 for degrees of freedom 20 – 2 = 18 is = 2.10.
- **Decision:** The calculated value of $|t| = 0.75$ $|t| = 0.75 \leq t_{0.05, 18}$

  So the null hypothesis is accepted i.e. there is no significant deviation among the hybrid fishes obtained through reciprocal crosses.

Example 2: The body weight (kg) of 8 adult males & of eight adult females is presented in respectively the first and second columns of table. Find out whether or not the mean weight of males is significantly higher than that of females.

Table Weight (kg)

Males (X_1)	*Females* (X_2)
50	49
58	52
60	51
55	56
59	55
56	53
54	52
64	48
456	416

[*C.U.* (*Zoology Hons.*) *2004*]

Solution:

- *Null Hypothesis*: Mean weights of males are not significantly higher than females.
- *Alternative Hypothesis*: Mean weights of males are significantly higher.
- Calculation:

Sl. No.	*Male* X_1	$X_1 - \bar{X}_1 = D_1$	D_1^2	*Female* X_2	$X_2 - \bar{X}_2 = D_2$	D_2^2
1.	50	50 – 57 = –7	49	49	49 – 52 = –3	09
2.	58	58 – 57 = +1	01	52	52 – 52 = 0	0
3.	60	60 – 57 = +3	09	51	51 – 52 = –1	01
4.	55	55 – 57 = –2	04	56	56 – 52 = +4	16
5.	59	59 – 57 = +2	04	55	55 – 52 = +3	09
6.	56	56 – 57 = –1	01	53	53 – 52 = +1	01
7.	54	54 – 57 = –3	09	52	52 – 52 = 0	00
8.	64	64 – 57 = +7	49	48	48 – 52 = –4	16
	456		$\Sigma D^2 = 126$	416		$\Sigma D_2^2 = 52$

$$\bar{X}_1 = \frac{456}{8} = 57 \qquad \bar{X}_2 = \frac{416}{8} = 52$$

$$n_1 = 8 \qquad n_2 = 8$$

$$\Sigma D_1^2 = 126 \qquad \Sigma D_2^2 = 52$$

$$SD = \sqrt{\frac{\Sigma (X_1 - \bar{X}_1)^2 + \Sigma (X_2 - \bar{X}_2)^2}{n_1 + n_2 - 2}}$$

$$= \sqrt{\frac{126 + 52}{8 + 8 - 2}} = \sqrt{\frac{178}{14}} = \sqrt{12.714} = 3.57$$

$$SE\ (X_1 - X_2) = 3.57\sqrt{\frac{1}{8} + \frac{1}{8}} = 3.57\sqrt{\frac{2}{8}} = 3.57\sqrt{0.25} = 3.57 \times 0.5 = 1.785$$

$$|t| = \frac{57 - 52}{1.785} = \frac{5}{1.785} = 2.80$$

- *Level of significance* = 5% *i.e.*, 0.05.
- *Critical value* = Tabulated value of t at 0.05 for df 16 – 2 = 14 is 2.14.
- *Decision*: The calculated value of $|t| = 2.80$

$$|t|\ 2.80 > t_{0.05},\ 2.14$$

So the null hypothesis rejected *i.e.*, mean weights of males are not significantly higher than females.

Example 3: Applications of fertilizers were tested for the yield of rice grown in 10 plots. Another seed of 10 plots of similar size & condition were taken as control. Test the effect of fertilizer.

Plot No.:	**1**	**2**	**3**	**4**	**5**	**6**	**7**	**8**	**9**	**10**
Fertilizer applied	**16**	**14**	**18**	**15**	**13**	**17**	**16**	**15**	**14**	**13**
Fertilizer not applied	**10**	**12**	**11**	**9**	**13**	**13**	**12**	**14**	**13**	**11**

Solution:

- *Null Hypothesis*: No significant effect of fertilizer on yield of rice grown.
- *Alternative Hypothesis*: Significant effect of fertilizer in yield of rice grown.
- *Calculation*:

Plot No.	*Fertilizer applied (X)*	$X - \bar{X} = D_1$	D_1^2	*Fertilizer not not applied* Y_2	$Y - \bar{Y} = D_2$	D_2^2
1.	16	16 – 15.1 = 0.90	0.81	10	10 – 11.8 = –1.8	3.24
2.	14	14 – 15.1 = –1.1	1.21	12	12 – 11.8 = +0.2	0.04
3.	18	18 – 15.1 = +2.9	8.41	11	11 – 11.8 = –0.8	0.64
4.	15	15 – 15.1 = –0.1	0.01	9	9 – 11.8 = +2.8	7.84
5.	13	13 – 15.1 = –2.1	4.41	13	13 – 11.8 = +1.2	1.44
6.	17	17 – 15.1 = +1.9	3.61	13	13 – 11.8 = +1.2	1.44
7.	16	16 – 15.1 = +0.9	0.81	12	12 – 11.8 = +0.2	0.04
8.	15	15 – 15.1 = –0.1	0.01	14	14 – 11.8 = 2.2	4.84
9.	14	14 – 15.1 = –1.1	1.21	13	13 – 11.8 = 1.2	1.44
10.	13	13 – 15.1 = –2.1	4.41	11	11 – 11.8 = –0.8	0.64
	$\Sigma X = 151$		$\Sigma D_1^2 = 24.90$	$\Sigma Y = 118$		$\Sigma D_2^2 = 21.60$

$$\bar{X} = \frac{151}{10} = 15.1 \qquad \bar{Y} = \frac{118}{10} = 11.8$$

$$n_1 = 10 \qquad n_2 = 10$$

$$\bar{X} = 15.1 \qquad \bar{Y} = 11.8$$

$$\Sigma D_1^2 = 24.9 \qquad \Sigma D_2^2 = 21.6$$

$$SD\,(X - Y) = \sqrt{\frac{\Sigma(X - \bar{X})^2 + \Sigma(Y - \bar{Y})^2}{n_1 + n_2 - 2}}$$

$$= \sqrt{\frac{\Sigma D_1^2 + \Sigma D_2^2}{10 + 10 - 2}} = \sqrt{\frac{24.9 + 21.6}{18}}$$

$$= \sqrt{\frac{46.5}{18}} = \sqrt{2.584} = 1.607$$

$$SE = 1.607\sqrt{\frac{1}{10} + \frac{1}{10}} = 1.607\sqrt{0.2}$$

$$= 1.607 \times 0.447 = 0.718$$

$$|t| = \frac{\bar{X} - \bar{Y}}{SE} = \frac{15.1 - 11.8}{0.718} = \frac{3.3}{0.718} = 4.596.$$

- *Level of significance*: 5% level *i.e.*, 0.05.
- *Critical value*: Tabulated value of t at 0.05 for df (10 + 10 – 2) = 18 is 2.10.
- *Decision*: The calculated value of $|t|$ is 4.596

$$|t|\ 4.59 > t_{0.05,\ 18} = 2.10$$

So the null hypothesis is rejected *i.e.*, there is significant difference between the control & fertilizer used rice plant growth.

Example 4: Body length of fishes of a species was obtained from two ponds. They were measured as follows (in cm):

Pond A:	**20**	**24**	**20**	**28**	**22**	**20**	**24**	**32**	**24**	**26**
Pond B:	**12**	**10**	**8**	**10**	**6**	**4**	**14**	**20**	**10**	**6**

Calculate the mean difference in total body length between the two ponds of fish is significant or not.

Solution:

- *Null Hypothesis*: There is no significant difference in total body length between the fishes of the two ponds.
- *Alternative Hypothesis*: Presence of significant differences in body length between the fishes of the two ponds.
- *Calculation*:

	Pond A (X)	$X - \bar{X} = D_1$	D_1^2	*Pond B (Y)*	$Y - \bar{Y} = D_2$	D_2^2
1.	20	20 – 24 = –4	16	12	12 – 10 = +2	04
2.	24	24 – 24 = 0	0	10	10 – 10 = 0	00
3.	20	20 – 24 = –4	16	08	8 – 10 = –2	04
4.	28	28 – 24 = +4	16	10	10 – 10 = 0	00
5.	22	22 – 24 = –2	04	06	06 – 10 = –4	16
6.	20	20 – 24 = –4	16	04	04 – 10 = –6	36
7.	24	24 – 24 = 0	0	14	14 – 10 = +4	16
8.	32	32 – 24 = +8	64	20	20 – 10 = +10	100
9.	24	24 – 24 = 0	0	10	10 – 10 = 0	00
10.	26	26 – 24 = +2	04	06	06 – 10 = –4	16
	$\Sigma X = 240$		136	$\Sigma Y = 100$		192

Pond A (X)	Pond B (Y)
$n_1 = 10$	$n_2 = 10$
$\bar{X} = \frac{240}{10} = 24$	$\bar{Y} = \frac{100}{10} = 10$
$\Sigma D_1^2 = 136$	$\Sigma D_2^2 = 192$

$$SD\,(X - Y) = \sqrt{\frac{\Sigma(X-\bar{X})^2 + \Sigma(Y-\bar{Y})^2}{n_1 + n_2 - 2}}$$

$$= \sqrt{\frac{D_1^2 + D_2^2}{10 + 10 - 2}} = \sqrt{\frac{136 + 192}{18}} = \sqrt{\frac{328}{18}} = \sqrt{18.22} = 4.27$$

$$SE = S.D.\sqrt{\frac{1}{n_1} + \frac{1}{n_2}} = 4.27\sqrt{\frac{1}{10} + \frac{1}{10}} = 4.27\sqrt{0.2} = 4.27 \times 0.45 = 1.92$$

$$|\,t\,| = \frac{\bar{X} - \bar{Y}}{SE} = \frac{24 - 10}{1.92} = \frac{14}{1.92} = 7.29$$

- *Level of significance*: 5% level *i.e.*, 0.05.
- *Critical value*: Tabulated value of t at 0.05 for *df* 20 – 2 = 18 is 2.10.
- *Decision*: The calculated value of $|\,t\,| = 7.29$

$$|\,t\,| = 7.29 > t_{0.05,\,18} = 2.10$$

So the null hypothesis is rejected.

Example 5: In order to find the effect of *Azolla* growth on the rice field & experimentally grown *Azolla* in 10 similar field plots before rice planting & other 10 similar plots were taken as control without *Azolla* growth. Rice was grown in all these plots & yields were noted.

Plot No.:	1	2	3	4	5	6	7	8	9	10
With Azolla:	15.3	15.8	16.1	17.0	15.5	16.5	16.2	15.5	17.1	16.3
Without Azolla:	14.5	13.8	15.9	13.9	14.8	14.9	15.2	15.0	14.1	13.7

Verify whether there is any significant effect of Azolla growth on the gain yield of rice.

Solution:

- *Null Hypothesis*: There is no significant effect of Azolla growth in grain yield of rice.
- *Alternative Hypothesis*: Azolla growth has some effect on the grain yield of rice.
- *Calculation*:

Plot No.	With Azolla (X)	$X - \bar{X} = D_1$	$(X - \bar{X})^2 = D_1^2$	Without Azolla (Y)	$Y - \bar{Y} = D_2$	$(Y - \bar{Y})^2 = D_2^2$
1.	15.3	15.3–16.12 = –0.82	0.6724	14.5	14.5–14.58 = –0.08	0.0064
2.	15.8	15.8–16.12 = –0.32	0.1024	13.8	13.5–14.58 = –1.08	1.1664
3.	16.1	16.1–16.12 = –0.02	0.0004	15.9	15.9–14.58 = +1.32	1.7424
4.	17.0	17.0–16.12 = +0.88	0.7744	13.9	13.9–14.58 = –0.68	0.4624
5.	15.5	15.5–16.12 = –0.62	0.3844	14.8	14.8–14.58 = +0.22	0.0484
6.	16.5	16.5–16.12 = +0.38	0.1444	14.9	14.9–14.58 = +0.32	0.1024
7.	16.2	16.2–16.12 = +0.08	0.0064	15.2	15.2–14.58 = +0.62	0.3844

Plot No.	With Azolla (X)	$X-\bar{X}=D_1$	$(X-\bar{X})^2 = D_1^2$	Without Azolla (Y)	$Y-\bar{Y}=D_2$	$(Y-\bar{Y})^2 = D_2^2$
8.	15.4	15.4–16.12 = –0.72	0.5184	15.0	15.0–14.58 = +0.42	0.1764
9.	17.1	17.1–16.12 = 0.98	0.9604	14.1	14.1–14.58 = –0.48	0.2304
10.	16.0	16.3–16.12 = +0.18	0.0324	13.7	13.7–14.58 = –0.88	0.7744
	$\Sigma X = 161.2$		$\Sigma D_1^2 = 3.596$	$Y = \Sigma 145.8$		$\Sigma D_2^2 = 5.0940$

$$\bar{X} = \frac{161.2}{10} = 16.12 \qquad \bar{Y} = \frac{145.8}{10} = 14.58$$

$$\bar{X} = 16.12 \qquad \bar{Y} = 14.58$$

$$n_1 = 10 \qquad n_2 = 10$$

$$\Sigma (X-\bar{X})^2 = \Sigma D_1^2 = 3.596 \qquad \Sigma (Y-\bar{Y})^2 = \Sigma D_2^2 = 5.0940$$

$$SD = \sqrt{\frac{D_1^2 + D_2^2}{n_1 + n_2 - 2}} = \sqrt{\frac{3.596 + 5.0940}{20-2}} = \sqrt{\frac{8.690}{18}} = \sqrt{0.482} = 0.694$$

$$SE = S.D\sqrt{\frac{1}{n_1} + \frac{1}{n_2}} = 0.694\sqrt{\frac{1}{10} + \frac{1}{10}} = 0.694\sqrt{\frac{2}{10}} = 0.694\sqrt{.2}$$

$$= 0.694 \times 0.45 = 0.312$$

$$|t| = \frac{\bar{X} - \bar{Y}}{SE} = \frac{16.12 - 14.58}{.312} = \frac{1.54}{312} = 4.94$$

- *Level of significance*: 5% level *i.e.*, 0.05.
- *Critical value*: Tabulated value of t at 0.05 for df (20 – 2) = 18 is 2.10
- *Decision*: The calculated value of $|t| = 4.09$

$$|t| = 4.94 > t_{0.05,\ 18} = 2.10$$

So the null hypothesis is rejected.

Example 6: The students of two schools were measured for their heights, one school was east coast & another in west coast where there is slight difference in weather. The sampling results are as follows.

East coast:	**43**	**45**	**48**	**49**	**51**	**52**			
West coast:	**47**	**49**	**51**	**53**	**54**	**55**	**55**	**56**	**57**

Find whether there is any impact of weather on height taking other variable constant. Apply suitable statistical method.

Solution:

- *Null Hypothesis*: Weather has no impact on the height of the students.
- *Alternative Hypothesis*: Weather has some impact on the heights of students.

• *Calculation*:

Sl. No.	East coast (X)	$X-\bar{X}=D_1$	$(X-\bar{X})^2 = D_1^2$	West coast (Y)	$Y-\bar{Y}=D_2$	$(Y-\bar{Y})^2 = D_2^2$
1.	43	43 – 48 = –5	25	47	47 – 53 = –6	36
2.	45	45 – 48 = –3	09	49	49 – 53 = –4	16
3.	48	48 – 48 = 0	00	51	51 – 53 = –2	04
4.	49	49 – 48 = +1	01	53	53 – 53 = 0	00
5.	51	51 – 48 = +3	09	54	54 – 53 = +1	01
6.	52	52 – 48 = +4	16	55	55 – 53 = +2	04
7.				55	55 – 53 = +2	04
8.				56	56 – 53 = +3	09
9.				57	57 – 53 = +4	16
	288		$\Sigma D_1^2 = 60$	447		$\Sigma D_2^2 = 90$

$$\bar{X} = \frac{288}{6} = 48 \qquad \bar{Y} = \frac{477}{9} = 53$$

$$n_1 = 6 \qquad n_2 = 9$$

$$\bar{X} = 48 \qquad \bar{Y} = 53$$

$$\Sigma D_1^2 = 90 \qquad \Sigma D_2^2 = 90$$

$$SD\ (X - Y) = \sqrt{\frac{\Sigma (y-\bar{y})^2 + \Sigma (X-\bar{X})^2}{n_1 + n_2 - 2}}$$

$$= \sqrt{\frac{\Sigma D_1^2 + \Sigma D_2^2}{n_1 + n_2 - 2}}$$

$$= \sqrt{\frac{60+90}{6+9-2}} = \sqrt{\frac{150}{13}}$$

$$= \sqrt{11.5} = 3.39$$

$$SE = SD\sqrt{\frac{1}{n_1} + \frac{1}{n_2}}$$

$$= 3.39 \times \sqrt{\frac{1}{6} + \frac{1}{9}}$$

$$= 3.39 \times \sqrt{\frac{5}{18}}$$

$$= 3.39 \times \sqrt{0.28} = 3.39 \times 0.52 = 1.76$$

$$|t| = \frac{\bar{X} - \bar{Y}}{SE} = \frac{48-53}{1.76} = \frac{-5}{1.76} = \frac{500}{176} = 2.840$$

• *Level of significance*: 5% level *i.e.*, 0.05.

• *Critical value*: Tabulated of t at 0.05 for df (9 + 6 – 2) = 13 is 2.16.

• *Decision*: The calculated value of $|t|$ is 2.840

$$|t|\ 2.840 > t_{0.05,\ 13} = 2.16.$$

So the null hypothesis is rejected.

Example 7: Find whether or not there is a significant difference between the mean inter-orbital widths (mm) of two groups of pigeons, consisting of 10 & 8 pigeons respectively.

Group 1 (X_1):	**11.9**	**11.4**	**11.9**	**11.4**	**11.2**	**12.2**	**12.6**	**12.2**	**12.7**	**12.5**
Group 2 (X_2):	**10.8**	**10.9**	**10.5**	**11.0**	**10.6**	**10.5**	**10.8**	**10.5**		

Solution:

- *Null Hypothesis*: No significant difference between the mean inter-orbital width of two group of pigeons.
- *Alternative Hypothesis*: Presence of significant difference between inter-orbital widths of two groups of pigeons.
- *Calculation*:

	Group I X_1	$X_1 - \bar{X}_1 = D_1$	$(X_1 - \bar{X}_1)^2 = D_1^2$	*Group II* (X_2)	$X_2 - \bar{X}_2 = D_2$	$(X_2 - \bar{X}_2)^2 = D_2^2$
1.	11.9	11.9 – 12 = –0.1	0.01	10.8	10.8 – 10.7 = +0.1	0.01
2.	11.4	11.4 – 12 = –0.6	0.36	10.9	10.9 – 10.7 = +0.2	0.04
3.	11.9	11.9 – 12 = –0.1	0.01	10.5	10.5 – 10.7 = –0.2	0.04
4.	11.4	11.4 – 12 = –0.6	0.36	11.0	11.0 – 10.7 = +0.3	0.09
5.	11.2	11.2 – 12 = –0.8	0.64	10.6	10.6 – 10.7 = –0.1	0.01
6.	12.2	12.2 – 12 = +0.2	0.04	10.5	10.5 – 10.7 = –0.2	0.04
7.	12.6	12.6 – 12 + +0.6	0.36	10.8	10.8 – 10.7 = +0.1	0.01
8.	12.2	12.2 – 12 = +0.2	0.04	10.5	10.5 – 10.7 = –0.2	0.04
9.	12.7	12.7 – 12 = +0.7	0.49			
10.	12.5	12.5 – 12 = +0.5	0.25			
	120.0	$\Sigma X_1^2 = 0.256$		85.6		$\Sigma X_2^2 = 0.28$

Gr I

$$n_1 = 10$$

$$\bar{X}_1 = \frac{120}{10} = 12 \text{ mm}$$

Gr II

$$n_2 = 8$$

$$\bar{X}_2 = \frac{85.6}{8} = 10.7 \text{ mm}$$

$S.D.\ (X_1 - X_2)$

$$= \sqrt{\frac{\Sigma(X_1 - \bar{X}_1)^2 + \Sigma(X_2 - \bar{X}_2)^2}{n_1 + n_2 - 2}} = \sqrt{\frac{D_1^2 + D_2^2}{n_1 + n_2 - 2}} = \sqrt{\frac{2.56 + 0.28}{10 + 8 - 2}}$$

$$= \sqrt{\frac{2.84}{16}} = \sqrt{0.1775} = 0.42 \text{ (approx)}$$

$$SE = SD\sqrt{\frac{1}{n_1} + \frac{1}{n_2}} = 0.42\sqrt{\frac{1}{10} + \frac{1}{8}} = 0.42\sqrt{\frac{4+5}{40}} = 0.42\sqrt{\frac{9}{40}} = 0.199$$

$$|t| = \frac{\bar{X}_1 - \bar{X}_2}{SE} = \frac{12.0 - 10.7}{0.199} = \frac{1.3}{0.199} = 0.6533$$

- *Level of significance*: 5% level *i.e.*, 0.05.
- *Critical value*: Tabulated value of $|t|$ at 0.05 for

$$df = 10 + 8 - 2 = 16 \text{ is } 2.12.$$

- *Decision*: The calculated value of $|t| = 6.533$

$$|t| = 6.533 > t_{0.05, 16} = 2.12$$

So the null hypothesis is rejected i.e. there is significant difference between the mean inter orbital width of two groups of pigeons.

Example 8: Ten rats were fed with rice in first months & body weights of the rats were recorded. In the next months they were fed with grams & their weights were measured again. The respective weights of ten rats in two months are as follows.

Weight in 1st months:	**50**	**60**	**58**	**52**	**51**	**62**	**58**	**55**	**50**	**65**
Weight in 2nd months:	**56**	**58**	**68**	**61**	**56**	**59**	**64**	**60**	**50**	**62**

Test the given data to find the impact of grams in rat's nutrition.

Solution:

- *Null Hypothesis*: Grams have no impact in rat's nutrition.
- *Alternative Hypothesis*: Grams have impact in rat's nutrition.
- Calculation:

Sl. No.	*Weight in 1st month* (X_1)	*Weight in 2nd month* (X_2)	*Difference* $X_1 - X_2 = D$	D^2
1.	50	56	50 − 56 = −6	36
2.	60	58	60 − 58 = +2	04
3.	58	68	58 − 68 = −10	100
4.	52	61	52 − 61 = −9	81
5.	51	56	51 − 56 = −5	25
6.	62	59	62 − 59 = +3	09
7.	58	64	58 − 64 = −6	36
8.	55	60	55 − 60 = −5	25
9.	50	50	50 − 50 = 0	00
10.	65	62	65 − 62 = +3	09
			$\Sigma D = -41 + 8 = -33$	$\Sigma D^2 = 325$

$n = 10 \qquad \bar{D} = \frac{\Sigma D}{n} = \frac{-33}{10} = -3.3$

$$S.D. = \sqrt{\frac{\Sigma D^2 - \frac{(\Sigma D)^2}{n}}{n-1}} = \sqrt{\frac{325 - \frac{(-33)^2}{10}}{10-1}} = \sqrt{\frac{325 - \frac{1089}{10}}{9}}$$

$$= \sqrt{\frac{325 - 108.9}{9}} = \sqrt{\frac{216.1}{9}} = \sqrt{24.1} = 4.9$$

$$SE = \frac{SD}{\sqrt{n}} = \frac{4.9}{\sqrt{10}} = \frac{4.9}{3.16} = 1.55$$

$$|t| = \frac{3.3}{1.5} = 2.13.$$

- *Level of significance*: 0.05 levels.
- *Critical value*: Tabulated value of $|t|$ at 0.05 levels at *df* (10 − 1) = 9 is 2.26.

- *Decision*: The calculated value of $|t|$ is $2.13 < |t|_{0.05,\,9} = 2.26$

So the null hypothesis accepted i.e. grams have no impact in rat nutrition.

Example 9: Albino rats were administered with an aurvedic medicine at the rate of 10 mg/ 10 kg day for 7days. Initial & final body weights of the rats were recorded as shown in the table. Determine whether the drug has any significant effect on the gain or loss of body weight of the rat.

Rat No.	**1**	**2**	**3**	**4**	**5**	**6**	**7**	**8**	**9**	**10**
Initial body wt.:	**110**	**115**	**102**	**98**	**112**	**110**	**97**	**120**	**102**	**110**
Final body wt.:	**109**	**116**	**100**	**95**	**108**	**112**	**98**	**115**	**98**	**111**

Solution:

- *Null Hypothesis*: Drug has no significant effect in gain or loss of body weight of rat.
- *Alternative Hypothesis*: Drug has some effect on gain or loss of body weight of rat.
- *Calculation*:

Sl. No.	*Initial weight* (x_1)	*Final weight* (x_2) *(after application of medicine)*	*Difference* $X_1 - X_2 = D$	D^2
1.	110	109	110 − 109 = +1	01
2.	115	116	115 − 116 = −1	01
3.	102	100	102 − 100 = +2	04
4.	98	95	98 − 95 = +3	09
5.	112	108	112 − 108 = +4	16
6.	110	112	110 − 112 = −2	4
7.	97	98	97 − 98 = −1	01
8.	120	115	120 − 115 = +5	25
9.	102	98	102 − 98 = +4	16
10.	110	111	110 − 111 = −1	01
			$\Sigma D = 14$	$\Sigma D^2 = 78$

$n = 10$ $\qquad \bar{D} = \frac{\Sigma D}{n} = \frac{14}{10} = 1.4$

$$SD = \sqrt{\frac{\Sigma D^2 - \frac{(\Sigma D)^2}{n}}{n-1}}$$

$$= \sqrt{\frac{78 - \frac{(14)^2}{10}}{10-1}} = \sqrt{\frac{78 - \frac{196}{10}}{9}} = \sqrt{\frac{78 - 19.6}{9}}$$

$$= \sqrt{\frac{58.4}{9}} = \sqrt{6.49} = 2.55$$

$$SE = \frac{SD}{\sqrt{n}} = \frac{2.55}{\sqrt{10}} = \frac{2.55}{3.16} = 0.806$$

$$|t| = \frac{\bar{D}}{SE} = \frac{1.4}{0.806} = 1.736 = 1.74$$

- *Level of significance*: 5% *i.e.*, 0.05
- *Critical value*: The critical value of '*t*' at 0.05 level for 10 – 1 = 9 *df i.e.*, $|t|_{0.05, 9} = 2.26$.
- *Decision*: Since calculated '*t*' = 1.74 < tabulated $t_{0.05, 09} = 2.26$, so the null hypothesis is accepted.

Example 10: Ten individuals each two strains (*A* & *B*) of a species were measured for a particular trait in the environment with the following results:

Strain *A*:	**10**	**12**	**8**	**14**	**12**	**9**	**7**	**7**	**13**	**16**
Strain *B*:	**13**	**9**	**14**	**12**	**10**	**13**	**15**	**11**	**14**	**12**

Is there significant difference in trait between 2 strains?

Solution:

- *Null Hypothesis*: There is no significant difference in trait.
- *Alternative Hypothesis*: There is significant difference in trait.
- *Calculation*:

	A (X)	$X - \bar{X} = D_1$	$(X - \bar{X})^2 = D_1^2$	$B = (Y)$	$Y - \bar{Y} = D$	$(Y - \bar{Y})^2$ D_2^2
1.	10	10 – 10.8 = 0.8	0.64	13	13 – 12.3 = 0.7	0.49
2.	12	12 – 10.8 = 1.2	1.44	9	09 – 12.3 = 3.3	10.89
3.	8	8 – 10.8 = 2.8	7.84	14	14 – 12.3 = 1.7	2.89
4.	14	14 – 10.8 = 3.2	10.24	12	12 – 12.3 = 0.3	0.09
5.	12	12 – 10.8 = 1.2	1.44	10	10 – 12.3 = 2.3	5.29
6.	9	9 – 10.8 = 1.8	3.24	13	13 – 12.3 = 0.7	0.49
7.	7	7 – 10.8 = 3.8	14.44	15	15 – 12.3 = 2.7	7.29
8.	7	7 – 10.8 = 3.8	14.44	11	11 – 12.3 = 1.3	1.69
9.	13	13 – 10.8 = 2.2	4.84	14	14 – 12.3 = 1.7	2.89
10.	16	16 – 10.8 = 5.2	27.04	12	12 – 12.3 = 0.9	0.09
	$\Sigma X = 108,\ n_1 = 10$		$\Sigma D_1^2 = 85.6$	$\Sigma Y = 123, n_2 = 10$		$\Sigma D_2^2 = 32$

$$\bar{X} = \frac{108}{10} = 10.8 \qquad = \frac{123}{10} = 12.3$$

$$SD\ (X - Y) = \sqrt{\frac{(X - \bar{X})^2 + (Y - \bar{Y})^2}{n_1 + n_2 - 2}} = \sqrt{\frac{85.6 + 32.1}{20 - 2}} = \sqrt{\frac{117.7}{18}}$$

$$= \sqrt{6.539} = 2.557$$

$$SE = SD\sqrt{\frac{1}{n_1} + \frac{1}{n_2}} = 2.557\sqrt{\frac{2}{10}} = 2.557\sqrt{0.2} = 2.557 \times 0.447 = 1.14$$

$$|t| = \frac{\bar{X} - \bar{Y}}{SE} = \frac{10.8 - 12.3}{1.14} = \frac{1.5}{1.14} = 1.31$$

- *Level of Significance*: 0.05
- *Critical Value*: Tabulated value of *t* for 18*df* at 0.05 is 2.10.
- *Decision*: The calculated value of *t* is 1.31 < $|t|_{0.05, 18} = 2.10$.

So the null hypothesis is accepted i.e. there is no significant difference in trait.

Example 11: Ten soldiers visit a rifle range for two consecutive weeks. For the first week their scores are 67, 24, 57, 55, 63, 54, 56, 68, 33, 43 and during second week they score in same order 70, 38, 58, 56, 67, 68, 72, 42, 38. Examine if there is any significant difference in the performance.

Solution:

- *Null Hypothesis*: There is no significant difference in the performance of soldiers.
- *Alternative Hypothesis*: Significant difference in the performance of soldiers.

	A (X)	$X - \bar{X} = D_1$	$(X - \bar{X})^2 = D_1^2$	$B = (Y)$	$Y - \bar{Y} = D_2$	$(Y - \bar{Y})^2 = D_2^2$
1.	67	67 – 52 = 15	225	70	70 – 56.7 = 13.3	176.89
2.	24	24 – 52 = –28	784	38	38 – 56.7 = –18.7	349.69
3.	57	57 – 52 = 05	25	58	58 – 56.7 1.3	1.69
4.	55	55 – 52 = 03	09	58	58 – 56.7 = 1.3	1.69
5.	63	63 – 52 = 11	121	56	56 – 56.7 = –0.7	0.49
6.	54	54 – 52 = 02	04	67	67 – 56.7 = 10.3	106.09
7.	56	56 – 52 = 04	16	68	68 – 56.7 = –11.3	127.69
8.	68	68 – 52 = 16	256	72	72 – 56.7 = 15.3	234.09
9.	33	33 – 52 = –19	361	42	42 – 56.7 = –14.7	216.09
10.	43	43 – 52 = –09	81	38	38 – 56.7 = –18.7	349.69
	$\Sigma X = 520$		$\Sigma D_1^2 = 1882$	$\Sigma Y = 567$		$\Sigma D_2^2 = 1564.1$

$$\bar{X} = \frac{520}{10} = 52 \qquad \bar{Y} = \frac{567}{10} = 56.7$$

$$\Sigma D_1^2 = 1882,\ n_1 = 10 \qquad \Sigma D_2^2 = 1564.1,\ n_2 = 10$$

$$SD(X - Y) = \sqrt{\Sigma \frac{(X - \bar{X})^2 + \Sigma (Y - \bar{Y})^2}{n_1 + n_2 - 2}} = \sqrt{\frac{1882 + 1564.1}{20 - 2}} = \sqrt{\frac{3446.1}{18}}$$

$$= \sqrt{191.45} = 13.84$$

$$SE = SD\sqrt{\frac{1}{n_1} + \frac{1}{n_2}} = 13.84\sqrt{\frac{2}{10}} = 13.84\sqrt{0.2} = 13.84 \times 0.447 = 6.186$$

$$|t| = \frac{\bar{X} - \bar{Y}}{SE} = \frac{52 - 56.7}{6.186} = \frac{-4.7}{6.186} = 0.759 = 0.76$$

- *Level of significance* = 0.05.
- *Critical value*: Tabulated value of t for df 18 at 0.05 is 2.10.
- *Decision*: The calculated value of (t) is 0.76 $|t| < t_{0.05,\,18}$ =2.10, so the null hypothesis is accepted *i.e.*, no significant difference in the performance of soldiers.

Example 12: Two strains of mice were grown under the same environment and fed with same ratios. Twenty adult mice in each strain of the same age and sex were then measured for body weight (in gms) and results were as follows:

Strain	Mean body weight	Variance
A	16	16
B	28	25

Is there any significant difference in body weight between two strains?

Solution: Since data is quantitative we shall verify it by using t test.

- *Null Hypothesis*: There is no significant difference in the mean body weight.
- *Alternation Hypothesis*: There is significant difference in the mean body weight.

• *Calculation*:

	Mean	N	Variance	Variance of Mean	Variance of mean $=(s^2 \bar{x}-\bar{y})$	$SE=\sqrt{s^2 \bar{x}-\bar{y}}$
A	16	20	16	$\frac{s^2x}{n}=\frac{16}{20}=0.8$	$(0.8+1.25)=2.05$	$\sqrt{2.05}$
B	28	20	25	$\frac{s^2y}{n}=\frac{25}{20}=1.25$		1.43

$$t=\frac{\text{Difference between sample mean}}{\text{Standard error of the difference between means}}=\frac{16-28}{1.43}$$

$$=\frac{1200}{143}=8.39.$$

• *Degrees of freedom*: 20 + 20 − 2 =38.
• *Level of significance*: 0.05.
• *Critical value*: Tabulated value of $|t|$ for 38*df* at 0.05 is 2.02 (approximate *i.e.*, 40*df*).
• *Decision*: Since the calculated value of $|t|$ is 8.37 > $t_{0.05,\,38}$ = 2.02, so the null hypothesis is rejected. The difference between mean body weight persists.

Example 13: The body weight of 10 fishes (*Labeo rohita*) of different ponds is given below:

	1	2	3	4	5	6	7	8	9	10
Pond A:	**85**	**75**	**70**	**90**	**80**	**75**	**80**	**80**	**90**	**85**
Pond B:	**55**	**75**	**80**	**65**	**60**	**70**	**70**	**55**	**60**	**80**

Find if there is significant difference between the mean body weights of the above mentioned two groups of fishes.

***t* scores:** $t_{0.05\,(.9)}$ **= 2.093** $t_{0.05\,(18)}$ **= 2.101** $t_{0.05(20)}$ **= 2.086,** $t_{0.05\,(17)}$ **= 2.110**

Solution:

• *Null Hypothesis*: No significant difference in mean body weight.
• *Calculation*:

Sl. No.	Pond A (X)	$X-\bar{X}=D_1$	D_1^2	Pond B (Y)	$Y-\bar{Y}=D_2$	$=D_2^2$
1.	85	85 − 81 = 4	16	55	55 − 67 = −12	149
2.	75	75 − 81 = −6	36	75	75 − 67 = 8	64
3.	70	70 − 81 = −11	121	80	80 − 67 = 13	169
4.	90	90 − 81 = 9	81	65	65 − 67 = −2	04
5.	80	80 − 81 = −1	01	60	60 − 67 = −7	49
6.	75	75 − 81 = −6	36	70	70 − 67 = +3	09
7.	80	80 − 81 = −1	01	70	70 − 67 = +3	09
8	80	80 − 81 = −1	01	55	55 − 67 = −12	144
9.	90	90 − 81 = +9	81	60	60 − 67 = −7	49
10.	85	85 − 81 = +4	16	80	80 − 67 = +13	169
	Σ810		390	670		810

Pond $A = X$ $n_1 = 10$ Pond $B = Y$ $n_2 = 10$

$$\bar{X} = \frac{810}{10} = 81 \qquad \bar{Y} = \frac{670}{10} = 67$$

$$\Sigma D_1^2 = 390 \qquad \Sigma D_2^2 = 810$$

$$SD\ (X - Y) = \sqrt{\frac{\Sigma D_1^2 + \Sigma D_2^2}{n_1 + n_2 - 2}} = \sqrt{\frac{390 + 810}{10 + 10 - 2}} = \sqrt{\frac{1200}{18}} = \sqrt{66.66} = 8.169$$

$$SE = \sqrt{\frac{1}{n_1} + \frac{1}{n_2}} = 8.16\sqrt{\frac{1}{10} + \frac{1}{10}} = 8.16 \times \sqrt{0.2} = 8.16 \times 0.44$$

$$= 3.59$$

$$|t|^2 = \frac{\bar{X} - \bar{Y}}{SE} = \frac{81 - 67}{3.59} = \frac{14}{3.59} = 3.89$$

$$df = n_1 + n_2 - 2 = 18$$

- *Level of Significance*: 5% level *i.e.*, 0.05.
- *Critical value*: Tabulated value at 0.05 for *df* 18 is 2,101.
- Decision: The calculated value of $|t| = 3.89$

  $|t| = 3.89 > t_{0.05\,(18)} = 2.101$

  So the null hypothesis is rejected. Therefore it is inferred that there is significant difference presence, between the mean body weights of fishes of two different ponds.

Example 14: Work out the *t* test to find whether or not the mean gill weights (mg) differ in the following two groups of samples of 10 fishes of *Tilapia mossambica* from two habitats.

	1	2	3	4	5	6	7	8	9	10
Group X	**100**	**80**	**65**	**75**	**110**	**75**	**95**	**78**	**80**	**92**
Group Y	**77**	**63**	**68**	**70**	**55**	**62**	**75**	**70**	**60**	**80**

Critical *t* scores: $t_{0.05\,(19)} = 2.093$ $t_{0.05(18)} = 2.101$ $t_{0.05(9)} = 2.262$ $t_{0.05(8)} = 2.306$

Solution:

- *Null Hypothesis*: No significant difference present in the mean gill weight of the two samples.

Sl. No.	*Habitat*	$X - \bar{X} = D_1$	D_1^2	*Habitat*	$Y - \bar{Y} = D_1$	$= D_2^2$
1.	100	100 – 85 = 15	225	77	77 – 66 = 11	121
2.	80	80 – 85 = –5	25	63	63 – 66 = –3	09
3.	65	65 – 85 = –20	400	68	68 – 66 = 2	04
4.	75	75 – 85 = –10	100	70	70 – 66 = –4	16
5.	100	110 – 85 = 15	225	55	55 – 66 = –10	100
6.	75	75 – 85 = –10	100	62	62 – 66 = –4	16
7.	95	95 – 85 = 10	100	75	75 – 66 = 9	81
8.	78	78 – 85 = –7	49	70	70 – 66 = 4	16
9.	80	80 – 85 = –5	25	60	60 – 66 = –6	36
10.	92	92 – 85 = 7	49	60	60 – 66 = –6	36
	850		1298	660		435

$$\sum X = 850 \quad \bar{X} = 85 \qquad \sum Y = 660 \quad \bar{Y} = 66$$

$$n_1 = 10 \qquad \sum D_1^2 = 1298 \qquad \sum D_2^2 = 435 \quad n_2 = 10$$

$$SD\ (X - Y) = \sqrt{\frac{\sum D^2 + \sum D_2}{n_1 + n_2 - 2}} = \sqrt{\frac{1298 + 435}{10 + 10 - 2}}\sqrt{\frac{1733}{18}} = \sqrt{96.28} = 9.81$$

$$SE = SD\sqrt{\frac{1}{x_1} + \frac{1}{x_2}} = 9.81\sqrt{\frac{1}{10} + \frac{1}{10}} = 9.8\sqrt{10.2} = 9.8 \times 0.44 = 4.31$$

$$|t| = \frac{\bar{X} - \bar{Y}}{SE} = \frac{85 - 66}{4.31} = \frac{19}{4.31} = 4.48.$$

- *Level of significance*: 5% level *i.e.*, 0.05.
- *Critical value*: The tabulated value of $|t|$ at 0.05 for *df* 18 is = 2.01.
- *Decision*: Since the calculated value of $|t|$ is 4.4 & the tabulated value of $|t|$ at 0.05 for *df* 18 is 2.01. So the null hypothesis is rejected.

The mean weight of gill differs.

Example 15: Workout *t* test to find whether or not the mean oxygen consumption (milliliter/hr) of 10 carp fishes varied before or after irradiation. The data are given below:

Fish No.	**1**	**2**	**3**	**4**	**5**	**6**	**7**	**8**	**9**	**10**
Before irradiation	**3.2**	**2.7**	**2.9**	**3.0**	**2.8**	**2.9**	**2.8**	**3.2**	**3.1**	**3.0**
After irradiation	**2.7**	**2.5**	**2.6**	**2.6**	**2.5**	**2.6**	**2.4**	**2.8**	**2.5**	**2.5**

$$t_{\text{scores}} = t_{0.05\ (18)} = 2.101 \qquad t_{0.05\ (19)} = 2.093$$

$$t_{0.05\ (17)} = 2.110 \qquad t_{0.05\ (9)} = 2.262$$

Solution:

- *Null Hypothesis*: Oxygen consumption remains constant before & after irradiation.
- *Calculation*:

Sl. No.	*Before irradiation* (X_1)	*After irradiation* X_2	$X_1 - X_2 = D$	D^2
1.	3.2	2.7	3.2 − 2.7 = +0.5	0.25
2.	2.7	2.5	2.7 − 2.5 = +0.2	0.04
3.	2.9	2.6	2.9 − 2.6 = +0.3	0.09
4.	3.0	2.6	3.0 − 2.6 = +0.4	0.16
5.	2.8	2.5	2.8 − 2.5 = +0.3	0.09
6.	2.9	2.6	2.9 − 2.6 = +0.3	0.09
7.	2.8	2.4	2.8 − 2.4 = 0.4	0.16
8.	3.2	2.8	3.2 − 2.8 = 0.4	0.16
9.	3.1	2.5	3.1 − 2.5 = 0.6	0.36
10.	3.0	2.5	3.0 − 2.5 = 0.5	0.25
			$\sum D = 3.9$	1.65

$$n = 10 \qquad \bar{D} = \frac{\sum D}{n} = \frac{3.9}{10} = 0.39$$

$$\sum D^2 = 1.65$$

$$SD = \sqrt{\frac{\sum D^2 - \frac{(\sum D)^2}{n}}{n-1}} = \sqrt{\frac{1.65 - \frac{(3.9)^2}{10}}{10-1}} = \sqrt{\frac{1.65 - \frac{15.21}{10}}{9}} = \sqrt{\frac{1.65 - 1.52}{9}}$$

$$= \sqrt{\frac{0.13}{9}} = \sqrt{0.0145} = 0.120$$

$$SE = \frac{SD}{\sqrt{n}} = \frac{0.120}{\sqrt{10}} = \frac{0.120}{3.16} = 0.0379$$

$$|t| = \frac{\bar{D}}{SF} = \frac{0.39}{0.0379} = 10.29.$$

- *Level of significance*: 5% level *i.e.*, 0.05.
- *Critical value*: Tabulated value of *f* for *df* 9 is 2.26 at 0.05 level.
- *Decision*: The calculated value of $|t|$ is 10.29. $|t| > t_{0.05\,(9)} = 2.26$.

So the null hypothesis is rejected. Therefore irradiation have impact on oxygen consumption.

Example 16: The fasting glucose level of seven (7) persons was 4, 6, 8, 5, 9, 3 & 7 m mol/L. After 75 gm of oral glucose, the blood sugar level was 8, 10, 13, 9, 8, 12 & 14 m mol/L respectively. Use appropriate statistical test whether oral glucose had significantly increased the blood sugar level or not.

Solution:

- *Null Hypothesis*: Oral glucose has no impact on blood sugar level.

Blood sugar fasting	*After 75 gm oral glucose*	*D*	D^2
4	8	4 − 8 = −4	16
6	10	6 − 10 = −4	16
8	13	8 − 13 = −5	25
5	9	5 − 7 = −2	04
9	8	9 − 8 = +1	01
3	12	3 − 12 = −9	81
7	14	7 − 14 = −7	49
		−30	192

$$\bar{D} = \frac{\sum D}{n} \qquad \therefore \bar{D} = \frac{-30}{7} = -4.285$$

$$n = 7$$

$$SD = \sqrt{\frac{\sum D^2 - \frac{(\sum D)^2}{n}}{n-1}}$$

$$= \sqrt{\frac{192 - \frac{(-30)^2}{7}}{6}} = \sqrt{\frac{192 - \frac{900}{7}}{6}} = \sqrt{\frac{192 - 128.57}{6}} = \sqrt{\frac{63.43}{6}}$$

$$= \sqrt{10.57} = 3.25$$

$$SE = \frac{SD}{\sqrt{n}} = \frac{3.25}{\sqrt{7}} = \frac{3.25}{2.64} = 1.23$$

$$|t| = \frac{\bar{D}}{SE} = \frac{4.28}{1.23} = 3.479 = 3.48.$$

- *Level of significance*: = 0.05 level
- *Critical value*: Tabulated value of t at 0.05 level for df 7 – 1 = 6 is = 2.45.
- *Decision*: The calculated value of $|t|$ = 3.48

It is $72.45 t_{0.05\,(6)}$ (2.45)

So the null hypothesis is rejected *i.e.*, oral glucose significantly increased blood sugar level.

11
CHAPTER
Z-TEST

Deviation from the mean in a normal distribution or curve is called relative or standard normal deviate and is given the symbol '**Z**'.

It is measured in terms of *SD* and indicates how much an observation is bigger or smaller than mean in units of *SD*. So *Z* will be a ratio *i.e.*,

$$Z = \frac{\text{Observation} - \text{Mean}}{SD} = \frac{X - \overline{X}}{SD}$$

- ***Z-test:*** When *Z* test is applied to the sampling variability, the difference observed between a sample estimate and that of population is expressed in terms of *SE* instead of *SD*. The score of the value of ratio between the observed difference and *SE* is called "*Z*'.
- ***Conditions for applying Z test:***
  (*i*) The sample size must be larger than 30.
  (*ii*) The sample must be randomly selected.
  (*iii*) The variable is assumed to follow normal distribution.
  (*iv*) The data must be quantitative.
- ***Characteristics:***
  (*i*) If the distance in terms of *SE* or *Z* scores falls within mean ±1.96 *SE i.e.*, in the zone of acceptance (*confidence limit* 95%) the null hypothesis (*Ho*) is accepted.
  (*ii*) The distance from the mean at which null hypothesis (*Ho*) is rejected is called *level of significance*. It falls in the zone of rejection for *Ho* and is denoted by letter '*P*'
  (*iii*) Greater the *Z* value, lesser will be the '*P*'.
  (*iv*) *P* at 5% level written as 0.05% and at 2.5% as 0.025 and so on.
- ***Z-test:***
  (*a*) Test of significance of difference between a sample mean (μ) and population mean ($\overline{X}$)

$$Z = \frac{\overline{X} - \mu}{SE(X)}$$

  (*b*) Test of significance of difference between two sample means

$$Z = \frac{\overline{X}_1 - \overline{X}_2}{SE(\overline{X}_1 - \overline{X}_2)}$$

- ***One – tailed test:*** It is a test of statistical hypothesis where either alternative hypothesis is one sided is called one tailed test or one sided test.
  (*i*) It may be right tailed or left tailed test.
  (*ii*) If we want to know one particular drug is than the other. It will be one tailed test.

- ***Two – tailed test:*** It is a test of statistical hypothesis based on rejected region represented by both sides of the standard normal curve.

Example: *I.Q.* of malnourished children is different from that of well nourished children.

- ***Working Procedure:***

(*i*) Set up – Null Hypothesis (*Ho*): There is no significant difference between sample means, population mean.

$$Ho = \bar{X} = \mu$$

Alternative Hypothesis (H_1): $H_1 = \bar{X} \neq \mu$

(*ii*) Computation of test of *Z* statistic:

(*a*) When the S.D. of the population is known.

Here $S.E.\ \bar{X}$ (standard error of mean) $= \dfrac{\sigma}{\sqrt{n}}$.

$$Z = \frac{\bar{X} - \mu}{SE(\bar{X})}$$

n = Sample $\sigma = S.D.$ μ = Sample mean

(*b*) When the *S.D.* of the population is not known: Let '*S*' be the *S.D.* of the sample, to calculate the standard error of mean *i.e.*,

$$SE(\bar{X}) = \frac{\sigma}{\sqrt{n}}$$

$$Z = \frac{\bar{X} - \mu}{SE(\bar{X})} = \frac{\bar{X} - \mu}{\frac{S}{\sqrt{n}}} = \frac{(\bar{X} - \mu)\sqrt{n}}{S}$$

(*iii*) Let the level of significance be α.

(*iv*) Find the critical value of *Z* *i.e.*, Z_α at the level of significance from the table.

Areas under the normal curve $-Z_\alpha$ values.

Critical value Z_α	*Level of significance*		
	1%	*5%*	*10%*
Two tailed test	$\lvert Z_\alpha \rvert = 2.58$	$\lvert Z_\alpha \rvert = 1.96$	$\lvert Z_\alpha \rvert = 1.645$
Right tailed test	$\lvert Z_\alpha \rvert = 2.33$	$\lvert Z_\alpha \rvert = 1.645$	$\lvert Z_\alpha \rvert = 1.28$
Left tailed test	$\lvert Z_\alpha \rvert = 2.33$	$\lvert Z_\alpha \rvert = 1.645$	$\lvert Z_\alpha \rvert = 1.28$

(*v*) (*a*) If computed value of $Z = \lvert Z \rvert <$ critical value Z_α at a level of significance then accept the null hypothesis (*Ho*).

(*b*) If computed value of $Z = \lvert Z \rvert >$ critical value Z_α then reject the null hypothesis (*Ho*) and accept the alternative hypothesis (H_1).

Q1. The mean *I.Q.* of a sample of 1600 children was 99. It is likely that this was a random sample from a population with mean *I.Q.* 100 and standard deviation 15.

Solution:

- *Null Hypothesis:* Sample has not been drawn from the population with mean *I.Q.* 100.
- *Alternative Hypothesis:* Sample has been drawn from the population hypothesis.
- *Calculation:*

Here, $n = 1600,\ \bar{X} = 99,\ \mu = 100,\ S.D.$ i.e. $\sigma = 15$

$$SE = \frac{\sigma}{\sqrt{n}} = \frac{15}{\sqrt{1600}} = \frac{15}{40} = 0.375$$

$$|Z| = \left|\frac{\bar{X} - \mu}{SE(\bar{X})}\right| = \left|\frac{99 - 100}{0.375}\right| = \left|\frac{-1}{0.375}\right| = |2.67|$$

- *Critical value:*

Since Z_α (level of significance) is not given so we take $\alpha = 0.05$.

- *Decision:* Calculated value of Z is 2.67. Since $|Z|$ is $2.67 > Z_\alpha$ 1.96, null hypothesis *Ho* is rejected *i.e.*, sample has not been drawn from population with mean 100 and σ 15.

Q2. A college conducts both day & night classes intended to be identical. A sample of 100 day students yields examination results as under.

$$\bar{X}_1 = \mathbf{72.4\ \&\ } \sigma_1 = 14.8$$

A sample of 200 night students yields examination results as under

$$\bar{X}_2 = \mathbf{73.9\ \&\ } \sigma_2 = 17.9$$

Are the two means statistically equal at 10% level?

Solution:

Here $\quad n_1 = 100,\ \ n_2 = 200$

$$\bar{X}_1 = 72.4,\ \bar{X}_2 = 73.9$$

$$\sigma_1 = 14.8,\ \sigma_2 = 17.9$$

- *Null Hypothesis Ho:* $\mu_1 = \mu_2$
- *Alternative Hypothesis* H_1: $\mu_1 \neq \mu_2$
- *Calculation:*

SE of means $\quad (\bar{X}_1 - \bar{X}_2) = \sqrt{\frac{\sigma_1^2}{n_1} + \frac{\sigma_2^2}{n_2}} = \sqrt{\frac{(14.8)^2}{100} + \frac{(17.9)^2}{200}}$

$$= \sqrt{\frac{219.04}{100} + \frac{320.14}{200}} = \sqrt{2.19 + 1.60}$$

$$= \sqrt{3.79} = 1.946 = 1.95$$

$$Z| = \left|\frac{\bar{X}_1 - \bar{X}_2}{SE(X_1 - X_2)}\right| = \left|\frac{72.4 - 73.9}{1.95}\right| = \left|\frac{-1.5}{1.95}\right| = 0.769$$

- *Level of significance:* Here $Z_\alpha = 0.10$
- *Critical value:* at 0.10 is $Z_\alpha = 1.645$
- *Decision:* Since calculated value of $Z = 0.769 <$ critical value $(\alpha = 0.10) = 1.645 =$ *Ho* is accepted.

Hence we conclude that two means are statistically equal.

Q3. A sample of 400 items is taken from a normal population whose mean is 4 and whose variance is also 4. If the sample mean is 4.45, can the sample mean be regarded as truly random sample?

Solution:

- *Null Hypothesis (Ho):* Sample can not be regarded as having been drawn from the population with mean 4.
- *Alternative Hypothesis* (H_1): Sample be regarded as having been drawn from population with mean 4.
- *Calculation:*

$$\text{Population mean } (\mu) = 4. \text{ Sample mean } (\bar{X}) = 4.45$$

$$\text{variance } (\sigma^2) = 4$$

$$S.D. = \sqrt{4} = 2$$

$$\text{Standard error of mean } SE = \frac{\sigma}{\sqrt{400}} = \frac{2}{\sqrt{400}} = \frac{2}{20} = 0.1$$

$$Z = \frac{\bar{X} - \mu}{SE(\bar{X})} = \frac{4.45 - 4}{0.1} = \frac{.45}{.1} = 4.5$$

- *Critical value:* Here level of significance is not given. So we take 5% level *i.e.*, 0.05.
- *Decision:* Since $|Z| = 4.5 > Z_\alpha$ 1.96 so H_0 is rejected at 5% level of significance *i.e.*, sample can not be regarded as having been drawn from the population with mean 4.

Q4. A simple sample of 1,000 members is found to have a mean 3.42 cm. Could it be reasonably regarded as a simple sample from a large population arose mean is 3.30 cm and *S.D* is 2.6 cm?

Solution:

- *Null Hypothesis* (H_0): Sample is drawn from a normal population with $\mu = 3.30$ and $\sigma = 2.6$
- *Alternate Hypothesis* (H_1): Sample has not been drawn from normal population with $\mu = 3.30$ and $\sigma = 2.6$
- *Calculation:*

$$\text{Here, } n = 1000 \quad \bar{X} = 3.42, \quad \mu = 3.30, \ \sigma = 2.6$$

$$|Z| = \frac{\bar{X} - \mu}{SE} = \frac{3.42 - 3.30}{\frac{2.6}{\sqrt{1000}}}$$

$$= \frac{.12}{2.6} \times \sqrt{1000} = \frac{12}{260} \times 31.6 = \frac{379.2}{260} = 1.458 = 1.46$$

- *Critical value:* Since Z_α is not given we take $\alpha = 0.05$
- *Decision:* Computed value of Z is 1.46
  $|Z|_{0.05} = 1.96 > Z = 1.46$. So Ho is accepted. So we say that the sample is drawn from normal population.

Q5. The standard deviation of the height of B.Sc. (Bio) student of Serampore College is 5.0 cm. Two samples are taken. The standard deviation of 100 students of 1st year is 4.5 cm and 100 students of 2nd year is 5.5 cm. Test the significance of difference of standard deviations of the samples.

Solution:

- *Null Hypothesis* (H_0): There is no significant difference between the standard deviation of two samples *i.e.,* H_0: $\sigma_1 = \sigma_2$
- *Alternative Hypothesis* (H_1): Presence of significant difference between the standard deviation *i.e.,* H_0: $\sigma_1 \neq \sigma_2$
- *Calculation:*

$$n_1 = 100,\ n_2 = 100$$
$$S_1 = 4.5,\ \ S_2 = 5.5$$
$$\sigma = 5.0$$

$$\text{Standard Error } (S.E)(S_1 - S_2) = \sqrt{\frac{\sigma_2^1}{2n_1} + \frac{\sigma_2^2}{2n_2}}$$

$$= \sqrt{\frac{(5)^2}{2 \times 100} + \frac{(5)^2}{2 \times 100}} = \sqrt{\frac{25}{200} + \frac{25}{200}}$$

$$= \sqrt{0.125 + 0.125} = \sqrt{.25} = .5$$

$$\text{Test Statistic } (Z) = \frac{S_1 - S_2}{SE(S_1 - S_2)} = \frac{4.5 - 5.5}{.5} = \frac{-1.0}{.5} = -2.0$$

$$|Z| = 2.0$$

- *Level of significance:* Since no significance is given, so we take $\alpha = 0.05$ level.
- *Critical value:* Critical value of Z at $\alpha = 0.05$ is 1.96
- *Decision:* Calculated value of Z is 2.0. Since $|Z|$ is $= 2.0 > Z_\alpha$ 1.96 the null hypothesis Ho is rejected.

  So we say there is significant difference between the two standard deviation.

Q6. Random samples drawn from two places (Serampore & Sheoraphuli) gave the following data. Relating to the wing length of *Anopheles* mosquitoes.

	Serampore (A)	***Sheoraphuli (B)***
Mean (in mm) ($\overline{X}_1$)	**3.60**	($\overline{X}_2$) **= 3.58**
Standard deviation (S_1)	**1.8**	S_2 **= 1.6**
Number of Sample (n_1)	**50**	n_2 **= 50**

Test at 5% level that the mean wing length is the same for mosquitoes at two places.

Solution:

- *Null Hypothesis* (*Ho*): There is no significant difference between mean wing length of the two samples *i.e.,* $Ho = \overline{X}_1 = \overline{X}_2$
- *Alternative Hypothesis* (H_1): $H_1 = \overline{X}_1 \neq \overline{X}_2$
- *Calculation:*

$$n_1 = 50, \quad n_2 = 50$$

$$\overline{X}_1 = 3.60, \quad \overline{X}_2 = 3.58$$

$$S_1 = 1.8, \quad S_2 = 1.6$$

Standard Error (*S.E*):

$$(S_1 - S_2) = \sqrt{\frac{\sigma_1^2}{n_1} + \frac{\sigma_2^2}{n_2}} = \sqrt{\frac{(1.8)^2}{50} + \frac{(1.6)^2}{50}} = \sqrt{\frac{3.24}{50} + \frac{2.56}{50}}$$

$$= \sqrt{0.0648 + 0.0512} = \sqrt{0.116} = 0.34$$

$$Z = \frac{\overline{X}_1 - \overline{X}_2}{SE} = \frac{3.60 - 3.58}{0.34} = \frac{0.02}{0.34} = 0.058$$

- *Level of significance:* Significance at 5% level *i.e.,* 0.05 is 1.96.
- *Critical value:* Here Z_α is 0.05 level *i.e.,* 1.96
- *Decision:* Since $| Z |$ is 0.058 < Z_α 1.96. So *Ho* is accepted *i.e.,* null hypothesis is accepted.

We can say the mean wing length of both the samples are equal.

12

CHAPTER

F-TEST OR FISHER'S–F TEST

This test was originated by the statistician **R.A. Fisher**. It is popularly known as **F-test**. It is also called variance ratio test as comparison of sample variance involves in this test.

- ***F-test:*** It is a test of hypothesis concerning two variances derived from two samples.
- ***F-statistic:*** It is the ratio of two independent "Unbiased estimators" of population variances & expressed as:

$$F = \frac{\sigma_1^2}{\sigma_2^2}$$

(*i*) $n_1 - 1$ the degrees of freedom for numerator & $n_2 - 1$ the degrees of freedom for denominator.

(*ii*) F-table gives variance ratio values at different levels of significance at $df = (n_1 - 1)$ given horizontally & degrees of freedom (df) = $(n_2 - 1)$ given vertically.

(*iii*) Generally, σ_1^2 is greater than σ_2^2 but if σ_2^2 is greater than σ_1^2, in such cases the two variances should be interchanged so that the value of '***F***' is always greater than 1.

(*iv*) If the ***F***-ratio value is smaller than the table value, the null hypothesis (*Ho*) is accepted. It indicates that the samples are drawn from the same population.

If the calculated ***F*** value is greater than the table value, the null hypothesis (*Ho*) is rejected and conclude that the standard deviations in the two populations are not equal.

- **Working Procedure:**

(*i*) Set up the null hypothesis (*Ho*) $\sigma_1^2 = \sigma_2^2$ and alternative hypothesis $(H_1) = \sigma_1^2 \neq \sigma_2^2$

(*ii*) Calculate the variances of two samples and then calculate the F statistic *i.e.*,

$$F = \frac{\sigma_1^2}{\sigma_2^2} \text{ if } \sigma_1^2 \geq \sigma_2^2$$

or $$F = \frac{\sigma_2^2}{\sigma_1^2} \text{ if } \sigma_2^2 \geq \sigma_1^2$$

(*iii*) Take level of significance at 0.05.

(*iv*) Compare the computed F-value with the table value and degrees of freedom $(n_1 - 1)$ horizontally and degrees of freedom $(n_2 - 1)$ vertically.

- **Assumption of F test:**

(*i*) The value in each group should be normally distributed.

(*ii*) The variation of each value around its own group mean *i.e.*, error should be independent of each value.

(*iii*) The variances within each group should be equal for all group *i.e.*, $\sigma_1^2 = \sigma_2^2 \ldots\ldots \sigma_n^2$.

- **Uses:**
  - (*i*) *F* test for equality of population variances.
  - (*ii*) To test the two independent samples (*x* & *y*) have been drawn from the normal populations with same variances (σ^2).
  - (*iii*) Whether the two independent estimates of the populations variances are homogeneous or not.

Q.1. Two samples are drawn from two normal populations. From the following data test whether the two samples have the same variances at 5% level.

Sample I:	**60**	**65**	**71**	**74**	**76**	**82**	**85**	**87**		
Sample II:	**64**	**66**	**67**	**85**	**78**	**88**	**86**	**85**	**63**	**91**

Solution:

- *Null Hypothesis: Ho:* $\sigma_1^2 = \sigma_2^2$ *i.e.,* two samples have the same variances.
- *Alternative Hypothesis:* $H_1: \sigma_1^2 \neq \sigma_2^2$
- *Calculation:*

Sample – I			Sample – II		
X	$X - \bar{X}$	$(X - \bar{X})^2$	Y	$Y - \bar{Y}$	$(Y - \bar{Y})^2$
60	60 – 75 = – 15	225	61	61 – 77 = – 16	256
65	65 – 75 = – 10	100	66	66 – 77 = – 11	121
71	71 – 75 = – 4	16	67	67 – 77 = – 10	100
74	74 – 75 = – 1	01	85	85 – 77 = + 8	64
76	76 – 75 = + 1	01	78	78 – 77 = + 1	01
82	82 – 75 = + 7	49	88	88 – 77 = + 11	121
85	85 – 75 = + 10	100	86	86 – 77 = + 9	81
87	87 – 75 = + 12	144	85	85 – 77 = + 8	64
			63	63 – 77 = – 14	196
			91	91 – 77 = + 14	196
$\sum X = 600$		$\sum (X-\bar{X})^2 = 636$	$\sum Y = 770$		$\sum (Y-\bar{Y})^2 = 1200$

$$\bar{X} = \frac{600}{8} = 75 \qquad \bar{Y} = \frac{770}{10} = 77$$

$$n_1 = 8 \qquad n_{II} = 10$$

$$\text{Variance of sample I} = \sigma_I^2 = \frac{\sum (X - \bar{X})^2}{n_I - 1} = \frac{636}{8-1} = \frac{636}{7} = 90.857$$

$$\text{Variance of sample II} = \sigma_{II}^2 = \frac{\sum (Y - \bar{Y})^2}{n_{II} - 1} = \frac{1200}{10-1} = \frac{1200}{9} = 133.33$$

$$\text{Test-statistic} \qquad F = \frac{\sigma_{II}^2}{\sigma_I^2} = \frac{133.33}{90.857} = 1.467 = 1.48$$

- *Critical value:* The table of *F* at 0.05 for degrees of freedom for 7 is $F_{0.05}$ 3.68.
- *Decision:* The computed value of $F = 1.48 <$ table value 3.68, the null Hypothesis is accepted. The samples (I & II) have the same variance.

Q.2. Two samples were drawn from two normal population of Birbhum district & their values are

A:	66	67	82	75	76	90	92	88	84		
B:	66	64	78	74	87	85	82	95	93	97	92

Test whether the two populations have the same variance at 5% level of significance.

Solution:

- *Null Hypothesis: Ho:* $\sigma_A^2 = \sigma_B^2$ *i.e.*, two samples have same variance.
- *Alternative Hypothesis:* H_1: $\sigma_A^2 \neq \sigma_B^2$

Sample – A			*Sample – B*		
X	$X - \bar{X}$	$(X - \bar{X})^2$	Y	$Y - \bar{Y}$	$(Y - \bar{Y})^2$
66	66 – 80 = – 14	196	66	66 – 83 = – 17	289
67	67 – 80 = – 13	169	64	64 – 83 = – 19	361
82	82 – 80 = + 02	04	78	78 – 83 = – 05	25
75	75 – 80 = – 05	25	74	74 – 83 = – 09	81
76	76 – 80 = – 04	16	87	87 – 83 = + 04	16
90	90 – 80 = + 10	100	85	85 – 83 = + 02	04
92	92 – 80 = + 12	144	82	82 – 83 = – 01	01
88	88 – 80 = + 08	64	95	95 – 83 = + 12	144
84	84 – 80 = + 04	16	93	93 – 83 = + 10	100
			97	97 – 83 = + 14	196
			92	92 – 83 = + 09	81
$\sum X = 720$		$\sum (X - \bar{X})^2 = 734$	$\sum Y = 913$		$\sum (Y - \bar{Y})^2 = 1298$

$n = 9 \qquad \bar{X} = \frac{720}{9} = 80 \qquad n = 11 \qquad \bar{Y} = \frac{913}{11} = 83$

$$\text{Variance of sample A} = \sigma_A^2 = \frac{\sum (X - \bar{X})^2}{n-1} = \frac{734}{9-1} = \frac{734}{8} = 91.75$$

$$\text{Variance of sample B} = \sigma_B^2 = \frac{\sum (Y - \bar{Y})^2}{n-1} = \frac{1298}{11-1} = \frac{1298}{10} = 129.8$$

$$\text{Test-statistic} \qquad F = \frac{\sigma_B^2}{\sigma_A^2} = \frac{129.8}{91.75} = 1.4147 = 1.415$$

- *Critical value:* The table value of F at 0.05 for the *df* (11 – 1) = 10 & (9 – 1) = 8 is 3.35.
- *Decision:* The computed value F is 1.415 < table value of $F_{0.05}$ 3.35 = the null hypothesis H_0 is accepted. The population from where the samples are taken has the same variances.

Q.3. For a random sample (*A*) of 10 albino rats fed on vitamins, the increase in weight in grams in 15 days and another sample of 12 albino rats fed on another vitamin, increase in weight. which are given in table.

A:	10	06	16	17	13	12	08	14	15	09		
B:	07	13	22	15	12	14	21	08	18	17	23	10

Test whether both the samples come from population having same variance. (The $F_{0.05}$ for $v_2 = 11$ $v_1 = 9$ is 3.11)

Solution:

- *Null Hypothesis:* There is no significant variance between the two populations *i.e.,* $H_0: \sigma_1^2 = \sigma_2^2$.
- *Alternative Hypothesis:* $H_1: \sigma_1^2 \neq \sigma_2^2$.
- *Calculation:*

Sample – A			Sample – B		
X	$X-\bar{X}$	$(X-\bar{X})^2$	Y	$Y-\bar{Y}$	$(Y-\bar{Y})^2$
10	10 – 12 = – 02	04	07	07 – 15 = – 08	64
06	06 – 12 = – 06	36	13	13 – 15 = – 02	04
16	16 – 12 = + 04	16	22	22 – 15 = + 07	49
17	17 – 12 = + 05	25	15	15 – 15 = 00	00
13	13 – 12 = + 01	01	12	12 – 15 = – 03	09
12	12 – 12 = 00	00	14	14 – 15 = – 01	01
08	08 – 12 = – 04	16	21	21 – 15 = + 06	36
14	14 – 12 = + 02	04	08	08 – 15 = – 07	49
15	15 – 12 = + 03	09	18	18 – 15 = + 03	09
09	09 – 12 = – 03	09	17	17 – 15 = + 02	04
			23	23 – 15 = + 08	64
			10	10 – 15 = – 05	25
$\Sigma X = 120$		$\Sigma(X-\bar{X})^2 = 120$	$\Sigma Y = 180$		$\Sigma(Y-\bar{Y})^2 = 314$

$n = 10 \quad \bar{X} = \frac{120}{10} = 12 \qquad n = 12 \quad \bar{Y} = \frac{180}{12} = 15$

Variance of sample A = $\sigma_1^2 = \frac{\Sigma(X-\bar{X})^2}{n-1} = \frac{120}{10-1} = \frac{120}{9} = 13.3$

Variance of sample B = $\sigma_2^2 = \frac{\Sigma(Y-\bar{Y})^2}{n-1} = \frac{314}{12-1} = \frac{314}{11} = 28.545 = 28.55$

Test-statistic $F = \frac{\sigma_2^2}{\sigma_1^2} = \frac{28.55}{13.3} = 2.14$

- *Critical value:* The table value of F at 0.05 for (10 – 1) 9 & (12 – 1)11 is 3.11.
- *Decision:* Since the calculated value of F is less than tabulated F so null hypothesis is accepted *i.e.,* variance of two populations are same.

Q.4. In a test given to two groups of students (*A & B*) drawn from south point school of Calcutta, the marks obtained were as follows.

Group A:	46	44	30	35	26	28	29		
Group B:	18	36	20	49	50	34	49	36	41

Examine at 5% level whether the two groups of school students have the same variance.

Solution:

- *Null Hypothesis:* There is no significant variance between the two groups of students of the school. $H_0: \sigma_1^2 = \sigma_2^2$.

- *Alternative Hypothesis:* H_1: $\sigma_1^2 \neq \sigma_2^2$.
- *Calculation:*

Group – A			Group – B		
X	$X - \bar{X}$	$(X - \bar{X})^2$	Y	$Y - \bar{Y}$	$(Y - \bar{Y})^2$
46	46 – 34 = + 12	144	18	18 – 37 = – 19	361
44	44 – 34 = + 10	100	36	36 – 37 = – 01	01
30	30 – 34 = – 04	16	20	20 – 37 = – 17	289
35	35 – 34 = + 01	01	49	49 – 37 = + 12	144
26	26 – 34 = – 08	64	50	50 – 37 = + 13	169
28	28 – 34 = – 06	36	34	34 – 37 = – 03	09
29	29 – 34 = – 05	25	49	49 – 37 = + 12	144
			36	36 – 37 = – 01	01
			41	41 – 37 = + 04	16
$\Sigma X = 238$		$\Sigma(X - \bar{X})^2 = 386$	$\Sigma Y = 333$		$\Sigma(Y - \bar{Y})^2 = 1134$

$n = 7 \quad \bar{X} = \frac{238}{7} = 34 \quad n = 9 \quad \bar{Y} = \frac{333}{9} = 37$

Variance of sample A = $\sigma_X^2 = \frac{386}{7-1} = \frac{386}{6} = 64.33$

Variance of sample B = $\sigma_Y^2 = \frac{1134}{9-1} = \frac{1134}{8} = 141.75$

Test-statistic $F = \frac{\sigma_Y^2}{\sigma_X^2} = \frac{141.75}{64.33} = 2.203$

- *Critical value:* The table value of 'F' at 5% level for *df* 8 & 6 is $F_{0.05} = 4.15$
- *Decision:* The computed value of $F = 2.203 <$ table value of F 4.15. So the null hypothesis is accepted *i.e.,* two populations have the same variance.

Q.5. The following data presents the yields of potato on ten sub divisions of equal area of two agricultural plots of Arambagh (Hooghly, West Bengal).

	1	2	3	4	5	6	7	8	9	10
Plot A:	6.2	5.9	5.8	5.9	5.8	6.0	5.9	6.0	6.3	6.2
Plot B:	5.5	5.8	5.7	5.9	6.2	6.0	5.8	5.7	5.6	5.8

Test whether the two samples taken from two random population have the same variance (5% for F for $v_1 = 9$ & $v_2 = 9$ is 3.18).

Solution:

- *Null Hypothesis:* There is no significance variance between the two population H_0: $\sigma_X^2 = \sigma_Y^2$.
- *Alternative Hypothesis:* H_1: $\sigma_X^2 \neq \sigma_Y^2$.

• *Calculation:*

	Plot – A			Plot – B		
S.No.	X	$X-\bar{X}$	$(X-\bar{X})^2$	Y	$Y-\bar{Y}$	$(Y-\bar{Y})^2$
1	6.2	6.2 – 6.0 = + 0.2	0.04	5.5	5.5 – 5.8 = – 0.3	0.09
2	5.9	5.9 – 6.0 = – 0.1	0.01	5.8	5.8 – 5.8 = 00	0.00
3	5.8	5.8 – 6.0 = – 0.2	0.04	5.7	5.7 – 5.8 = – 0.1	0.01
4	5.9	5.9 – 6.0 = – 0.1	0.01	5.9	5.9 – 5.8 = + 0.1	0.01
5	5.8	5.8 – 6.0 = – 0.2	0.04	6.2	6.2 – 5.8 = + 0.4	0.16
6	6.0	6.0 – 6.0 = 00	00	6.0	6.0 – 5.8 = + 0.2	0.04
7	5.9	5.9 – 6.0 = – 0.1	0.01	5.8	5.8 – 5.8 = 00	00
8	6.0	6.0 – 6.0 00	00	5.7	5.7 – 5.8 = – 0.1	0.01
9	6.3	6.3 – 6.0 = + 0.3	0.09	5.6	5.6 – 5.8 = – 0.2	0.04
10	6.2	6.2 – 6.0 = + 0.2	0.04	5.8	5.8 – 5.8 = 00	00
	$\sum X=60.0$		$\sum(X-\bar{X})^2=0.28$	$\sum Y=58.0$		$\sum(Y-\bar{Y})^2=0.37$

$$n = 10 \quad \bar{X} = \frac{60.0}{10} = 6.0 \quad n = 10 \quad \bar{Y} = \frac{58.0}{10} = 5.8$$

$$\text{Variance of sample A} - \sigma_1^2 = \frac{0.28}{10-1} = \frac{0.28}{9} = 0.03$$

$$\text{Variance of sample B} = \sigma_2^2 = \frac{0.37}{10-1} = \frac{0.37}{9} = 0.04$$

Since $\sigma_1^2 > \sigma_1^2$

$$F = \frac{\sigma_1^2}{\sigma_2^2} = \frac{0.03}{0.04} = 0.75$$

• *Critical value:* The table value of F at 0.05 for 929 df is 3.18.

• *Decision:* Since the calculated value of F is less than table value F, so the null hypothesis is accepted.

Q.6. The standard deviations calculated from two random samples of sizes 9 & 13, 2.1 & 1.8 respectively. Test, whether the samples are drawn from the normal population with the same S.D.? (The 5%) value of F from tables with df 8 & 12 is $F_{0.05}$ = 2.85.

Solution:

• *Null Hypothesis:* $\sigma_1^2 = \sigma_2^2$ *i.e.,* samples taken from the population with same S.D.

• *Alternative Hypothesis:* $\sigma_1^2 \neq \sigma_2^2$ *i.e.,* sample taken from the population with the different S.D.

• *Calculation:*

Here $n_1 = 9$ *S.D.* $(S_1) = 2.1$ $n_2 = 13$ *S.D.* $(S_2) = 1.8$

$$\therefore \quad \sigma_1^2 = \frac{S_1^2 n_1}{n_1 - 1} \qquad \sigma_2^2 \; \frac{S_2^2 n^2}{n_2 - 1}$$

$$\sigma_1^2 = \frac{(2.1)^2 \times 9}{9-1} = 4.96 \qquad \sigma_2^2 = \frac{(1.8)^2 \times 13}{13-1} = \frac{(1.8)^2 \times 13}{12} = 3.51$$

$$F_1 = \frac{\sigma_1^2}{\sigma_2^2} = \frac{4.96}{3.51} = 1.41$$

- *Critical value:* The table value of F at 0.05 for degrees of freedom 8 *i.e.*, 9 – 1 & 12 *i.e.*, 13 – 1 is 2.85.
- *Decision:* Since calculated value of F is 1.41 is less than the tabulated value at 0.05 for 8, 12 degrees of freedom. $F_{0.05}$ 1.41 ≤ $F_{0.05}$ 2.85, we accept the null hypothesis *Ho i.e.*, the sample may be drawn from normal population with same standard deviation.

Q.7. In a laboratory experiment two samples gave the following results.

Sample	Size	Sample mean	Sum of squares of deviation from the mean
1	10	15	90
2	12	14	108

Test the equality of sample variances at 5% level of significance.

Solution:

- *Null Hypothesis:* $H_0 = \sigma_1^2 = \sigma_2^2$
- *Alternative Hypothesis:* $\sigma_1^2 \neq \sigma_2^2$
- *Calculation:*

$$\text{Variances of sample } I \ (\sigma_1^2) = \frac{(X-\overline{X})^2}{n_1 - 1} = \frac{90}{10-1} = \frac{90}{9} = 10$$

$$\text{Variances of sample } II \ (\sigma_2^2) = \frac{(Y-\overline{Y})^2}{n_2 - 1} = \frac{108}{12-1} = \frac{108}{11} = 9.818 = 9.82$$

$$\text{Test statistic } F = \frac{\sigma_1^2}{\sigma_2^2} = \frac{10}{9.82} = 1.018$$

- *Critical value:* The table value of F at 0.05 for 10 – 1= 9 *df* & 12 – 1 = 11 *df* is 2.90.
- *Decision:* Since the computed value of F = 1.018 < table value of $F_{0.05}$ = 2.90. So the null hypothesis is accepted *i.e.*, the two population have the same variance.

Q.8. In a microbiology laboratory experiment two samples gave the following results.

Sample	Size	Sum of squares of deviation from the mean
1	08	84.4
2	10	102.6

Test whether this difference is significant at 5% level (Given that F for 7, 9 is 3.29).

Solution:

- *Null Hypothesis (Ho)*: $\sigma_1^2 = \sigma_2^2$ the sample is homogeneous.
- *Calculation:*

$n_1 = 8 \quad \sigma_1^2 = 84.4 \quad n_2 = 10 \quad \sigma_2^2 = 102.6$

$$\text{Variance of Sample (1) is} = \frac{84.4}{8-1} = \frac{84.4}{7} = 12.057$$

$$\text{Variance of Sample (2) is} = \frac{102.6}{10-1} = \frac{102.6}{9} = 11.4$$

Since $\sigma_1^2 > \sigma_2^2$

$$F = \frac{12.06}{11.4} = 1.057$$

- *Critical value:* The table value of F at 7 & 9 *df* is 3.29.
- *Decision:* Since calculated value of F 1.057 < table value of $F_{0.05}$ = 3.29 the null hypothesis is accepted.

Q.9. The following results were obtained from a laboratory of Pune from two samples.

Sample	*Size*	*Sample average*	*Standard deviation*
A	**10**	**2.01**	**0.050**
B	**10**	**1.96**	**0.031**

Does the result indicate they are identical (F = 9.9 *df* 3.18).

Solution:

- *Null Hypothesis* (*Ho*): The sample population were identical $Ho: \sigma_1^2 = \sigma_2^2$
- *Calculation:*

$A: n_1 = 10$ $\quad B: n_2 = 10$

$\bar{X} = 2.01$ $\quad \bar{Y} = 1.96$

$\sigma_A = 0.50$ $\quad \sigma_B = 0.031$

$$\sigma_1^2 A = \frac{n_1 \times \sigma_1^2}{n_1 - 1} = \frac{10 \times (0.050)^2}{10 - 1} = \frac{10 \times .0025}{9} = \frac{0.025}{9} = 0.0028$$

$$\sigma_2^2 B = \frac{n_1 \times \sigma_2^2}{n_2 - 1} = \frac{10 \times (0.031)^2}{10 - 1} = \frac{10 \times .00096}{9} = \frac{0.0096}{9} = 0.0010$$

Since $\sigma_2^2 B > \sigma_1^2 A$

$$F = \frac{0.0010}{0.0028} = 0.357 = 0.38$$

- *Critical value:* The tabulated value of F at 0.05 level is (9, 9) *df* is 3.18.
- *Decision:* Since the calculated value of F is 0.38 < table value of F *i.e.*, 3.18 at *df* (9, 9) so the null hypothesis is accepted.

Q.10. The following results were obtained from two independent samples (*Labeo rohita*).

Sample	*Sample size*	*Sample average*	*Standard deviation*
A	**6**	**29**	**4.0**
B	**5**	**25**	**2.1**

Test whether the two samples may be regarded as a drawn from the same pond of natural population. (Given $F_{0.05}$ at 5, 4 *df* is 6.26)

Solution:

- *Null Hypothesis*: (*Ho*): The samples are drawn from the same population $\sigma_1^2 = \sigma_1^2$
- *Calculation:*

Population A $\quad$ Population B

$n = 6$ $\quad \bar{X} = 29$ $\quad n = 5$ $\quad \bar{X} = 25$

$\sigma_1 = 4.0$ $\quad \sigma_2 = 2.1$

$$\text{Variances of } A \text{ i.e., } \sigma_1^2 = \frac{n \times (\sigma_1)^2}{n_1 - 1} = \frac{6 \times (4)^2}{6 - 1} = \frac{6 \times 16}{5} = \frac{96}{5} = 19.2$$

$$\text{Variances of } B \text{ i.e., } \sigma_2^2 = \frac{n \times (\sigma_2)^2}{n_2 - 1} = \frac{5 \times (2.1)^2}{5 - 1} = \frac{5 \times 4.41}{4} = \frac{22.05}{4} = 5.51$$

Since $\sigma_1^2 > \sigma_2^2$

$$F = \frac{\sigma_1^2}{\sigma_2^2} = \frac{19.2}{5.51} = 3.48$$

- *Critical value:* The tabulated value of F at 0.05 level in 5.4 *df* is 6.26.
- *Decision:* Since the calculated value of F is 3.48 < table value of F *i.e.,* 6.26 at *df* (5, 4) so the null hypothesis is accepted.

13

CHAPTER

CORRELATION

The statistical tool for measuring the degree of relationship between the two variables *i.e.*, a change in one variable results a positive or negative change in the other and also a greater change in one variable results in corresponding greater or smaller change in other variable is known as correlation.

- **Co-efficient of Correlation:**

The extent or degree of relationship between the two variables is measured in terms of another parameter called co-efficient of correlation.

It is denoted by 'r' *i.e.*, $-1 \leq r \leq 1$.

- **Properties of Co-efficient of Correlation.**

I. It is a measure of the closeness between the two variables.

II. It lies between -1 and $+1$ *i.e.*, $-1 \leq r \leq 1$.

III. The correlation is perfect and positive if $r = 1$ and it is perfect and negative if $r = -1$.

IV. If, $r = 0$ then there is no correlation between the two variables and said to be independent.

TYPES OF CORRELATION:

(*a*) Perfect Positive Correlation:

I. Here two variables denoted by letter X and Y are directly proportional and fully correlated with each other.

II. The correlation co-efficient $(r) = +1$ *i.e.*, both the variables increases and decreases in the same proportion.

III. Perfect correlation is not usually found in nature but approaching to that extent.

Example: Height and weight, age and weight up to a certain age.

IV. The graph forms a straight line originating from the lower ends of X and Y axes.

(*b*) Perfect Negative Correlation:

I. In this two variables (X & Y) are inversely proportional to each other *i.e.*, when one rises, the other falls in the same proportion.

II. Here correlation co-efficient $r = -1$.

III. It is not usually available in nature but some approaching to that extent.

Examples:

(*i*) Mean weekly temperature and intensity of cold in winter.

(*ii*) Pressure and volume of gas at a particular temperature.

IV. The graph will contain all the observations on a straight line starting from either of the extreme ends because one variable rises and other falls in a fixed proportion or $r = -1$.

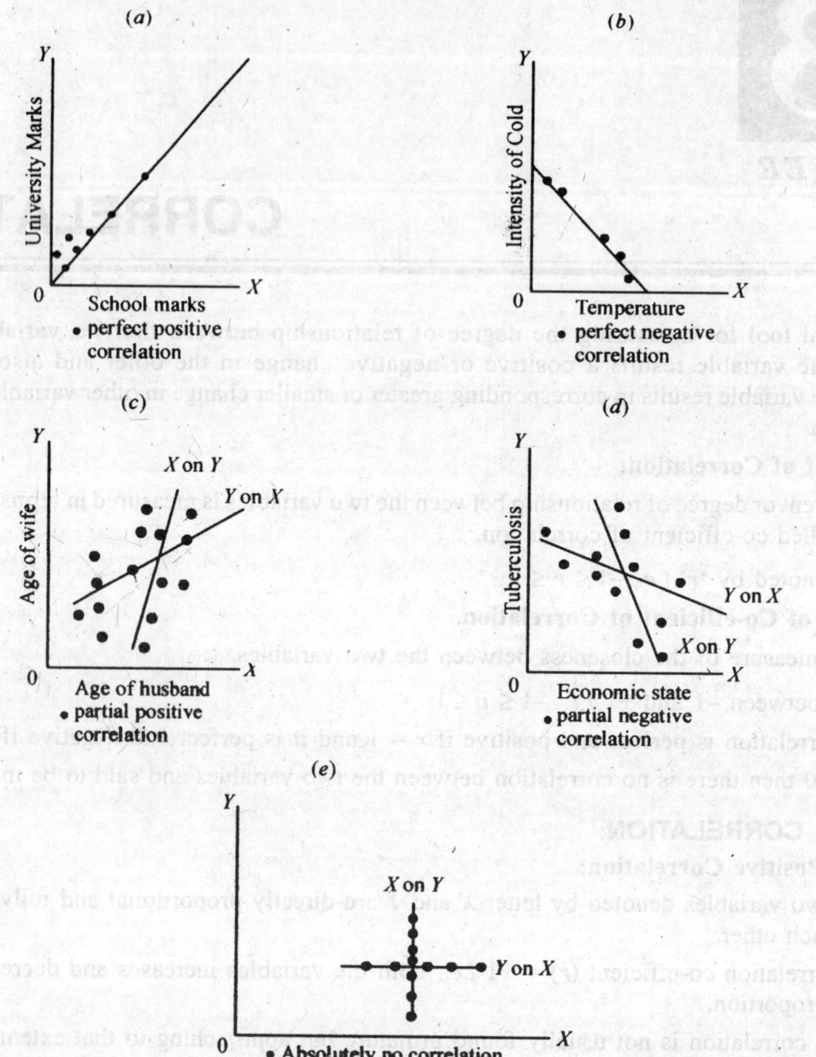

Fig. 13.1 The diagrams are taken from hypothetical numbers to show differnet types of correlation.

(c) Moderately (partial) Positive Correlation:

I. Here the non zero values of correlation co-efficient (r) lies between 0 and +1 *i.e.*, $0 < r < 1$.

Examples:

(*i*) Infant mortality rate and overcrowding.

(*ii*) Age of husband and age of wife.

(*iii*) Temperature and pulse rate.

II. Here scatter diagram will be, around an imaginary mean line rising from lower extreme values of both variables.

(d) Moderately (partial) Negative Correlation:

I. In this case, the non zero values of correlation co-efficient lies between −1 and 0 *i.e.*, $-1 < r < 0$.

Examples:

(*i*) Income and infant mortality rate.

(*ii*) Age and vital capacity in adult.

II. In this case the scatter diagram will be of the same type but the mean imaginary line will rise from extreme values of one variable.

(*e*) Absolutely No Correlation:

I. Here the value of correlation coefficient is zero, indicating that no linear relationship exists between the two variables.

II. There is no mean or imaginary line indicating trend of correlation.

III. Here the variable '*X*' is completely independent of variable '*Y*'.

Example: Height and pulse rate.

IV. Here the points are scattered, so that no imaginary line can be drawn, the correlation will be zero.

(*f*) Perfect and imperfect Correlation:

I. In perfect correlation, the dots lie exactly on a straight line.

II. Here the changes in the corresponding values of the two variables are proportional directly or inversely. The degree of imperfect correlation lies between perfect correlation and no or zero correlation.

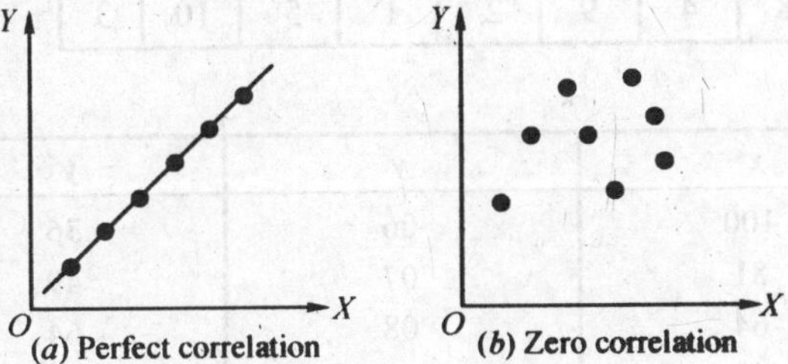

Fig. 13.2 (*a* & *b*) Showing (*a*) Perfect corrélator & (*b*) Zero correlation.

- **Simple and multiple Correlation:**

I. It deals with the study of two variables.

II. The relationship between this two variable is called simple correlation.

Example: Trunk length and wing length in a sample of cockroach.

In multiple correlations we study more than two variables simultaneously.

Example: Oxygen consumption and the tracheal ventilation & atmospheric oxygen tension.

- **Partial and Total Correlation:**

The study of two variables partialling out (excluding) some other variables is called partial correlation.

Example: Oxygen consumption and tracheal veritilation excluding partialling out atmospheric oxygen tension.

In total correlation all the facts are taken into account.

- **Liner and non linear correlation:**

If the ratio of change between two variables is uniform, then there will be linear correlation between them.

Example: Body weight and gill weight in a sample of fishes.

Here the amount of change in one variable does not bear constant ratio of the amount of change in the other variables.

Example: Initial velocity of an enzyme action and the corresponding substrate concentration, it forms a rectangular hyperbola.

- **Methods of Studying Correlation:**
  1. Karl Pearson's coefficient of correlation.
  2. Rank correlation (*Spearman's and Kendall's coefficient*).
  3. Scatter diagram.
- **Computation of 'r' without using deviation from mean:**

When the values of variables (x_i & y_i) are small the value of 'r' can be made directly without using deviations from means.

$$r = \frac{\sum xy - \frac{\sum x \sum y}{n}}{\sqrt{\sum x^2 - \frac{(\sum x)^2}{n}} \times \sqrt{\sum y^2 - \frac{(\sum y)^2}{n}}}$$

Example 1. Calculate the coefficient of correlation between x and y.

x	10	9	8	7	6	5	4	3	2	1
y	6	7	8	4	9	2	1	5	10	3

Solution:

x	x^2	y	y^2	xy
10	100	06	36	60
09	81	07	49	63
08	64	08	64	64
07	49	04	16	28
06	36	09	81	54
05	25	02	04	10
04	16	01	01	04
03	09	05	25	15
02	04	10	100	20
01	01	03	09	03
55	385	55	385	321

$\sum x = 55 \quad \sum x^2 = 385 \quad \sum y = 55 \quad \sum y^2 = 385 \quad \sum xy = 321 \quad n = 10$

$$r = \frac{321 - \frac{55 \times 55}{10}}{\sqrt{385 - \frac{(55)^2}{10}} \times \sqrt{385 - \frac{(55)^2}{10}}} = \frac{321 - 302.5}{\sqrt{385 - 302.5} \times \sqrt{385 - 302.5}}$$

$$= \frac{18.5}{\sqrt{82.5 \times 82.5}} = \frac{18.5}{82.5} = 0.224$$

Example 2. The following data refers to recombination values obtained from the two regions *viz.* R_1 & R_2 on a chromosome of six different strains (subspecies) of *Anopheles* mosquitoes.

Strain:	**1**	**2**	**3**	**4**	**5**	**6**
Region 1 (R_1):	**36.7**	**20.0**	**18.3**	**34.5**	**20.8**	**30.8**
Region 2 (R_2):	**16.2**	**24.3**	**32.3**	**10.4**	**36.6**	**15.4**

Comment on the result.

Solution:

	$R_1(X)$	X^2	Y	Y^2	XY
1	36.7	1346.89	16.2	262.44	594.54
2	20.0	400.00	24.3	590.49	486.00
3	18.3	334.89	32.3	1043.29	591.09
4	34.5	1190.25	10.4	108.16	358.80
5	20.8	432.64	36.6	1339.56	761.28
6	30.8	984.64	15.4	237.16	474.32
	161.1	4653.31	135.2	3581.1	3266.03

$$r = \frac{\sum XY - \frac{\sum X \sum Y}{n}}{\sqrt{\sum X^2 - \frac{\left(\sum X\right)^2}{n}} \times \sqrt{\sum Y^2 - \frac{\left(\sum Y\right)^2}{n}}}$$

$$= \frac{3266.03 - \frac{161.1 \times 135.2}{6}}{\sqrt{4653.31 - \frac{(161.1)^2}{6}} \times \sqrt{3581.1 - \frac{(135.2)^2}{6}}}$$

$$= \frac{3266.03 - \frac{21780.72}{6}}{\sqrt{4653.31 - \frac{25953.21}{6}} \times \sqrt{3581.1 - \frac{18279.04}{6}}}$$

$$= \frac{3266.03 - 3630.12}{\sqrt{(4653.31 - 4325.53) \times (3581.1 - 3046.50)}} = \frac{-364.09}{\sqrt{327.78 \times 534.6}}$$

$$= \frac{-364.09}{\sqrt{175231.18}} = \frac{-364.09}{418.606} = -0.869 = -0.87$$

This negative correlation means that high recombination value in one region results in lower recombination value in the adjacent region & vice versa.

- **Computation of 'r' directly by using deviation from mean**

Direct Method: If X and Y are two variates having their means $\overline{X}$ and $\overline{Y}$ respectively. then

$$r(XY) = \frac{\sum dxdy}{\sqrt{\sum dx^2 \times \sum dy^2}}$$

where $dx = x_1 - \overline{x}$

$dy = y_1 - \overline{y}$

$dx^2 = \left(x_1 - \overline{x}\right)^2$

$dy^2 = (y_1 - \overline{y})^2$

It also can be written as

$$r_{xy} = \frac{\sum dxdy}{n\sigma_x \times \sigma_y}$$ [When the deviation of the item is taken from the actual mean]

where n = number of observation in x or y series

σ_x = standard deviation of x

σ_y = standard deviation of y

Working Procedure:

I. Denote one series by X and the other series by Y.

II. Calculate $\bar{X}$ (mean) and $\bar{Y}$ mean of the x and y series respectively.

III. Take the deviations of the observations in X series from $\bar{X}$ and written as $\sigma_x = X - \bar{X}$. Take the deviations of the observations in Y series from $\bar{Y}$ and written it under the column headed by $\sigma_y = Y - \bar{Y}$.

IV. Square the deviations and written them under the columns headed by dx^2 and dy^2

V. Multiply the respective dx and dy and write it under the column headed by $dx\ dy$.

VI. Apply the following formula to calculate r or r_{xy} (the coefficient of correlation)

$$r = \frac{\sum dxdy}{\sqrt{\sum dx^2 \times \sum dy^2}} \text{ or } r = \frac{\sum dxdy}{n\sigma_x \times \sigma_y}$$

n = number of observation

σ_x = S.D. of X

σ_y = S.D. of Y

Significance of r:

(*i*) The computed r can be transformed into t score for interpretation by using standard error (Sr) of r.

(*ii*) The degrees of freedom df of computed t should be taken as $n - 2$.

$$r = Sr = \sqrt{\frac{t - r^2}{n - 2}} \quad t = \frac{r}{Sr} \quad df = n - 2$$

(*iii*) The computed r is considered significant at or below the level of significance whose critical t either equal or lower than computed $t\ (p \le \alpha)$.

Example 1. Find the coefficient correlation between the heights of fathers and daughters both from the following Ganguli family members.

Height of father (in cm)	64	65	66	67	68	69	70
Height of daughters (in cm)	66	67	68	69	70	71	72

Solution: Let the height of father denoted by X & height of daughters by Y, then

$$\bar{X} = \frac{64 + 65 + 66 + 67 + 68 + 69 + 70}{7} = \frac{469}{7} = 67$$

$$\bar{Y} = \frac{66 + 67 + 68 + 69 + 70 + 71 + 72}{7} = \frac{483}{7} = 69$$

Let us prepare the table.

Height of father (X)	dx X – 67	dx^2 $(X-67)^2$	Height of daughter (Y)	dy Y – 69	dy^2 $(Y-69)^2$	dxdy
64	64 – 67 = –3	9	66	66 – 69 = –3	9	9
65	65 – 67 = –2	4	67	67 – 69 = –2	4	4
66	66 – 67 = –1	1	68	68 – 69 = –1	1	1
67	67 – 67 = 0	0	69	69 – 69 = 0	0	0
68	68 – 67 = +1	1	70	70 – 69 = +1	1	1
69	69 – 67 = +2	4	71	71 – 69 = +2	4	4
70	70 – 67 = +3	9	72	72 – 69 = +3	9	9
$\overline{x} = 67$	$\sum dx = 0$	$\sum dx^2 = 28$	$\overline{y} = 69$	$\sum dy = 0$	$\sum dy^2 = 28$	$\sum dx \sum dy = 28$

Now

$$r = \frac{\sum dxdy}{\sqrt{\sum dx^2 \times \sum dy^2}}$$

$$= \frac{28}{\sqrt{28 \times 28}}$$

$$= \frac{28}{28} = 1$$

Example 2. Calculate the correlation co-efficient between *X* and *Y* from the following data.

X	5	9	13	17	21
Y	12	20	25	33	35

Solution: Let us prepare the table.

X	dx X – 13	dx^2 $(X-13)^2$	Y	dy Y – 25	dy^2 $(Y-25)^2$	dxdy
05	5 – 13 = – 8	64	12	12 – 25 = – 13	169	104
09	9 – 13 = – 4	16	20	20 – 25 = – 5	25	20
13	13 – 13 = 0	0	25	25 – 25 = 0	0	0
17	17 – 13 = + 4	16	33	33 – 25 = + 8	64	32
21	21 – 13 = + 8	64	35	35 – 25 = + 10	100	80
65 $\overline{X} = \frac{65}{5} = 13$	$\sum dx = 0$	$\sum dx^2 = 160$	125 $\overline{Y} = \frac{125}{5} = 25$	$\sum dy = 0$	$\sum dy^2 = 358$	$\sum dxdy = 236$

Now

$$r = \frac{\sum dxdy}{\sqrt{\sum dx^2 \times \sum dy^2}}$$

$$= \frac{236}{\sqrt{160 \times 358}} = \frac{236}{\sqrt{57280}}$$

$$= \frac{236}{239.33} = 0.986\,(approx)$$

Example 3. Calculate correlation co-efficient between X & Y for the following data.

X	1	2	3	4	5	6	7	8	9
Y	10	11	12	14	13	15	16	17	18

Solution: Let us prepare the table.

X	dx X – 5	dx^2 $(X-5)^2$	Y	dy Y – 14	dy^2 $(Y-14)^2$	dxdy
1	1 – 5 = – 4	16	10	10 – 14 = – 4	16	16
2	2 – 5 = – 3	9	11	11 – 14 = – 3	9	9
3	3 – 5 = – 2	4	12	12 – 14 = – 2	4	4
4	4 – 5 = – 1	1	14	14 – 14 = 0	0	0
5	5 – 5 = 0	0	13	13 – 14 = – 1	1	0
6	6 – 5 = + 1	1	15	15 – 14 = + 1	1	1
7	7 – 5 = + 2	4	16	16 – 14 = + 2	4	4
8	8 – 5 = + 3	9	17	17 – 14 = + 3	9	9
9	9 – 5 = + 4	16	18	18 – 14 = + 4	16	16
$\sum x = 45$ $\overline{x} = 5$	$\sum dx = 0$	$\sum dx^2$ $= 60$	$\sum y = 126$ $\overline{y} = 14$	$\sum dy = 0$	$\sum y^2$ $= 60$	$\sum dxdy$ $= 59$

$$r = \frac{\sum dxdy}{\sqrt{\sum dx^2 \times \sum dy^2}}$$

$$= \frac{59}{\sqrt{60 \times 60}} = \frac{59}{60} = .98\,(approx)$$

- **Short Cut Method or Assume Mean Method:**

I. When the terms of series 'x' and 'y' are big and calculation of $\overline{X}$ (mean) and Y (mean) becomes difficult.

II. The means of X and Y are not integers.

The following formula is applied.

$$r_{xy} = \frac{\sum dxdy - \left(\frac{\sum dx \sum dy}{n}\right)}{\sqrt{\sum dx^2 - \frac{\left(\sum dx\right)^2}{n}} \times \sqrt{\sum dy^2 - \frac{\left(\sum dy\right)^2}{n}}}$$

Where $dx = x - a$

$dy = y - b$

a = assumed mean of x series.

b = assumed mean of y series.

n = number of observation of x or y.

- **Working Procedure:**

I. Take any term 'a' (Preferably the middle one) of X series as assumed mean and any term 'b' (*preferably middle one*) as assumed mean of Y.

II. Take deviations of the observation in 'X' series from 'a' *i.e.,* $dx = X - a$.

Take deviations of the observation in 'Y' series from 'b' *i.e.,* $dy = Y - b$.

There are written under the columns *dx* and *dy* respectively.

III. Square the deviations and write them under the columns headed by dx^2 and dy^2.

IV. Multiply the respective '*dx*' and '*dy*' and write it under the column *dx.dy*.

V. Apply the following formula to calculate *r* or r_{xy} (correlation coefficient).

$$r = \frac{\sum dxdy - \left(\frac{\sum dx \sum dy}{n}\right)}{\sqrt{\sum dx^2 - \frac{\left(\sum dx\right)^2}{n}} \times \sqrt{\sum dy - \frac{\left(\sum dy\right)^2}{n}}}$$

Example: Calculate coefficient-correlation between *X* and *Y* for the following data.

X	1	3	4	5	7	8	10
Y	2	6	8	10	14	16	20

Solution: Let 5 be the assumed mean for the values of *x* and 14 be the assumed for the values of *y*.

X	*dx* *X* – 5	dx^2 $(X-5)^2$	*Y*	*dy* *Y* – 14	dy^2 $(Y-14)^2$	*dxdy*
1	1 – 5 = – 4	16	2	2 – 14 = – 12	144	48
3	3 – 5 = – 2	4	6	6 – 14 = – 8	64	16
4	4 – 5 = – 1	1	8	8 – 14 = – 6	36	6
5	5 – 5 = 0	0	10	10 – 14 = – 4	16	0
7	7 – 5 = + 2	4	14	14 – 14 = 0	0	0
8	8 – 5 = + 3	9	16	16 – 14 = + 2	4	6
10	10 – 5 = + 5	25	20	20 – 14 = + 6	36	30
	$\sum dx = 3$	$\sum dx^2 = 59$		$\sum dy = -30 + 8 = -22$	$\sum dy^2 = 300$	$\sum dxdy = 106$

$$_{xy} = \frac{\sum dxdy - \left(\frac{\sum dx \sum dy}{n}\right)}{\sqrt{\sum dx^2 - \frac{\left(\sum dx\right)^2}{n}} \times \sqrt{\sum dy^2 - \frac{\left(\sum dy\right)^2}{n}}}$$

$$= \frac{106 - \left(\frac{3.-22}{7}\right)}{\sqrt{59 - \frac{(3)^2}{7}} \times \sqrt{300 - \frac{(-22)^2}{7}}} = \frac{106 + \frac{66}{7}}{\sqrt{59 - \frac{9}{7}} \times \sqrt{300 - \frac{484}{7}}}$$

$$= \frac{\frac{742 + 66}{7}}{\sqrt{\frac{413 - 9}{7}} \times \sqrt{\frac{2100 - 484}{7}}} = \frac{\frac{808}{7}}{\sqrt{\frac{404}{7} \times \frac{1616}{7}}} = \frac{\frac{808}{7}}{\frac{808}{7}} = 1$$

• Covariance Method or Product Moment Method

The most widely used mathematical method of measuring correlation is "Pearson co-efficient correlation" due to *Karl Pearson.*

r = Product moment correlation co-efficient.

$$= \frac{\text{covariance of } x \text{ and } y}{(S.D\, of\, X)(S.D\, of\, Y)} = \frac{\sum(X-\bar{X})(Y-\bar{Y})}{N\sigma_x\sigma_y}$$

$$\left[\text{cov.}(X.Y) = \frac{\sum(X-\bar{X})(Y-\bar{Y})}{N}\right]$$

σ_X = S.D. of X
σ_Y = S.D. of Y
N = Number of observation

Example 1. Marks of 10 students in Mathematics and Statistics are given below.

Mathematics (X)	32	38	48	43	40	22	41	69	35	64
Statistics (Y)	30	31	38	43	33	11	27	76	40	59

Calculate correlation co-efficient between X & Y using product moment formula.

Solution: Let 43 be the assumed mean for the values of X & 40 be the assumed mean for the values of Y.

	X	dx X – 43	dx^2 $(X-43)^2$	Y	dy Y – 40	dy^2 $(Y-40)^2$	dxdy
1	32	32 – 43 = – 11	121	30	30 – 40 = – 10	100	110
2	38	38 – 43 = – 5	25	31	31 – 40 = – 9	81	45
3	48	48 – 43 = + 5	25	38	38 – 40 = – 2	04	–10
4	43	43 – 43 = 0	0	43	43 – 40 = + 3	09	0
5	40	40 – 43 = – 3	9	33	33 – 40 = – 7	49	21
6	22	22 – 43 = – 21	441	11	11 – 40 = – 29	841	609
7	41	41 – 43 = – 2	4	27	27 – 40 = – 13	169	26
8	69	69 – 43 = + 26	676	76	76 – 40 = + 36	1296	936
9	35	35 – 43 = – 8	64	40	40 – 40 = 0	0	0
10	64	64 – 43 = + 21	441	59	59 – 40 = + 19	361	399
		$\Sigma dx = 2$	$\Sigma dx^2 = 1806$		$\Sigma dy = -12$	$\Sigma dy^2 = 2910$	2136

$$r = \frac{\sum(X-\bar{X})(Y-\bar{Y})}{N\sigma_x\sigma_y}.$$ It may be written as

$$r = \frac{n\sum dxdy - (\sum dx)(\sum dy)}{\sqrt{n\sum dx^2 - \left(\sum dx\right)^2} \times \sqrt{n\sum dy^2 - \left(\sum dy\right)^2}} \quad (Proved)$$

Here $\sum dx = 2 \quad \sum dy = -12 \quad n = 10$

$\sum dx^2 = 1806 \quad \sum dy^2 = 2910 \quad \sum dxdy = 2136$

$$r = \frac{10.2136 - 2.(-12)}{\sqrt{10.1806 - (2)^2} \times \sqrt{10.2910 - (-12)^2}}$$

$$= \frac{10.2136 + 24}{\sqrt{10.1806 - 4} \times \sqrt{10.2910 - 144}}$$

$$= \frac{21360 + 24}{\sqrt{18060 - 4}\sqrt{29100 - 144}} = \frac{21384}{\sqrt{18056}\sqrt{28956}}$$

$$= \frac{21384}{134.3 \times 170.16} = \frac{21384}{22852} = 0.935\,(approx)$$

Example 2. Calculate Pearson's coefficient of correlation from the following data and interpret the result.

A	**104**	**111**	**104**	**114**	**118**	**117**	**105**	**108**	**106**	**100**	**104**	**105**
B	**57**	**55**	**47**	**45**	**45**	**50**	**64**	**63**	**66**	**62**	**69**	**61**

Solution:

A(X)	$X - \bar{X} = dx$ $\bar{X} = 100$	dx^2	B(Y)	$Y - \bar{Y} = dy$ $\bar{Y} = 50$	dy^2	$dx \times dy$
104	104 – 100 = 4	16	57	57 – 50 = 7	49	4 × 7 = 28
111	111 – 100 = 11	121	55	55 – 50 = 5	25	11 × 5 = 55
104	104 – 100 = 4	16	47	47 – 50 = –3	09	4 × –3 = –12
114	114 – 100 = 14	196	45	45 – 50 = –5	25	14 × –5 = –70
118	118 – 100 = 18	324	45	45 – 50 – –5	25	18 × –5 = –90
117	117 – 100 = 17	289	50	50 – 50 = 0	00	17 × 0 = 0
105	105 – 100 = 5	25	64	64 – 50 = 14	196	5 × 14 = 70
108	108 – 100 = 8	64	63	63 – 50 = 13	169	6 × 16 = 96
106	106 – 100 = 6	36	66	66 – 50 = 16	256	0 × 12 = 0
100	100 – 100 = 0	00	62	62 – 50 = 12	144	4 × 19 = 76
104	104 – 100 = 4	16	69	69 – 50 = 19	361	5 × 11 = 55
105	105 – 100 = 5	25	61	61 – 50 = 11	121	
	$\sum dx = 96$	$\sum dx^2 = 1128$		$\sum dy = 84$	$\sum dy^2 = 1380$	$\sum dxdy = 312$

$$r_{xy} = \frac{\sum dxdy - \left(\frac{\sum dx \sum dy}{n}\right)}{\sqrt{\sum dx^2 - \frac{\left(\sum dx\right)^2}{n}} \times \sqrt{\sum dy^2 - \frac{\left(\sum dy\right)^2}{n}}}$$

$\sum dx = 96$ $\quad \sum dx^2 = 1128$

$\sum dy = 84$ $\quad \sum dy^2 = 1380$

$\sum dxdy = 312$

$$r_{xy} = \frac{312 - \left(\frac{96 \times 84}{12}\right)}{\sqrt{1128 - \frac{(96)^2}{12}} \times \sqrt{1380 - \frac{(84)^2}{12}}} = \frac{312 - \frac{8064}{12}}{\sqrt{1128 - 768} \times \sqrt{1380 - 588}}$$

$$r_{xy} = \frac{312 - 672}{\sqrt{360} \times \sqrt{792}} = \frac{-360}{\sqrt{285120}} = \frac{-360}{533.96} = 0.6742 = 0.67\,(approx)$$

The result *i.e.,* the value of correlation coefficient indicates that there is negative correlation between the two variables.

Example 3. Ten students of Zoology of Serampore college obtained marks in Taxonomy and Animal physiology are given below. Calculate Karl Pearson's correlation coefficient between the marks in Taxonomy and Animal physiology.

Marks in Taxonomy	10	25	13	25	22	11	12	25	21	20
Names in Animal Physiology	12	22	16	15	18	18	17	23	24	17

Solution:

Marks in Taxonomy (X)	Assume mean 18 $X - \bar{X} = dx$	dx^2	Marks in Animal Phsiology (Y)	Assume mean 18 $Y - \bar{Y} = dy$	dy^2	dx dy
10	10 – 18 = – 8	64	12	12 – 18 = – 6	36	– 8 × 6 = 48
25	25 – 18 = + 7	49	22	22 – 18 = + 4	16	7 × 4 = 28
13	13 – 18 = – 5	25	16	16 – 18 = – 2	04	– 5 × – 2 = 10
25	25 – 18 = + 7	49	15	15 – 18 = – 3	09	7 × – 3 = – 21
22	22 – 18 = + 4	16	18	18 – 18 = 0	00	4 × 0 = 0
11	11 – 18 = – 7	49	18	18 – 18 = 0	00	–7 × 0 = 0
12	12 – 18 = – 6	36	17	17 – 18 = – 1	01	–6 × –1 = 6
25	25 – 18 = + 7	49	23	23 – 18 = + 5	25	7 × 5 = 35
21	21 – 18 = + 3	09	24	24 – 18 = + 6	36	3 × 6 = 18
20	20 – 18 = + 2	04	17	17 – 18 = – 1	01	2 × –1 = –2
184	$\sum dx = 4$	350	182	$\sum dy = 2$	$\sum dy^2 = 128$	$\sum dxdy = 122$

$$r_{xy} = \frac{\sum dxdy - \left(\frac{\sum dx \sum dy}{n}\right)}{\sqrt{\sum dx^2 - \frac{\left(\sum dx\right)^2}{n}} \times \sqrt{\sum dy^2 - \frac{\left(\sum dy\right)^2}{n}}}$$

$\sum dx = 4$ $\quad \sum dx^2 = 350$

$\sum dy = 2$ $\quad \sum dy^2 = 128$

$\sum dxdy = 122$ $\quad n = 10$

$$= \frac{122 - \left(\frac{4 \times 2}{10}\right)}{\sqrt{350 - \frac{(4)^2}{10}} \times \sqrt{128 - \frac{(2)^2}{10}}} = \frac{122 - 0.8}{\sqrt{350 - 1.6} \times \sqrt{128 - 0.4}}$$

$$= \frac{121.2}{\sqrt{340.4} \times \sqrt{127.6}} = \frac{121.2}{\sqrt{44455.84}} = \frac{121.2}{210.8455}$$

$$= \frac{121.2}{210.85} = 0.5748$$

Example 4. Calculate Karl Pearson's co-efficient of correlation from the following data.

Serial No. of Student:	1	2	3	4	5	6	7	8	9	10
Marks in Zoology:	15	18	21	24	27	30	36	39	42	48
Marks in Chemistry:	25	25	27	27	31	33	35	41	41	45

Solution:

Sl. No.	Marks in Zoology (X)	$X - \bar{X} = dx$	dx^2	Marks in Chemistry (Y)	$Y - \bar{Y} = dy$	dy^2	$dx.dy$
1	15	15 – 30 = – 15	225	25	25 – 33 = – 8	64	120
2	18	18 – 30 = – 12	144	25	25 – 33 = – 8	64	96
3	21	21 – 30 = – 09	81	27	27 – 33 = – 6	36	54
4	24	24 – 30 = – 06	36	27	27 – 33 = – 6	36	36
5	27	27 – 30 = – 03	09	31	31 – 33 = – 2	04	06
6	30	30 – 30 = 00	00	33	33 – 33 = 00	00	00
7	36	36 – 30 = + 06	36	35	35 – 33 = + 02	04	12
8	39	39 – 30 = + 09	81	41	41 – 33 = + 08	64	72
9	42	42 – 30 = + 12	144	41	41 – 33 = + 08	64	96
10	48	48 – 30 = + 18	324	45	45 – 33 = + 12	144	216
	$\sum x = 300$		$\sum dx^2 = 1080$	330		$\sum dy^2 = 480$	708

$$\bar{X} = \frac{\sum x}{N} = \frac{300}{10} = 30 \quad \bar{Y} = \frac{\sum y}{N} = \frac{330}{10} = 33 \quad \sum dxdy = 708$$

$$\therefore \quad r(xy) = \frac{\sum dxdy}{\sqrt{\sum dx^2 \times \sum dy^2}} = \frac{708}{\sqrt{1080 \times 480}}$$

$$= \frac{708}{\sqrt{9 \times 12 \times 48 \times 100}} = \frac{708}{\sqrt{9 \times 3 \times 4 \times 3 \times 16 \times 10^2}}$$

$$= \frac{708}{\sqrt{9^2 \times 8^2 \times 10^2}} = \frac{708}{9 \times 8 \times 10} = \frac{708}{720} = 0.9833 = 0.98$$

Example 5. The rainfall and the output of wheat per acre for a farm under Kalyani agricultural institute was as follows.

Rain fall (in cms):	40	20	32	35	40	45	43	30	25	50
Wheat production (in quintals):	120	120	145	150	100	120	120	135	130	140

Find the correlation coefficient between the rainfall and wheat production.

Solution:

Rain fall (X)	$X - \bar{X} = dx$	dx^2	Wheat production (Y)	$Y - \bar{Y} = dy$	dy^2	$dx.dy$
40	40 – 36 = + 04	16	120	120 – 130 = – 10	100	– 40
20	20 – 36 = – 16	256	120	120 – 130 = – 10	100	160
32	32 – 36 = – 04	16	145	145 – 130 = + 15	225	– 60
35	35 – 36 = – 01	01	150	150 – 130 = + 20	400	– 20
40	40 – 36 = + 04	16	100	100 – 130 = – 30	900	– 120
45	45 – 36 = + 09	81	120	120 – 130 = – 10	100	– 90
43	43 – 36 = + 07	49	120	120 – 130 = – 10	100	– 70
30	30 – 36 = – 06	36	155	155 – 130 = + 25	625	– 150
25	25 – 36 = – 11	121	130	130 – 130 = 00	00	00
50	50 – 36 = + 14	196	140	140 – 130 = + 10	100	140
$\sum x = 360$		$\sum dx^2 = 788$	$\sum y = 1300$		$\sum dy^2 = 2650$	–250

$$\bar{X} = \frac{360}{10} = 36 \quad \bar{Y} = \frac{1300}{10} = 130$$

$$r(xy) = \frac{\sum dxdy}{\sqrt{\sum dx^2 \times \sum dy^2}} = \frac{-250}{\sqrt{788 \times 2650}}$$

$$= \frac{-250}{\sqrt{2088200}} = \frac{-250}{1445} = -0.173$$

Example 6. Find out the Karl Pearson's coefficient of correlation of the following data.

Runs of Team A:	14	19	21	26	22	15	20	19	24
Runs of Team B:	31	36	37	50	45	33	41	39	48

Solution:

Team A Runs (X)	$X - \bar{X} = dx$	dx^2	Team B Runs (Y)	$Y - \bar{Y} = dy$	dy^2	dx.dy
14	14 – 20 = – 06	36	31	31 – 40 = – 09	81	54
19	19 – 20 = – 01	01	36	36 – 40 = – 04	16	04
21	21 – 20 = + 01	01	37	37 – 40 = – 03	09	03
26	26 – 20 = + 06	36	50	50 – 40 = + 10	100	60
22	22 – 20 = + 02	04	45	45 – 40 = + 05	25	10
15	15 – 20 = – 05	25	33	33 – 40 = – 07	49	35
20	20 – 20 = 00	00	41	41 – 40 = + 01	01	00
19	19 – 20 = – 01	01	39	39 – 40 = – 01	01	01
24	24 – 20 = + 04	16	48	48 – 40 = + 08	64	32
$\sum x = 180$		$\sum dx^2 = 120$	$\sum y = 360$		$\sum dy^2 = 346$	$\sum dxdy = 193$

$$\sum X = 180 \qquad \sum y = 360$$

$$\bar{X} = \frac{180}{9} = 20 \quad \bar{Y} = \frac{360}{9} = 40 \quad n = 9$$

$$\sum dx^2 = 120 \; \sum dy^2 = 346 \; \sum dxdy = 193$$

$$r_{xy} = \frac{\sum dxdy}{\sqrt{\sum dx^2 \times \sum dy^2}} = \frac{193}{\sqrt{120 \times 346}}$$

$$= \frac{193}{\sqrt{41520}} = \frac{193}{203.76} = 0.947 = 0.95$$

Example 7. Find the coefficient of correlation between the height of the brothers and sisters from the following Banerjee family members.

Heights of brother (X) (in cm):	65	66	67	68	69	70	71
Heights of sisters (Y) (in cm):	67	68	60	69	72	72	69

Solution:

Height of brother (X)	$\bar{X}$ = 68 X − $\bar{X}$ = dx	dx^2	Height of sisters (Y)	$\bar{Y}$ = 69 Y − $\bar{Y}$ = dy	dy^2	dx.dy
65	65 − 68 = − 03	09	67	67 − 69 = − 02	04	06
66	66 − 68 = − 02	04	68	68 − 69 = − 01	01	02
67	67 − 68 = − 01	01	66	66 − 69 = − 03	09	03
68	68 − 68 = 00	00	69	69 − 69 = 00	00	00
69	69 − 68 = + 01	01	72	72 − 69 = + 03	09	03
70	70 − 68 = − 02	04	72	72 − 69 = − 03	09	06
71	71 − 68 = + 03	09	69	69 − 69 = 00	00	00
$\sum x = 476$		$\sum dx^2 = 28$	$\sum y = 483$		$\sum dy^2 = 32$	$\sum dxdy = 20$

$$\sum X = 476 \quad n = 7 \quad \sum Y = 483$$

$$\bar{X} = \frac{476}{7} = 68 \quad \bar{Y} = \frac{483}{7} = 69$$

$$\sum dx^2 = 28 \quad \sum dy^2 - 32 \quad \sum dxdy = 20$$

$$r_{xy} = \frac{\sum dxdy}{\sqrt{\sum dx^2 \times \sum dy^2}} = \frac{20}{\sqrt{28 \times 32}}$$

$$= \frac{20}{\sqrt{896}} = \frac{20}{29.9} = 0.6688 = 0.67$$

Example 8. Calculate the Pearson's coefficient correlation between the ages of husband & wife of some tribal members.

Age of husband (X):	**38**	**34**	**35**	**20**	**40**	**43**	**56**
Age of wife (Y):	**33**	**30**	**32**	**20**	**31**	**32**	**53**

Solution:

Age of husband (X)	$\bar{X}$ = 38 X − $\bar{X}$ = dx	dx^2	Age of wife (Y)	$\bar{Y}$ = 33 Y − $\bar{Y}$ = dy	dy^2	dx.dy
38	38 − 38 = 00	00	33	33 − 33 = 00	00	00
34	34 − 38 = − 04	16	30	30 − 33 = − 03	09	12
35	35 − 38 = − 03	09	32	32 − 33 = − 01	01	03
20	20 − 38 = − 18	324	20	20 − 33 = − 13	169	234
40	40 − 38 = + 02	04	31	31 − 33 = − 02	04	−04
43	43 − 38 = + 05	25	32	32 − 33 = − 01	01	−05
56	56 − 38 = + 18	324	53	53 − 33 = + 20	400	360
$\sum x = 266$		$\sum dx^2 = 702$	$\sum y = 231$		$\sum dy^2 = 584$	$\sum dxdy = 600$

$$\bar{X} = \frac{266}{7} = 38 \quad \bar{Y} = \frac{231}{7} = 33$$

$$\sum dx^2 = 702 \sum dy^2 = 584 \quad \sum dxdy = 600$$

$$r_{xy} = \frac{\sum dxdy}{\sqrt{\sum dx^2 \times \sum dy^2}} = \frac{600}{\sqrt{702 \times 584}}$$

$$= \frac{600}{\sqrt{409968}} = \frac{600}{640.287} = 0.937$$

$$\therefore \quad r = 0.937$$

Example 9. Find out the Pearson's correlation coefficient between shell height and shell breadth of the snails *Physa acuta*.

Serial No.	1	2	3	4	5	6	7	8	9
Shell height (in mm):	9.1	9.2	9.5	9.7	5.8	6.9	7.0	5.5	9.3
Shell breadth (in mm):	2.5	3.0	3.6	3.5	2.4	2.7	3.0	2.3	4.0

Comment on the computed correlation coefficient value.

Solution:

Sl. No.	Shell height (X)	$\bar{X} = 8$ $X - \bar{X} = dx$	dx^2	Shell breadth (Y)	$\bar{Y} = 3$ $Y - \bar{Y} = dy$	dy^2	dx.dy
1	9.1	9.1 – 8 = 1.1	1.21	2.5	2.5 – 3 = –0.5	0.25	–.55
2	9.2	9.2 – 8 = 1.2	1.44	3.0	3.0 – 3 = 0	00	00
3	9.5	9.5 – 8 = 1.5	2.25	3.6	3.6 – 3 = 0.6	0.36	.90
4	9.7	9.7 – 8 = 1.7	2.89	3.5	3.5 – 3 = 0.5	0.25	.85
5	5.8	5.8 – 8 = – 2.2	4.84	2.4	2.4 – 3 = –0.6	0.36	1.32
6	6.9	6.9 – 8 = – 1.1	1.21	2.7	2.7 – 3 = – 0.3	0.09	.33
7	7.0	7.0 – 8 = – 1.0	1.00	3.0	3.0 – 3 = 00	00	00
8	5.5	5.5 – 8 = – 2.5	6.25	2.3	2.3 – 3 = – 0.7	0.49	1.75
9	9.3	9.3 – 8 = 1.3	1.69	4.0	4.0 – 3 = 1.0	1.00	1.30
	72.0		22.78	27.0		2.80	6.45

$$\sum X = 72.0 \quad n = 9 \quad \sum Y = 27.0$$

$$\bar{X} = \frac{72}{9} = 8 \quad \bar{Y} = \frac{27.0}{9} = 3.0$$

$$\sum dx^2 = 22.78 \quad \sum dy^2 = 2.8 \quad \sum dxdy = 6.45$$

$$r_{xy} = \frac{\sum dxdy}{\sqrt{\sum dx^2 \times \sum dy^2}} = \frac{6.45}{\sqrt{22.78 \times 2.8}}$$

$$= \frac{6.45}{\sqrt{63.784}} = \frac{6.45}{7.986} = \frac{6.45}{7.99} = 0.807\,(approx)$$

Positive correlation exist between shell height and shell breadth.

Example 10. Calculate Karl Pearson's coefficient correlation for the following data.

X:	78	89	96	69	59	79	68	62
Y:	125	137	156	112	107	136	123	104

Solution:

(X)	$\bar{X} = 75$ $X - \bar{X} = dx$	dx^2	(Y)	$\bar{Y} = 125$ $Y - \bar{Y} = dy$	dy^2	$dx.dy$
78	78 – 75 = 03	09	125	125 – 125 = 00	00	00
89	89 – 75 = 14	196	137	137 – 125 = 12	144	168
96	96 – 75 = 21	441	156	156 – 125 = 31	961	651
69	69 – 75 = –06	36	112	112 – 125 = –13	169	78
59	59 – 75 = –16	256	107	107 – 125 = –18	324	288
79	79 – 75 = 04	16	136	136 – 125 = 11	121	44
68	68 – 75 = –07	49	123	123 – 125 = –02	04	14
62	62 – 75 = –13	169	104	104 – 125 = –21	441	273
600		1172	1000		2164	1516

$\sum x = 600 \quad n = 8 \quad \sum y = 1000$

$$\bar{X} = \frac{600}{8} = 75 \quad \bar{Y} = \frac{1000}{8} = 125 \qquad \sum dx^2 = 1172 \quad \sum dy^2 = 2164 \quad \sum dxdy = 1516$$

$$r_{xy} = \frac{\sum dxdy}{\sqrt{\sum dx^2 \times \sum dy^2}} = \frac{1516}{\sqrt{1172 \times 2164}}$$

$$= \frac{1516}{\sqrt{2536208}} = \frac{1516}{1592.547} = \frac{1516}{1592.5} = 0.9519 = 0.952\,(approx)$$

Example 11. In order to find the correlation coefficient between two variables X and Y from 12 pairs of observation, the following calculation were made.

$$\sum x = 30 \quad \sum y = 5 \quad \sum x^2 = 670 \quad \sum y^2 = 285 \quad \sum xy = 334$$

On subsequent verification, it was found that the pair (11, 4) was copied wrongly, the correct values being (10, 14). Find the correct value of correlation coefficient.

Solution:

$\sum x$ -given $\sum x$ -incorrect + corrected value.

$$\text{Correct} \sum x = \sum x - 11 + 10 = 30 - 11 + 10 = 29$$

$$\text{Similarly corrected} \sum y = \sum y - 4 + 14 = 5 - 4 + 14 = 15$$

$$\sum x^2 = 670 - (11)^2 + (10)^2 = 670 - 121 + 100 = 649$$

$$\sum y^2 = 285 - (4)^2 + (14)^2 = 285 - 16 + 196 = 465$$

$$\sum xy = 334 - (11 \times 4) + (10 \times 14) = 334 - 44 + 140 = 430$$

Therefore the correct value of coefficient correlation (r) is

$$r = \frac{\sum xy - \frac{\sum x \sum y}{n}}{\sqrt{\sum x^2 - \frac{\left(\sum x\right)^2}{n}} \times \sqrt{\sum y^2 - \frac{\left(\sum y\right)^2}{n}}}$$

$$= \frac{430 - \frac{29 \times 15}{12}}{\sqrt{649 - \frac{(29)^2}{12}} \times \sqrt{465 - \frac{(15)^2}{12}}} = \frac{430 - \frac{435}{12}}{\sqrt{649 - \frac{841}{12}} \times \sqrt{465 - \frac{225}{12}}}$$

$$= \frac{430 - 36.25}{\sqrt{649 - 70.08} \times \sqrt{465 - 18.75}} = \frac{393.75}{\sqrt{578.92} \times \sqrt{446.25}}$$

$$= \frac{393.75}{24.061 \times 21.1246} = \frac{393.75}{24.061 \times 21.13}$$

$$= \frac{393.75}{508.27} = 0.7746 = 0.7747$$

$\therefore$ The correct value of $r = 0.77$

Example 12. A computer while calculating the correlation coefficient between two variables X and Y form 25 pairs of observation obtained the following calculations.

$$n = 25 \sum x = 125 \sum y = 100 \sum x^2 = 650 \sum y^2 = 460 \text{ and } \sum xy = 508$$

It was however discovered at the time of checking that two pairs of observation were not correctly copied. They were taken as (6, 14) and (8, 6) while the correct values were (8, 12) and (6, 8). Prove that the correct value of the correlation coefficient should be $\frac{2}{3}$.

Solution:

$\sum x$ given *i.e.,* 125

$$\sum \text{corrected } x = \sum x - \text{Sum of incorrect values} + \text{Sum of correct value}$$

$$\sum x = 125 - (6 + 8) + (8 + 6) = 125$$

$$\sum \text{corrected } y = \sum y - \text{Incorrect value} + \text{correct value}$$

$$\sum y = 100 - (14 + 6) + (12 + 8) = 100$$

$$\sum x^2 = \sum x^2 - (6^2 + 8^2) + (8^2 + 6^2)$$

$$= 650 - (36 + 64) + (64 + 36) = 650$$

$$\sum y^2 = \sum y^2 - (14^2 + 6^2) + (12^2 + 8^2)$$

$$= 460 - (196 + 36) + (144 + 64)$$

$$= 460 - 232 + 208 = 436$$

$$\sum xy = 508 - (6 \times 14 + 8 \times 6) + (8 \times 12 + 6 \times 8)$$

$$= 508 - (84 + 48) + (96 + 48)$$

$$= 508 - 132 + 144 = 520$$

$\therefore$ Corrected value of coefficient correlation (r) is

$$r_{xy} = \frac{\sum xy - \frac{\sum x \sum y}{n}}{\sqrt{\sum x^2 - \frac{(\sum x)^2}{n}} \times \sqrt{\sum y^2 - \frac{(\sum y)^2}{n}}}$$

$$= \frac{520 - \frac{125 \times 100}{25}}{\sqrt{650 - \frac{(125)^2}{25}} \times \sqrt{436 - \frac{(100)^2}{25}}} = \frac{520 - 500}{\sqrt{650 - 625} \times \sqrt{436 - 400}}$$

$$= \frac{20}{\sqrt{25} \times \sqrt{36}} = \frac{20}{5 \times 6} = \frac{20}{30} = \frac{2}{3}$$

Example 13. A computer obtained the following data

$$n = 30 \sum x = 120 \sum y = 90 \sum x^2 = 600 \sum y^2 = 250 \sum xy = 356$$

Later it was found that pairs $\begin{array}{c|c|c} x & 8 & 12 \\ \hline y & 10 & 7 \end{array}$ **are wrong while the correct values are** $\begin{array}{c|c|c} x & 8 & 10 \\ \hline y & 12 & 8 \end{array}$

Find the correct values of coefficient correlation.

Solution:

Corrected $\sum x = \sum x$ given – Sum of two incorrect values of x + Sum of two correct values of x

$= 120 - (8 + 12) + (8 + 10) = 120 - 20 + 18 = 118$

Corrected $\sum y = \sum y$ given – Sum of two incorrect values of y

+ Sum of two correct values of y

$= 90 - (10 + 7) + (12 + 8) = 90 - 17 + 20 = 93$

Similarly

Corrected $\sum x^2 = 600 - (8^2 + 12^2) + (8^2 + 10^2) = 600 - (64 + 144) + (64 + 100)$

$= 600 - 208 + 164 = 556$

Corrected $\sum y^2 = 250 - (10^2 + 7^2) + (12^2 + 8^2) = 250 - (100 + 49) + (144 + 64)$

$= 250 - 149 + 208 = 309$

Corrected $\sum xy = 356 - (8 \times 12 + 10 \times 7) + (8 \times 12 + 10 \times 8)$

$= 356 - (80 + 84) + (96 + 80)$

$= 356 - 164 + 176 = 356 + 12 = 368$

$$(r) = \frac{\sum xy - \frac{\sum x \sum y}{n}}{\sqrt{\sum x^2 - \frac{\left(\sum x\right)^2}{n}} \times \sqrt{\sum y^2 - \frac{\left(\sum y\right)^2}{n}}}$$

$$= \frac{368 - \frac{118 \times 93}{30}}{\sqrt{556 - \frac{(118)^2}{30}} \times \sqrt{309 - \frac{(93)^2}{30}}} = \frac{368 - \frac{118 \times 31}{10}}{\sqrt{556 - \frac{118 \times 118}{30}} \times \sqrt{309 - \frac{93 \times 93}{30}}}$$

$$= \frac{\frac{3680-3658}{10}}{\sqrt{\frac{16680-13924}{30}} \times \sqrt{9270-8649}} = \frac{\frac{22}{10}}{\sqrt{\frac{2756}{30}} \times \sqrt{\frac{621}{30}}}$$

$$= \frac{2.2}{\sqrt{91.87} \times \sqrt{20.7}} = \frac{2.2}{9.58 \times 4.549} = \frac{2.2}{9.58 \times 4.55} = \frac{2.2}{43.58}$$

$= 0.0504$

Example 14. Find the correlation coefficient between earthworm density (X) and soil P^H (Y) using the following results obtained from 15 experiments.*CU. (M Sc. (Zoology) 2007*

$\sum X = 106.4 \sum X^2 = 755.95 \sum XY = 2058.4 \sum Y = 290 \sum Y^2 = 5696$

Test whether the above correlation coefficient is significant at 5% level [(Given $t_{0.05(13)} = 2.16$].

Solution:

$$r = \frac{n\sum XY - \sum X \sum Y}{\sqrt{\left[n\sum X^2 - \left(\sum X\right)^2\right]\left[n\sum Y^2 - \left(\sum Y\right)^2\right]}}$$

$$= \frac{15 \times 2058.4 - 106.4 \times 290}{\sqrt{\left[15 \times 755.95 - (106.4)^2\right] \times \left[15 \times 5696 - (290)^2\right]}}$$

$$= \frac{30876 - 30856}{\sqrt{(11339.25 - 11320.96) \times (85440 - 84100)}}$$

$$= \frac{20}{\sqrt{18.29 \times 1340}} = \frac{20}{\sqrt{24508.6}}$$

$$= \frac{20}{156.5522} = \frac{20}{156.56} = 0.12775 = 0.128$$

$$r = \frac{\sum XY - \frac{\sum X \sum Y}{n}}{\sqrt{\sum X^2 - \frac{\left(\sum X\right)^2}{n}} \times \sqrt{\sum Y^2 - \frac{\left(\sum Y\right)^2}{n}}}$$

$$= \frac{2058.4 - \frac{106.4 \times 290}{15}}{\sqrt{755.95 - \frac{(106.4)^2}{15}} \times \sqrt{5696 - \frac{(290)^2}{15}}}$$

$$= \frac{2058.4 - \frac{30856}{15}}{\sqrt{755.95 - \frac{11320.96}{15}} \times \sqrt{5696 - \frac{84100}{15}}}$$

$$= \frac{2058.4 - 2057.06}{\sqrt{755.95 - 754.73} \times \sqrt{5696 - 5606.47}} = \frac{1.34}{\sqrt{1.22} \times \sqrt{89.33}}$$

$$= \frac{1.34}{1.10 \times 9.45} = \frac{1.34}{10.39}$$

$r = 0.128$

$$S.E.\ of\ r\ i.e. = \sqrt{\frac{1-r^2}{n-2}} = \sqrt{\frac{1-(0.128)^2}{15-2}} = \sqrt{\frac{1-0.016}{13}} = \sqrt{\frac{0.984}{13}} = \sqrt{0.07569} = 0.275$$

$$t = \frac{r}{S_r} = \frac{0.127}{0.275} = 0.465$$

$$df = n - 2 = 15 - 2 = 13 \quad \alpha = 0.05 \quad \text{critical } t_{0.05(13)} = 2.16.$$

As the value of computed t less than the critical $t_{0.05}$. P is considered too high $(P > \alpha)$ so the Ho is consequently retained and the computed 'r' has no significance.

Example 15. The following data gives the marks obtained by 10 students in Accountancy & Statistics.

Accountancy: x; Statistics y;

x:	**45**	**70**	**65**	**30**	**90**	**40**	**50**	**75**	**85**	**60**
y:	**35**	**90**	**70**	**40**	**95**	**40**	**60**	**80**	**80**	**50**

Calculate the r value

Solution:

(x)	$dx = (x - \bar{x})$	dx^2	(y)	$dy = (y - \bar{y})$	dy^2	$dx.dy$
45	45 – 61 = 16	256	35	35 – 64 = – 29	841	464
70	70 – 61 = 09	81	90	90 – 64 = 26	676	234
65	65 – 61 = 04	16	70	70 – 64 = 06	36	24
30	30 – 61 = – 31	961	40	40 – 64 = – 24	576	744
90	90 – 61 = 29	841	95	95 – 64 = 31	961	899
40	40 – 61 = – 21	441	40	40 – 64 = – 24	576	504
50	50 – 61 = – 11	121	60	60 – 64 = – 04	16	44
75	75 – 61 = 19	196	80	80 – 64 = 16	256	224
85	85 – 61 = 24	576	80	80 – 64 = 16	256	384
60	60 – 61 = – 01	01	50	50 – 64 = – 14	196	14
$N = 10$	$\sum dx = 0$	$\sum dx^2 = 3490$		$\sum dy = 0$	$\sum dy^2 = 3490$	$\sum dxdy = 3535$

$$\sum dx = 0, \sum dx^2 = 3490; \sum dy = 0, \sum dy^2 = 4390 \ \& \sum dxdy = 3535$$

$$\bar{x} = \frac{\sum dx}{N} = \frac{610}{10} = 61$$

$$\bar{y} = \frac{640}{10} = 64$$

$$r = \frac{\sum dxdy}{\sqrt{\sum dx^2 \times \sum dy^2}} = \frac{3535}{\sqrt{3490 \times 4390}} = \frac{3535}{\sqrt{15321100}}$$

$$= \frac{3535}{3914.2176} = 0.903$$

Example 16. From the following data. Calculate the r value.

Weight of father (kgs):	**65**	**66**	**67**	**67**	**68**	**69**	**71**	**73**
Weight of son (kgs):	**67**	**68**	**67**	**68**	**72**	**70**	**69**	**70**

Solution:

Weight of Father (x)	$X - \bar{X} = dx$	dx^2	Weight of Son (y)	$Y - \bar{Y} = dy$	dy^2	dx.dy
65	65 – 68.25 = – 3.25	10.56	67	67–69 = – 2	4	6.5
66	66 – 68.25 = – 2.25	5.06	68	68–69 = – 1	1	2.25
67	67 – 68.25 = – 1.25	1.56	67	67–69 = – 2	4	2.5
67	67 – 68.25 = – 1.25	1.56	68	68–69 = –1	1	1.25
68	68 – 68.25 = – .25	.0625	72	72–69 = 3	9	– .75
69	69 – 68.25 = .75	.5625	70	70–69 = 1	1	.75
71	71 – 68.25 = 2.75	7.56	69	69–69 = 0	0	0
73	73 – 68.25 = 4.75	22.56	70	70–69 = 1	1	4.75
	$\sum dx = 0$	$\sum dx^2 = 49.49$			$\sum dy^2 = 21$	$\sum dxdy = 17.25$

$$\therefore \bar{x} = \frac{\sum x}{N} = \frac{546}{8} = 68.25$$

$$\sum dx = 0 \ \& \ \sum dx^2 = 49.49$$

$$\bar{y} = \frac{\sum y}{N} = \frac{551}{8} = 69$$

$$\sum dy^2 = 21 \& \sum dxdy = 17.25$$

So, $$r = \frac{\sum dxdy}{\sqrt{\sum dx^2 \times \sum dy^2}}$$

$$= \frac{17.25}{\sqrt{49.49 \times 21}} = \frac{17.25}{32.238} = .535$$

Example 17. Find coefficient of correlation from the data given below using rank difference method & comment on the result. ***[CU (Edu) : 2001]***

Students:	**A**	**B**	**C**	**D**	**E**	**F**	**G**	**H**	**I**	**J**
No. in Eng:	**48**	**72**	**54**	**51**	**37**	**43**	**83**	**74**	**58**	**61**
No in Beng:	**77**	**81**	**86**	**52**	**51**	**76**	**94**	**79**	**56**	**58**

Solution:

Eng		Beng			
Marks	Rank (R_1)	Marks	Rank(R_2)	$D = (R_1 - R_2)$	D^2
48	08	77	05	3	09
72	03	81	03	0	00
54	06	86	02	4	16
51	07	52	09	–2	04
37	10	51	10	0	00
43	09	76	06	3	09
83	01	94	01	0	00
74	02	79	04	–2	04
58	05	56	08	–3	09
61	04	58	07	–3	09
				0	60

$$R = 1 - \frac{6\sum D^2}{N(N^2-1)} \quad [n = 10]$$

$$= 1 - \frac{6 \times 60}{10(10^2-1)} = 1 - \frac{360}{10(100-1)}$$

$$= 1 - \frac{360}{10 \times 99} = 1 - \frac{360}{990} = 1 - .3636 = .6364$$

Example 18. Find the r value between earthworm density (x) & soil pH (y) using the following results obtained from 15 experiments.

$$\sum x = 106.4 \sum x^2 = 755.96 \sum xy = 2058.4 \sum y = 290 \sum y^2 = 5696$$

[C.U (M.Sc. Zoo.)2000]

Solution: $n = 15$

$$\text{Now Correlation coefficient } (r) = \frac{\sum dxdy}{\sqrt{\sum dx^2 \times \sum dy^2}}$$

$$\sum dxdy = \sum xy - \frac{\sum x.\sum y}{n}$$

$$= 2058.4 - \frac{106.4 \times 290}{15}$$

$$= 2058.4 - 2058 = 1.4$$

$$\sum dx^2 = \sum x^2 - \frac{\left(\sum x\right)^2}{n}$$

$$= 755.96 - \frac{(106.4)^2}{15}$$

$$= 755.96 - 754.73$$

$$= 1.23$$

$$\sum dy^2 = \sum y^2 - \frac{\left(\sum y\right)^2}{n}$$

$$= 5696 - \frac{(290)^2}{15}$$

$$= 5696 - 5606.67$$

$$= 89.33$$

$$r = \frac{1.4}{\sqrt{1.23 \times 89.33}} = \frac{1.4}{10.48} = 0.13356$$

Example 19. From 25 observations on the pair of varieties (x, y) the following data were obtained.

$$\sum x = 125 \sum x^2 = 650 \sum xy = 533 \sum y = 10 \sum y^2 = 484$$

Find the correlation coefficient. *[C.U. (M.Sc. Zoo.) 1997]*

Solution: It is given that $n = 25$

$$\text{Now Correlation coefficient } (r) = \frac{\sum dxdy}{\sqrt{\sum dx^2 \times \sum dy^2}}$$

$$\sum dxdy = \sum xy - \frac{\sum x.\sum y}{n}$$

$$= 533 - \frac{125 \times 10}{25}$$

$$= 533 - 50 = 483$$

$$\sum dx^2 = \sum x^2 - \frac{\left(\sum x\right)^2}{n}$$

$$= 650 - \frac{(125)^2}{25}$$

$$= 650 - 625$$

$$= 25$$

$$\sum dy^2 = \sum y^2 - m\frac{(10)^2}{25}$$

$$= 484 - 4$$

$$= 480$$

$$r = \frac{483}{\sqrt{25 \times 480}} = \frac{483}{\sqrt{12000}} = \frac{483}{109.54} = 4.41$$

Example 20. Calculate the correlation between the following two series of test scores by any one of the methods:

Individuals:	**1**	**2**	**3**	**4**	**5**	**6**	**7**	**8**	**9**	**10**
x:	**85**	**103**	**88**	**95**	**76**	**74**	**58**	**97**	**76**	**60**
y:	**110**	**98**	**118**	**104**	**112**	**124**	**119**	**95**	**94**	**97**

Solution:

x	x^2	y	y^2	xy
85	7225	110	12100	9350
103	10609	98	9604	10094
88	7744	118	13924	10384
95	9025	104	10816	9880
76	5776	112	12544	8512
74	5476	124	15376	9176
58	3364	119	14161	6902
97	9409	95	9025	9215
76	5776	94	8836	7144
60	3600	97	9409	5820
$\Sigma x = 812$	$\Sigma x^2 = 68004$	$\Sigma y = 1071$	$\Sigma y^2 = 115795$	$\Sigma xy = 86477$

Now $$r = \frac{\sum dx.dy}{\sqrt{\sum dx^2.\sum dy^2}}$$

So, $$\sum dx.dy = \sum xy - \frac{\sum x.\sum y}{n}$$

$$= 86477 - \frac{812 \times 1071}{10}$$

$$= 86477 - 86965.2 = -488.2$$

$$\sum dx^2 = \sum x^2 - \frac{\left(\sum x\right)^2}{n}$$

$$= 68004 - \frac{(812)^2}{10}$$

$$= 68004 - 65934.4 = 2069.6$$

$$\sum dy^2 = \sum y^2 - \frac{\left(\sum y\right)^2}{n}$$

$$= 115796 - \frac{(1071)^2}{10}$$

$$= 115796 - 114704.1 = 1091.9$$

$$r = \frac{-488.2}{\sqrt{2069.6 \times 1091.9}} = \frac{-488.2}{1503.26} = -0.32$$

Example 21. Calculate by any method, the correlation coefficient between the following two sets of scores.

Pupils:	**A**	**B**	**C**	**D**	**E**	**F**	**G**	**H**	**I**	**J**	**K**	**L**
***x*:**	**48**	**50**	**54**	**60**	**64**	**58**	**70**	**66**	**50**	**·50**	**46**	**63**
***y*:**	**22**	**32**	**29**	**33**	**30**	**36**	**40**	**36**	**21**	**36**	**26**	**43**

Solution:

x	$dx = (x - \bar{x})$	dx^2	y	$dy = (y - \bar{y})$	dy^2	$dx.dy$
48	–8.58	73.61	22	–10	100	85.8
50	-6.58	43.3	32	00	00	00
54	–2.58	6.65	29	–03	09	7.74
60	3.42	11.7	33	01	01	3.42
64	7.42	55.05	30	–02	04	–14.84
58	1.42	2.01	36	04	16	5.68
70	13.42	180	40	08	64	107.36
66	9.42	88.73	36	04	16	37.68
50	–6.58	43.3	21	–11	121	72.38
50	6.58	43.3	36	04	16	–26.32
46	–10.58	111.93	26	06	36	63.48
63	–6.42	41.21	43	11	121	70.62
		$\sum dx^2 = 700.79$			$\sum dy^2 = 504$	$\sum dxdy = 413.32$

$$\bar{x} = \frac{\sum x}{n} = \frac{679}{12} = 56.58$$

$$\bar{y} = \frac{\sum y}{n} = \frac{384}{12} = 32$$

$$r = \frac{\sum dx.dy}{\sqrt{\sum dx^2.\sum dy^2}} = \frac{413.32}{\sqrt{700.79 \times 504}}$$

$$= \frac{413.32}{\sqrt{353198.16}} = \frac{413.32}{594.30} = .6954$$

Example 22. From the following data do you consider that there exist correlation between the variables.

	1	2	3	4	5	6	7	8	9	10
X:	10	15	20	22	30	35	40	45	50	55
Y:	12	16	20	25	35	40	45	50	52	60

Solution: Let assume mean of x = 30 and y = 35.

	X	$x - \bar{x} = dx$ $x - 30$	dx^2	Y	$y - \bar{y} = dy$	dy^2	dydx
1	10	10 – 30 = –20	400	12	12 – 35 = –23	529	–20 × –23 = 460
2	15	15 – 30 = –15	225	16	16 – 35 = –19	361	–15 × –19 = 285
3	20	20 – 30 = –10	100	20	20 – 35 = –15	225	–10 × –15 = 150
4	22	22 – 30 = –08	64	25	25 – 35 = –10	100	–08 × –10 = 80
5	30	30 – 30 = 0	0	35	35 – 35 = 0	0	0 × 0 = 00
6	35	35 – 30 = +05	25	40	40 – 35 = +05	25	5 × 5 = 25
7	40	40 – 30 = +10	100	45	45 – 35 = +10	100	10 × 10 = 100
8	45	45 – 30 = +15	225	50	50 – 35 = +15	225	15 × 15 = 225
9	50	50 – 30 = +20	400	52	52 – 35 = +17	289	20 × 17 = 340
10	55	55 – 30 = +25	625	60	60 – 35 = +25	625	25 × 25 = 625
		+22	2164		+5	2479	2290

$$r = \frac{n\sum dxdy - \left(\sum dx \sum dy\right)}{\sqrt{n\sum dx^2 - \left(\sum dx\right)^2} \times \sqrt{n\sum dy^2 - \left(\sum dy\right)^2}}$$

$$= \frac{10 \times 2290 - 22 \times 5}{\sqrt{10 \times 2164 - (22)^2} \times \sqrt{10 \times 2479 - (5)^2}}$$

$$= \frac{22900 - 110}{\sqrt{21640 - 484} \times \sqrt{24790 - 25}} = \frac{22790}{\sqrt{21156} \times \sqrt{24765}}$$

$$= \frac{22790}{145.45 \times 157.37} = \frac{22790}{22889.46} = 0.9956 = 0.996$$

Calculated value is higher in 9 *df* at 5% level. So it is significantly correlated.

• Rank Correlation:

The product moment correlation coefficient (r) is calculated on the basis of 'value' of the variables. But quite often, situation arise in which data can not be measured quantitatively. In this case Rank correlation method is applied.

For example, "intelligence, beauty, honesty and sales efficiency" we may require to rank the individual variables in order of merit by using 1, 2, 3 etc. for both variables. The individuals are then said to be ranked and the number allotted to individual is called rank. Then we find difference (R) in ranks for each pair & apply the formula.

$$R = 1 - \frac{6\sum d^2}{n^3 - n}$$

Working Procedure:

I. Assign rank to each item of both the variables.

II. Calculate the difference of ranks of X from the ranks of Y and write it under the column headed by 'D'.

III. Square the difference 'D' and write it under the column 'D^2'

IV. Apply the formula.

$$R = 1 - \frac{6\sum D^2}{n^3 - n} = 1 - \frac{6\sum D^2}{n(n^3 - 1)}$$ n = number of observation.

	Range of r value	Relation
I.	± 0	No correlation
II.	± 0.00 to ± 0.20	Indifferent or negligible
III.	± 0.20 to ± 0.40	Low or slight correlation
IV.	± 0.40 to ± 0.70	Substantial or marked correlation
V.	± 0.70 to ± 1.00	High to very correlation

Type 1: When ranks are given

• Working Procedure

I. Compute D, the difference of the rank

II. Compute D^2 and get the sum $\sum D^2$

III. Substitute the value in the formula.

Example: Following are the ranks obtained by 10 students in two subjects, Biostatistics and Biophysics. To what extent the knowledge of the students in the two subjects is related?

Biostatistics:	**1**	**2**	**3**	**4**	**5**	**6**	**7**	**8**	**9**	**10**
Biophysics:	**2**	**4**	**1**	**5**	**3**	**9**	**7**	**10**	**6**	**8**

Solution:

Biostatistics Rank (R_1)	Biophysics Rank (R_2)	$D = R_1 - R_2$	D^2
1	2	$1 - 2 = -1$	1
2	4	$2 - 4 = -2$	4
3	1	$3 - 1 = +2$	4
4	5	$4 - 5 = -1$	1
5	3	$5 - 3 = +2$	4
6	9	$6 - 9 = -3$	9
7	7	$7 - 7 = 0$	0
8	10	$8 - 10 = -2$	4
9	6	$9 - 6 = +3$	9
10	8	$10 - 8 = +2$	4
			40

$$R = 1 - \frac{6\sum D^2}{N(N^2-1)} \qquad N = 10$$

$$R = 1 - \frac{6 \times 40}{10[(10)^2 - 1]}$$

$$= 1 - \frac{240}{10 \times 99} = \frac{990 - 240}{990} = \frac{750}{990} = +0.7575 = +0.76$$

Type 2: When ranks are not given

- **Working Procedure:**

I. When no ranks are given but actual data are given then we should assign the ranks.

II. We can give ranks by taking the highest as 1 or the lowest value as 1.

II. Next to the highest (lowest) as 2 and follow the same procedure for both the variables.

IV. Compute D, D^2 & sum D^2 and substitute the values in the formula.

Example: The marks obtained by the students in Zoology and Botany are as follows

Marks in Zoology:	**35**	**23**	**47**	**17**	**10**	**43**	**09**	**06**	**28**
Marks in Botany:	**30**	**33**	**45**	**23**	**08**	**49**	**12**	**04**	**31**

Compute their ranks in the two subjects and coefficient of correlation ranks.

Solution:

Zoology		*Botany*			
Marks	*Ranks* (R_1)	*Marks*	*Ranks* (R_2)	$D = R_1 - R_2$	D^2
35	3	30	5	3 – 5 = –2	4
23	5	33	3	5 – 3 = +2	4
47	1	45	2	1 – 2 = –1	1
17	6	23	6	6 – 6 = 0	0
10	7	08	8	7 – 8 = –1	1
43	2	49	1	2 – 1 = +1	1
09	8	12	7	8 – 7 = +1	1
08	9	04	9	9 – 9 = 0	0
28	4	31	4	4 – 4 = 0	0
$N = 9$					$\sum D^2 = 12$

$$R = 1 - \frac{6\sum D^2}{N(N^2-1)}$$

$$= 1 - \frac{6 \times 12}{9(81-1)} = 1 - \frac{72}{9 \times 80}$$

$$= 1 - \frac{72}{720} = \frac{720 - 72}{720} = \frac{648}{720} = 0.9$$

Example 1. In an examination 10 students obtained the following marks in Mathematics and Physics. Find the coefficient of rank correlation.

Mathematics	**90**	**30**	**82**	**45**	**32**	**65**	**40**	**88**	**73**	**66**
Physics	**85**	**42**	**75**	**68**	**45**	**63**	**60**	**90**	**62**	**58**

Solution:

Math		Physics			
Marks	*Ranks (R_1)*	*Marks*	*Ranks (R_2)*	$D = R_1 - R_2$	D^2
90	1	85	2	1 − 2 = −1	1
30	10	42	10	10 − 10 = 0	0
82	3	75	3	3 − 3 = 0	0
45	7	68	4	7 − 4 = 3	9
32	9	45	9	9 − 9 = 0	0
65	6	63	5	6 − 5 = 1	1
40	8	60	7	8 − 7 = 1	1
88	2	90	1	2 − 1 = 1	1
73	4	62	6	4 6 = −2	4
66	5	58	8	5 − 8 = −3	9
N = 10				0	26

$$R = 1 - \frac{6\sum D^2}{N(N^2-1)}$$

$$= 1 - \frac{6.26}{10(10^2-1)}$$

$$= 1 - \frac{156}{10.99}$$

$$= 1 - \frac{156}{990}$$

$$= 1 - \frac{990-156}{990} = \frac{834}{990} = .84$$

Example 2. Ten students got the following percentage of marks in Mathematics & Statistics. ind the rank correlation coefficient.

Maths	**08**	**36**	**98**	**25**	**75**	**82**	**92**	**62**	**65**	**35**
Statistics	**84**	**51**	**91**	**60**	**68**	**62**	**86**	**58**	**35**	**49**

Solution:

Math (Y)	R_1	*Stat. (Y)*	R_2	$D = R_1 - R_2$	D^2
08	10	84	3	10 − 3 = 7	49
36	7	51	8	7 − 8 = −1	1
98	1	91	1	1 − 1 = 0	0
25	9	60	6	9 − 6 = 3	9
75	4	68	4	4 − 4 = 0	0
82	3	62	5	3 − 5 = −2	4
92	2	86	2	2 − 2 = 0	0
62	6	58	7	6 − 7 = −1	1
65	5	35	10	5 − 10 = −5	25
35	8	49	9	8 − 9 = −1	1
N = 10					90

$$R = 1 - \frac{6\sum D^2}{N(N^2 - 1)}$$

$$= 1 - \frac{6 \times 90}{10(10^2 - 1)}$$

$$= 1 - \frac{6 \times 90}{10 \times 99}$$

$$= 1 - \frac{6}{11} = 1 - .545 = .455$$

Example 3. Calculate Karl Pearson's correlation co-efficient between x and y for the following data.

X	43	54	59	68	76
Y	105	98	84	63	50

Solution:

	X	R_1	Y	R_2	$D = R_1 - R_2$	D^2
1	43	5	105	1	5 − 1 = 4	16
2	54	4	98	2	4 − 2 = 2	4
3	59	3	84	3	3 − 3 = 0	0
4	68	2	63	4	2 − 4 = −2	4
5	76	1	50	5	1 − 5 = −4	16
$N = 5$						40

$$R = 1 - \frac{6\sum D^2}{N(N^2 - 1)}$$

$$= 1 - \frac{6 \times 40}{5(25 - 1)} = 1 - \frac{6 \times 40}{5 \times 24} = 1 - 2 = -1$$

Example 4. Ten students get the following marks in Ecology and Genetics.

Students (Roll No.):	**1**	**2**	**3**	**4**	**5**	**6**	**7**	**8**	**9**	**10**
Marks in Ecology:	**78**	**36**	**98**	**25**	**75**	**82**	**90**	**62**	**65**	**39**
Marks in Genetics:	**84**	**51**	**91**	**60**	**68**	**62**	**86**	**58**	**53**	**47**

Calculate the correlation coefficient.

Solution:

Roll	Ecology		Genetics		Rank difference	
No.	Marks	Ranks (R_1)	Marks	Ranks (R_2)	$D = R_1 - R_2$	D^2
1	78	4	84	3	4 − 3 = 1	1
2	36	9	51	9	9 − 9 = 0	0
3	98	1	91	1	1 − 1 = 0	0
4	25	10	60	6	10 − 6 = 4	16
5	75	5	68	4	5 − 4 = 1	1
6	82	3	62	5	3 − 5 = −2	4
7	90	2	86	2	2 − 2 = 0	0
8	62	7	58	7	7 − 7 = 0	0
9	65	6	53	8	6 − 8 = −2	4
10	39	8	47	10	8 − 10 = −2	4
$N = 10$						$\sum D^2 = 30$

$$R = 1 - \frac{6\sum D^2}{N(N^2 - 1)}$$

$$= 1 - \frac{6 \times 30}{10(100 - 1)} = 1 - \frac{6 \times 30}{990}$$

$$= 1 - \frac{180}{990} = 1 - \frac{2}{11} = \frac{11 - 2}{11} = \frac{9}{11} = 0.818 = 0.82$$

Example 5. The coefficient of rank correlation between marks in Biostatistics and Biophysics obtained by a certain group of P.G students of Serampore college is $\frac{2}{3}$ and the sum of the square of the differences in ranks is 55. Find the number of students in the group.

Solution:

$\sum$ Spearman's formula for rank correlation coefficient is $R = 1 - \frac{6\sum D^2}{N(N^2 - 1)}$

Here $R = \frac{2}{3}$ $\sum D^2 = 55$ $N = ?$

Therefore

$$\frac{2}{3} = 1 - \frac{6 \times 55}{N(N^2 - 1)}$$

or $$\frac{2}{3} = 1 - \frac{330}{N(N^2 - 1)}$$

or $$\frac{330}{N(N^2 - 1)} = 1 - \frac{2}{3} = \frac{3 - 2}{3} = \frac{1}{3}$$

or $$N(N^2 - 1) = 3 \times 330 = 990$$

$$N(N + 1)(N - 1) = 990 = 10 \times 11 \times 9$$

Hence $N = 10$ *i.e.*, the number of student is 10.

Example 6. The coefficient rank correlation of marks obtained by 10 students in Molecular Biology and Genetics was found to be 0.5. It was later discovered that the difference in ranks in the two subjects obtained by one of the two students was wrongly taken as 3 instead of 7. Find the correct coefficient of rank correlation.

Solution:

According to spearman's formula $R = 1 - \frac{6\sum D^2}{N(N^2 - 1)}$

Substituting the value, $R = 0.5$ $N = 10$

$$0.5 = 1 - \frac{6\sum D^2}{10(10^2 - 1)}$$

or $$0.5 = 1 - \frac{6\sum D^2}{990}$$

or $$1 - 0.5 = \frac{6\sum D^2}{990}$$

or $$6\sum D^2 = 0.5 \times 990 = 495$$

$$\sum D^2 = \frac{495}{6} = 82.5$$

Corrected value of

$$\sum D^2 = 82.5 - 3^2 + 7^2 = 122.5$$

The corrected value of

$$R = 1 - \frac{6\sum D^2}{10(100-1)} = 1 - \frac{6 \times 122.5}{990} = 1 - \frac{735}{990}$$

$$= 1 - 0.742 = 0.258 = 0.26.$$

Example 7. Marks obtained by nine students of RPM college in Botany and Zoology are given below.

Botany:	**35**	**23**	**47**	**17**	**10**	**43**	**09**	**06**	**28**
Zoology:	**30**	**33**	**45**	**23**	**08**	**49**	**12**	**04**	**31**

Calculate Spearman's coefficient of rank correlation and interpret.

Solution:

Botany (X)		*Zoology (Y)*		*Rank Difference*	
Marks	*Ranks (R_1)*	*Marks*	*Ranks (R_2)*	$D = R_1 - R_2$	D^2
35	3	30	5	3 − 5 = −2	4
23	5	33	3	5 − 3 = +2	4
47	1	45	2	1 − 2 = −1	1
17	5	23	6	5 − 6 = −1	1
10	7	08	8	7 − 8 = −1	1
43	2	49	1	2 − 1 = +1	1
09	8	12	7	8 − 7 = +1	1
06	9	04	9	9 − 9 = 0	0
28	4	31	4	4 − 4 = 0	0
$N = 9$					$\sum D^2 = 13$

According *to Spearman's* formula

$$R = 1 - \frac{\sum D^2}{N(N^2-1)}$$

$$= 1 - \frac{13}{9(81-1)} = 1 - \frac{13}{720} = \frac{720-13}{720} = \frac{707}{720} = 0.9819$$

$R = 0.9$

Hence the correlation coefficient shows a very high relationship. It means that students who are good Botany are also good in Zoology and vice versa.

Example 8. Find the spearman's rank correlation for the following data of marks obtained by 10 P.G students of Bidhan Nagar Govt. College Calcutta in Evolution and Biostatistics.

Students (Roll No.):	1	2	3	4	5	6	7	8	9	10
Marks in Evolution:	80	38	95	30	74	84	91	60	66	40
Marks in Biostat:	85	50	92	58	70	65	88	56	52	46

Solution:

Roll	Evolution (X)		Biostatistics (Y)		Rank difference	
No.	Marks	Ranks (R_1)	Marks	Ranks (R_2)	$D = R_1 - R_2$	D^2
1	80	4	85	3	4 – 3 = 1	1
2	38	9	50	9	9 – 9 = 0	0
3	95	1	92	1	1 – 1 = 0	0
4	30	10	58	6	10 – 6 = 4	16
5	74	5	70	4	5 – 4 = 1	1
6	84	3	65	5	3 – 5 = –2	4
7	91	2	88	2	2 – 2 = 0	0
8	60	7	56	7	7 – 7 = 0	0
9	66	6	52	8	6 – 8 = –2	4
10	40	8	46	10	8 – 10 = –2	4
N = 10						$\sum D^2 = 30$

$$R = 1 - \frac{6\sum D^2}{N(N^2-1)}$$

$$= 1 - \frac{6 \times 30}{10(100-1)} = 1 - \frac{6 \times 30}{990} = 1 - \frac{180}{990}$$

$$= \frac{990 - 180}{990} = \frac{810}{990} = \frac{81}{99} = \frac{9}{11} = 0.818 = 0.82$$

Example 9. The following table gives the two kinds of assessment in practical classes of 10 post graduate students of Serampore colleges.

Students:	1	2	3	4	5	6	7	8	9	10
Internal Assesement:	45	62	66	32	12	38	47	67	42	85
External Assesement:	39	48	65	32	20	35	45	77	30	62

Find spearman's rank correlation coefficient & interpret the result.

Solution:

Students	Internal Assesement		External Assesement		Rank difference	
	Marks	Ranks (R_1)	Marks	Ranks (R_2)	$D = R_1 - R_2$	D^2
1	45	6	39	6	6 – 6 = 0	0
2	62	4	48	4	4 – 4 = 0	0
3	66	3	65	2	3 – 2 = 1	1
4	32	9	32	8	9 – 8 = 1	1
5	12	10	20	10	10 – 10 = 0	0
6	38	8	35	7	8 – 7 = 1	1
7	47	5	45	5	5 – 5 = 0	0
8	67	2	77	1	2 – 1 = 1	1
9	42	7	30	9	7 – 9 = –2	4
10	85	1	62	3	1 – 3 = –2	4
N = 10						$\sum D^2 = 12$

$$R = 1 - \frac{6\sum D^2}{N(N^2-1)}$$

$$= 1 - \frac{6 \times 12}{10(100-1)} = 1 - \frac{72}{990}$$

$$= \frac{990-72}{990} = \frac{918}{990} = 0.927 = 0.93$$

A very high correlation. It means that student who scored high marks in internal assessment also scored high marks in external assessment and vice versa.

Example 10. The coefficient of ranks correlation between the marks in Zoology and Biochemistry obtained by a certain group of Delhi universities P.G students is $\frac{1}{3}$ and the sum of the square of the differences in ranks is 56. Find the number of students in the group.

Solution:

Sepearman's formula for rank correlation coefficient is $R = 1 - \frac{6\sum D^2}{N(N^2-1)}$

Here $R = \frac{1}{3}$ $N = ?$

$$\sum D^2 = 56$$

Therefore

$$\frac{1}{3} = 1 - \frac{6 \times 56}{N(N^2-1)}$$

or

$$\frac{1}{3} = 1 - \frac{336}{N(N^2-1)}$$

or

$$1 - \frac{1}{3} = \frac{336}{N(N^2-1)}$$

or

$$\frac{2}{3} = \frac{336}{N(N^2-1)}$$

or

$$N(N^2-1) = 336 \times \frac{3}{2}$$

$$N(N^2-1) = 168 \times 3 = 504$$

$$N(N+1)(N-1) = 8 \times 9 \times 7$$

Hence $N = 8$, *i.e.*, the number of students is 8.

Example 11. In the following table recorded data showing the test scores made by 10 sales men in an intelligence test and their weekly sales.

Salesmen	**1**	**2**	**3**	**4**	**5**	**6**	**7**	**8**	**9**	**10**
Test scores	**50**	**70**	**50**	**60**	**80**	**50**	**90**	**50**	**60**	**60**
Sales (Rs)	**25**	**60**	**45**	**50**	**45**	**20**	**55**	**30**	**45**	**30**

Calculate rank correlation coefficient between intelligence & sales efficiencies.

Solution:

Sales man	Intelligence		Sales		$D = R_1 - R_2$	D^2
	Test score	Ranks (R_1)	Amount	Ranks (R_2)		
1	50	8.5	25	9	8.5 – 9 = –0.5	.25
2	70	3	60	1	3 – 1 = 2	04
3	50	8.5	45	5	8.5 – 5 = 3.5	12.25
4	60	5	50	3	5 – 3 = 2	04
5	80	2	45	5	2 – 5 = –3	09
6	50	8.5	20	10	8.5 – 10 = –1.5	2.25
7	90	1	55	2	1 – 2 = –1	1
8	50	8.5	30	7.5	8.5 – 7.5 = 1	1
9	60	5	45	5	5 – 5 = 0	0
10	60	5	30	7.5	5 – 7.5 = –2.5	6.25
					0	$\sum D^2 = 40$

N.B. = [Three salesmen - (4, 9 & 10) with scores 60 each tie for the 4th place & hence given the average rank $\left(\frac{4+5+6}{3} = 5\right)$ 5, Again 4 salesman (1, 3, 6 & 8) tie for the 7th place with score 50 each & are given average rank $\left(\frac{7+8+9+10}{4}\right) = 8.5$. Similarly in sales 3 salesmen tie for 4th place $\left(\frac{4+5+6}{3} = 5\right)$ & 2 sales man tie for the 7th place $\left(\frac{7+8}{2} = 7.5\right)$ where average rank is given.]

Modified formula

$$R = 1 - \frac{6\sum D^2 + \frac{\sum(t^3 - t)}{12}}{n^3 - n}$$

$$R = 1 - \frac{6(40 + 9.5)}{10^3 - 10}$$

$$= 1 - \frac{6 \times 49.5}{10(99)} = 1 - \frac{297}{990} = 1 - .3 = .70$$

In the two series together there are in all 4 ties viz. 4, 3 & 3, 2 = 12

$$\frac{\sum(t^3 - t)}{12} = \frac{4^3 - 4}{12} + \frac{3^3 - 3}{12} + \frac{3^3 - 3}{12} + \frac{2^3 - 2}{12}$$

$$= 5 + 2 + 2 + .5 = 9.5$$

Example 12. Write down Spearman's formula for determining rank coefficient *R*. For the two series we have $\sum D^2 = 30$ and $N = 10$ find *R*.

Solution:

It is a non parametric measure of association between the two series (*X* & *Y*) of rank is called "Rank correlation coefficient". It is given by the formula.

$$R = 1 - \frac{6\sum D^2}{N(N^2 - 1)}$$

$\sum D^2 = 30 \quad N = 10$

$$R = 1 - \frac{6\sum 30}{10(100-1)} = 1 - \frac{180}{990} = 1 - \frac{2}{11} = \frac{11-2}{11} = \frac{9}{11} = 0.818 = 0.82$$

Example 13. Find the rank correlation coefficient of the following data.

Series X:	**115**	**109**	**112**	**87**	**98**	**120**	**98**	**100**	**98**	**118**
Series Y:	**75**	**73**	**85**	**70**	**76**	**82**	**65**	**73**	**68**	**80**

Solution:

Series X		*Series Y*		*Rank Difference*	
Score	*Ranks (R_1)*	*Score*	*Ranks (R_2)*	$D = R_1 - R_2$	D^2
115	3	75	5	3 – 5 = –2	4
109	5	73	6.5	5 – 6.5 = –1.5	2.25
112	4	85	1	4 – 1 = 3	9
87	10	70	8	10 – 8 = 2	4
98	8	76	4	8 – 4 = 4	16
120	1	82	2	1 – 2 = –1	1
98	8	65	10	8 – 10 = –2	4
100	6	73	6.5	6 – 6.5 = –0.5	0.25
98	8	68	9	8 – 9 = –1	1
118	2	80	3	2 – 3 = –1	1
$N = 10$					$\sum D^2 = 42.50$

N.B:

1. In series X, the highest score is 120 and therefore its rank is 1. In this series the rank of 87 is 7 and it occurs thrice. As there is a tie for the 7th place. We find the average rank $\frac{7+8+9}{3} = \frac{24}{3} = 8$. We assign 8 as rank to each of the three 98.
2. The score next to 98 is 87 and its rank is 10.
3. Similarly in series Y, the rank of 73 is 6 and it occurs twice. As there is tie for 6th place, we find the average rank $\frac{6+7}{2} = 6.5$. We assign 6.5 as rank to each of two 73.

According to the Spearman's formula

$$R = 1 - \frac{6\left[\sum D^2 + \frac{\sum t^3 - t}{12}\right]}{N(N^2-1)}$$ t = Number of individuats involved in a tie.

$N = 10,\ t = 3, 2$

$$\sum \frac{t^3 - t}{12} = \frac{3^3 - 3}{12} + \frac{2^3 - 2}{12} = \frac{27-3}{12} + \frac{8-2}{12} = \frac{24}{12} + \frac{6}{12} = 2 + 0.5 = 2.5$$

$$R = 1 - \frac{6(42.50 + 2.5)}{10(100-1)} = 1 - \frac{6 \times 45}{990}$$

$$= 1 - \frac{270}{990} = \frac{99.27}{99} = \frac{72}{99} = 0.7227 = 0.73$$

Example 14. Find out the rank correlation coefficient between the heights of M. Sc (Zoology) first year & M. Sc. (Zoology) second year students of Serampore College.

Height of 1st yr M.Sc student:	65	66	67	67	68	69	70	72
Height of 2nd yr M.Sc student:	**67**	**68**	**65**	**68**	**72**	**72**	**69**	**71**

Solution:

Series X		*Series Y*		*Rank Difference*	
score	*Ranks* (R_1)	*Score*	*Ranks* (R_2)	$D = R_1 - R_2$	D^2
65	8	67	7	8 − 7 = 1	1
66	7	68	5.5	7 − 5.5 = 1.5	2.25
67	5.5	65	8	5.5 − 8 = −2.5	6.25
67	5.5	68	5.5	5.5 − 5.5 = 0	0
68	4	72	1.5	4 − 1.5 = 2.5	6.25
69	3	72	1.5	3 − 1.5 = 1.5	2.25
70	2	69	4	2 − 4 = −2	4
72	1	71	3	1 − 3 = −2	4
N = 8					$\sum D^2 = 26$

N.B:

1. In X-series, the rank of 67 is 5 & it occurs twice. As there is tie for 5th place, we find the average rank $= \frac{5+6}{2} = \frac{11}{2} = 5.5$.

2. In Y series, the rank of 72 is 2 & it occurs twice, therefore its average rank is $\frac{1+2}{2} = \frac{3}{2} = 1.5$.

Similarly the rank of 18 is 5 but it occurs twice. its average rank will be $\frac{5+6}{2} = 5.5$.

According to Spearman's formula

$$R = 1 - \frac{6\left[\sum D^2 + \frac{\sum (t^3 - t)}{12}\right]}{N(N^2 - 1)}$$

	X series	*Y series*
t	2	2,2
N	8	8

The correction factor for X series

$$\frac{(t^3 - t)}{12} = \frac{2^3 - 2}{12} = \frac{8-2}{12} = \frac{6}{12} = \frac{1}{2} = 0.5$$

The correction factor for Y series

$$\frac{(t^3 - t)}{12} = \frac{(2^3 - 2)}{12} = \frac{(2^3 - 2)}{12} = \frac{6}{12} + \frac{6}{12} = 0.5 + 0.5 = 1$$

Total correction factor = 0.5 + 1 = 1.5

$$R = 1 - \frac{6\left[\sum D^2 + \frac{\sum (t^3 - t)}{12}\right]}{N(N^2 - 1)}$$

$$= 1 - \frac{6(26 + 1.5)}{8(64 - 1)} = 1 - \frac{6 \times 27.5}{8 \times 63}$$

$$= 1 - \frac{165}{504} = \frac{504 - 165}{504} = \frac{339}{504} = 0.6726 = 0.673$$

Example 15. Marks of 8 students of Microbiology of R.K. Mission Vidyamandira in Biochemistry & Biostatistics are given below.

Marks in Biochemistry:	**15**	**20**	**28**	**12**	**40**	**60**	**20**	**80**
Marks in Biostat:	**40**	**30**	**50**	**30**	**20**	**10**	**30**	**60**

Calculate Spearman's coefficient of rank correlation.

Solution:

Biochemistry (X)		*Biostatistics (Y)*		*Rank Difference*	
Marks	*Ranks (R_1)*	*Marks*	*Ranks (R_2)*	*$D = R_1 - R_2$*	*D^2*
15	7	40	3	7 – 3 = 4	16
20	5.5	30	5	5.5 – 5 = 0.5	.25
28	4	50	2	4 – 2 = 2	4
12	8	30	5	8 – 5 = 3	9
40	3	20	7	3 – 7 = –4	16
60	2	10	8	2 – 8 = –6	36
20	5.5	30	5	5.5 – 5 = 0.5	.25
80	1	60	1	1 – 1 = 0	0
					$\sum D^2 = 81.5$

Note:

1. In X series the rank of 20 is 5 & it occurs twice so its average rank is $\frac{5+6}{2} = 5.5$.

2. Similarly in Y series the rank of 30 is 4 and it occurs thrice, so its average rank is $\frac{4+5+6}{3} = 5$.

According to Spearman's formula

$$R = 1 - \frac{6\sum D^2 + \frac{\sum (t^3 - t)}{12}}{N(N^2 - 1)}$$

	X series	*Y series*
$t =$	2	3
$N =$	8	8

The correction factor for X series

$$\frac{t^3 - t}{12} = \frac{2^3 - 2}{12} = \frac{6}{12} = 0.5$$

The correction factor for Y series

$$\frac{t^3 - t}{12} = \frac{3^3 - 3}{12} = \frac{27 - 3}{12} = \frac{24}{12} = 2$$

The total correction factor = 0.5 + 2 = 2.5

$$R = 1 - \frac{6\sum D^2 + 2.5}{8(64-1)} = 1 - \frac{6(81.5 + 2.5)}{8 \times 63}$$

$$= 1 - \frac{6(81.5 + 2.5)}{8 \times 63} = 1 - \frac{6 \times 84}{8 \times 63} = 1 - \frac{504}{504} = 1 - 1 = 0$$

Example 16. The Mathematical aptitude score of 10 computer programmers of R. K. Mission Vidyamandira with job performance rating is given below.

Mathematical aptitude score:	**2**	**5**	**0**	**4**	**3**	**1**	**6**	**8**	**7**	**9**
Job performing rating:	**8**	**16**	**8**	**9**	**5**	**4**	**3**	**17**	**8**	**12**

Calculate coefficient of rank correlation and state whether those who have aptitude for mathematics are likely to be better programmers.

Solution:

Mathematical apptitude (X)		*Job performing rating (Y)*		*Rank Difference*	
Marks	*Ranks (R_1)*	*Marks*	*Ranks (R_2)*	$D = R_1 - R_2$	D^2
2	8	8	6	8 − 6 = 2	04
5	5	16	2	5 − 2 = 3	09
0	10	8	6	10 − 6 = 4	16
4	6	9	4	6 − 4 = 2	04
3	7	5	8	7 − 8 = −1	01
1	9	4	9	9 − 9 = 0	00
6	4	3	10	4 − 10 = −6	36
8	2	17	1	2 − 1 = 1	1
7	3	8	6	3 − 6 = −3	9
9	1	12	3	1 − 3 = −2	4
$N = 10$					$\sum D^2 = 84$

According to Spearman's formula for rank correlation coefficient (tied rank).

$$R = 1 - \frac{6\sum D^2 + \frac{\sum(t^3 - t)}{12}}{N(N^2 - 1)} \quad N = 10,\ t = 3$$

The correction factor for $(Y) = \frac{3^3 - 3}{12} = \frac{27 - 3}{12} = \frac{24}{12} = 2$

$$R = 1 - \frac{6(84 + 2)}{10(100 - 1)} = 1 - \frac{6 \times 86}{990}$$

$$= 1 - \frac{516}{990} = \frac{990 - 516}{990} = \frac{474}{990} = 0.478 \simeq 0.48$$

There is marked relationship & therefore those who have aptitude for mathematics are likely to be better computer programmer.

Example 17. The final position of 12 clubs in a football league of Calcutta and the average attendance at their home matches were as follows.

Club:	A	B	C	D	E	F	G	H	I	J	K	L
Position:	1	2	3	4	5	6	7	8	9	10	11	12
Attendance:	27	30	18	25	32	12	19	11	32	12	12	15

Calculate coefficient of correlation by ranks and comment on your result.

Solution:

Club Position (X)		*Attendance (Y)*		*Rank Difference*	
Position	*Rank (R_1)*	*Presence of Spectator*	*Rank (R_2)*	*$D = R_1 - R_2$*	D^2
1	1	27	4	1 − 4 = −3	09
2	2	30	3	2 − 3 = −1	01
3	3	18	7	3 − 7 = −4	16
4	4	25	5	4 − 5 = −1	01
5	5	32	1.5	5 − 1.5 = 3.5	12.25
6	6	12	10	6 − 10 = −4	16
7	7	19	6	7 − 6 = 1	1
8	8	11	12	8 − 12 = −4	16
9	9	32	1.5	9 − 1.5 = 7.5	56.25
10	10	12	10	10 − 10 = 0	00
11	11	12	10	11 − 10 = 1	1
12	12	15	8	12 − 8 = 4	16
$N = 12$					$\sum D^2 = 145.5$

$$R = 1 - \frac{6\sum D^2 + \frac{\sum(t^3 - t)}{12}}{N(N^2 - 1)}$$

Correction factor (Y) 2, 3

$$= \frac{(t^3 - t)}{12} = \frac{2^3 - 2}{12} + \frac{3^3 - 3}{12} = \frac{8 - 2}{12} + \frac{27 - 3}{12} = \frac{6}{12} + \frac{9}{12} = .5 + 2 = 2.5$$

$$R = 1 - \frac{6(145.5 + 2.5)}{12(144 - 1)} = 1 - \frac{6 \times 148}{12 \times 143}$$

$$= 1 - \frac{888}{1716} = \frac{1716 - 888}{1716} = \frac{828}{1716} = .482 = 0.48$$

I. There is a marked relationship between the position of the clubs in the league & the number of spectators. It is not high relationship.

II. Although club 9 position, where maximum number of spectators witnesed the match.

• KENDALL'S RANK CORRELATION COEFFICIENT

It is a nonparametric correlation coefficient for simple linear correlation between the ranks of the individuals or cases with respect to the two corresponding variables.

I. Professor **M.G Kendall** introduced this. He denoted this rank by Greek letter τ **(tau)**.
II. It is less powerful than product moment r but more powerful than spearman's **rho.**.
III. The value of **tau** range from –1.00 to + 1.00. The value 0.00 indicates absence of correlation.

Working Procedure:

I. Ranks are first assigned in an ascending order of magnitude to the scores of each variables (Continuous or discontinuous) separately.
II. The ranks (R_1) of the variable with no tied score are arranged in an ascending order along a column. And each such R_1 rank is paired in the adjoining column with the rank (R_2) of the other variable in the same individual.
III. If neither or each of the variables has tied scores the ranks of any of them are arranged as R_1 ranks in the ordered manner pairing them with the ranks (R_2) of the other variable in the respective individuals.
IV. Moving down word from the top of the column of the paired R_2 ranks of the second variables, each R_2 rank is used in turn as the pivotal rank for comparing with every successive subsequent rank following the pivotal rank.
V. Each subsequent R_2 rank is counted here as +1, 0 or –1 according as it exceeds. Equals (tied with) or lower than the particular pivotal rank.
VI. Such counts of subsequent ranks are entered in a third column and totalled score as $S\ ({}^NC_2)$.
VII. Here ranks are assigned to N individuals range from 1 to N.

Kendall's rank correlation (τ) *i.e.*, tau is given by

$$\text{tau } (\tau) = \frac{\text{Total Score}}{\text{Maximum Possible Score}} = \frac{S}{{}^NC_2} = \frac{S}{\frac{N(N-1)}{2}} = \frac{2S}{N(N-1)}$$

Significance:

1. For small sample ($N \leq 10$) the computed tau is compared with critical tau value for given sample.
2. In case of larger sample ($N \geq 10$) tau has symmetric & nearly normal sampling distribution.

Example 1. The marks obtained by seven PG students of Zoology of Serampore College in Biostatistics and Biophysics are given below.

	Students						
Subject:	**A**	**B**	**C**	**D**	**E**	**F**	**G**
Biostatistics:	**56**	**44**	**53**	**77**	**69**	**60**	**40**
Biophysics:	**33**	**42**	**46**	**62**	**60**	**60**	**32**

Calculate Kendall's coefficient of ranks us. Comment on your result.

Solution:

I. Let us first find the ranks of marks of seven students in Biostatistics & Biophysics.
II. Calculation of ranks

Students	*Biostatistics*		*Biophysics*	
	Marks	*Ranks*	*Marks*	*Ranks*
A	56	4	33	6
B	44	6	42	5
C	53	5	46	4
D	77	1	62	1
E	69	2	60	2.5
F	60	3	60	2.5
G	40	7	32	7

As there is tie for the 2nd & 3rd places in marks of Biophysics. We find the average *i.e.*,

$$\frac{2+3}{2} = \frac{5}{2} = 2.5.$$

Here, $N = 7$

R_1	R_2	*Count of*	*Subsequent Ranks*
1	1	+ 1 + 1 + 1 + 1 + 1 + 1	= 6
2	2.5	+ 0 + 1 + 1 + 1 + 1	= 4
3	2.5	+ 1 + 1 + 1 + 1	= 4
4	6	− 1 − 1 + 1	= − 1
5	4	+ 1 + 1	= 2
6	5	+ 1	= 1
7	7		
			16

Hence the Kendell's coefficient of rank correlation is

$$\tau \text{ (tau)} = \frac{2S}{N(N-1)} = \frac{2 \times 16}{7(7-1)} = \frac{2 \times 16}{7 \times 6} = \frac{16}{21} = 0.76$$

The result $\tau = 0.76$ indicates that there is high positive correlation between the marks of Biostatistics and Biophysics.

Example 2. Marks obtained by 10 students of R. K. Mission Vidyamandira in Microbiology & Chemistry are given below.

	Students									
Subject:	**A**	**B**	**C**	**D**	**E**	**F**	**G**	**H**	**I**	**J**
Microbiology:	**80**	**88**	**76**	**74**	**68**	**65**	**40**	**43**	**27**	**36**
Chemistry:	**72**	**84**	**90**	**66**	**54**	**50**	**54**	**38**	**30**	**43**

Calculate Kendall's coefficient of rank correlation between the performances in two subjects for these condition.

Solution:

Let us first find the ranks of ten students in Microbiology & Chemistry.

Students	*Microbiology*		*Chemistry*	
	Marks	*Ranks*	*Marks*	*Ranks*
A	80	2	72	3
B	88	1	84	2
C	76	3	90	1
D	74	4	66	4
E	68	5	54	5.5
F	65	6	50	7
G	40	8	54	5.5
H	43	7	38	9
I	27	10	30	10
J	36	9	43	8

As there is tie for the 5th & 6th places in marks of chemistry. We find the average *i.e.*, $\frac{5+6}{2} = 5.5$.

Here $N = 10$

R_1	R_2	Count of	Subsequent Ranks
1	2	+ 1 − 1 + 1 + 1 + 1 + 1 + 1 + 1 + 1	+ 7
2	3	− 1 + 1 + 1 + 1 + 1 + 1 + 1 + 1	+ 6
3	1	+ 1 + 1 + 1 + 1 + 1 + 1 + 1	+ 7
4	4	+ 1 + 1 + 1 + 1 + 1 + 1	+ 6
5	5.5	+ 1 + 0 + 1 + 1 + 1	+ 4
6	7	+ 1 − 1 + 1 + 1	+ 2
7	9	− 1 − 1 + 1	− 1
8	5.5	+ 1 + 1	+ 2
9	8	+ 1	+ 1
10	10		
			+ 34

Total score = 34

Here $N = 10$

Hence the required Kendall's coefficient of rank correlation *i.e.*,

$$\tau = \frac{2S}{N(N-1)} = \frac{2 \times 34}{10 \times 9} = \frac{34}{45} = 0.7555 = 0.76$$

Example 3. The ranks of 10 individuals at the start and on the finish of course training are given below.

Individuals:	**A**	**B**	**C**	**D**	**E**	**F**	**G**	**H**	**I**	**J**
Rank before:	**6**	**3**	**5**	**4**	**1**	**2**	**7**	**8**	**10**	**9**
Rank after:	**3**	**4**	**6**	**7**	**1**	**5**	**8**	**9**	**2**	**10**

Calculate Kendall's coefficient of rank correlation.

Solution: Let us first find the ranks of 10 individuals.

Calculation of ranks:

Rank before	Rank after	Count of Subsequent Ranks	
E = 1	1	+ 1 + 1 + 1 + 1 + 1 + 1 + 1 + 1 + 1	+ 9
F = 2	5	− 1 + 1 + 1 − 1 + 1 + 1 + 1 − 1	+ 2
B = 3	4	+ 1 + 1 − 1 + 1 + 1 + 1 − 1	+ 3
D = 4	7	− 1 − 1 + 1 + 1 + 1 − 1	0
C = 5	6	− 1 + 1 + 1 + 1 − 1	+ 1
A = 6	3	+ 1 + 1 + 1 − 1	+ 2
G = 7	8	+ 1 + 1 − 1	+ 1
H = 8	9	+ 1 − 1	0
J = 9	10	− 1	− 1
I = 10	2		
			+ 17

$N = 10$

$$\tau = \frac{2S}{N(N-1)} = \frac{2 \times 17}{10 \times 9} = \frac{17}{45} = 0.377 = 0.38$$

Example 4. The eight students of Charusila Bose Balika Vidyalaya (W.B) obtained the following marks in Physics & Mathematics.

Subject:	A	B	C	D	E	F	G	H
Mathematics:	60	70	65	55	75	40	30	35
Physics:	45	50	45	55	54	60	40	37

Calculate Kendall's coefficient of rank correlation. & comment on the result.

Solution:

I. Let us first find the ranks of marks of eight students in Mathematics & Physics.

II. *Calculation of ranks:*

Students	*Mathematics*		*Physics*	
	Marks	*Ranks*	*Marks*	*Ranks*
A	60	4	45	5.5
B	70	2	50	4
C	65	3	45	5.5
D	55	5	55	2
E	75	1	54	3
F	40	6	60	1
G	30	8	40	7
H	35	7	37	8

As tie for the 5th & 6th place in the marks of Physics. We find the average $\frac{5+6}{2} = 5.5$.

Here $N = 8$

R_1	R_2	*Count of Subsequent Ranks*	
1	3	+ 1 + 1 + 1 − 1 − 1 + 1 + 1	= 5
2	4	+ 1 + 1 − 1 − 1 + 1 + 1	= 2
3	5.5	0 − 1 − 1 + 1 + 1	= 0
4	5.5	− 1 − 1 + 1 + 1	= 0
5	2	− 1 + 1 + 1	= 1
6	1	+ 1 + 1	= 2
7	8	− 1	= − 1
8	7		
N = 8		Total score	= 9

The Kendall's coefficient of rank correlation in $\tau = \frac{2S}{N(N-1)} = \frac{2 \times 9}{8 \times 9} = \frac{2}{8} = 0.25$

Low positive correlation between the marks of Math & Physics.

Example 5. Work out Kendall's coefficient of rank correlation between gill weight and body weight of sample of 10 carp fishes.

Fish:	**1**	**2**	**3**	**4**	**5**	**6**	**7**	**8**	**9**	**10**
Gill weight (gm):	**0.15**	**0.11**	**0.19**	**0.30**	**0.13**	**0.20**	**0.25**	**0.28**	**0.16**	**0.22**
Body weight (gm):	**11.20**	**8.10**	**13.20**	**14.50**	**8.45**	**12.20**	**14.35**	**14.00**	**11.25**	**9.50**

Solution: Let us first find the ranks of weights (gill & body) of 10 carp fishes.

Fish	*Gill*		*Body*	
	Weight	*Ranks*	*Weight*	*Ranks*
1	0.15	3	11.20	4
2	0.11	1	8.10	1
3	0.19	5	13.20	7
4	0.30	10	14.50	10
5	0.13	2	8.45	2
6	0.20	6	12.20	6
7	0.25	8	14.35	9
8	0.28	9	14.00	8
9	0.16	4	11.25	5
10	0.22	7	9.50	3

Here $N = 10$

R_1	R_2	*Count of Subsequent Ranks*	
1	1	+ 1 + 1 + 1 + 1 + 1 + 1 + 1 + 1 + 1	= 9
2	2	+ 1 + 1 + 1 + 1 + 1 + 1 + 1 + 1	= 8
3	4	+ 1 + 1 + 1 − 1 + 1 + 1 + 1	= 5
4	5	+ 1 + 1 − 1 + 1 + 1 + 1	= 4
5	7	− 1 − 1 + 1 + 1 + 1	= 1
6	6	− 1 + 1 + 1 + 1	= 2
7	3	− 1 + 1	= 0
8	9	+ 1	= 1
9	8		
10	10		
$N = 10$		Total score	= 30

Kendall's coefficient of ranks correlation is $\tau = \dfrac{2S}{N(N-1)} = \dfrac{2 \times 30}{10 \times 9} = \dfrac{6}{9} = \dfrac{2}{3} = 0.666 = 0.67$

• Scatter Diagram or Dot Diagram Method:

It is the simplest graphical device of showing the correlation between the two variables (X & Y). Here statistical data are plotted against each pair of values of two variables by dots and then cluster of dots make a shape. Thus the diagrammatic representation of bivariate data is known as **scatter diagram.**

The scatter diagram may indicate both degree and the type of correlation.

• Procedure:

I. Here the values of one variable (X) and other variable (Y) are plotted along the X-axis and Y-axis.

II. If there are 'n' pairs of observation, finally the graph paper will contain 'n' points.

- **Observations:**

I. If the pattern of points (*dots*) on the scatter diagram shows a linear path diagonally across the graph paper from the bottom left hand corner to the top right, the correlation will be positive *i.e.*, $r = +1$.

II. If the pattern of dots are lying on straight line rising from the upper left hand corner to the lower right hand corner of the diagram, the correlation is said to be perfectly negative *i.e.*, $r = -1$.

III. When the dots do not indicate any straight line tendency, but a swarm or concentration around a curved line, correlation is small. Infact, if no straight line tendency is noticed, correlation will be zero, $r = 0$.

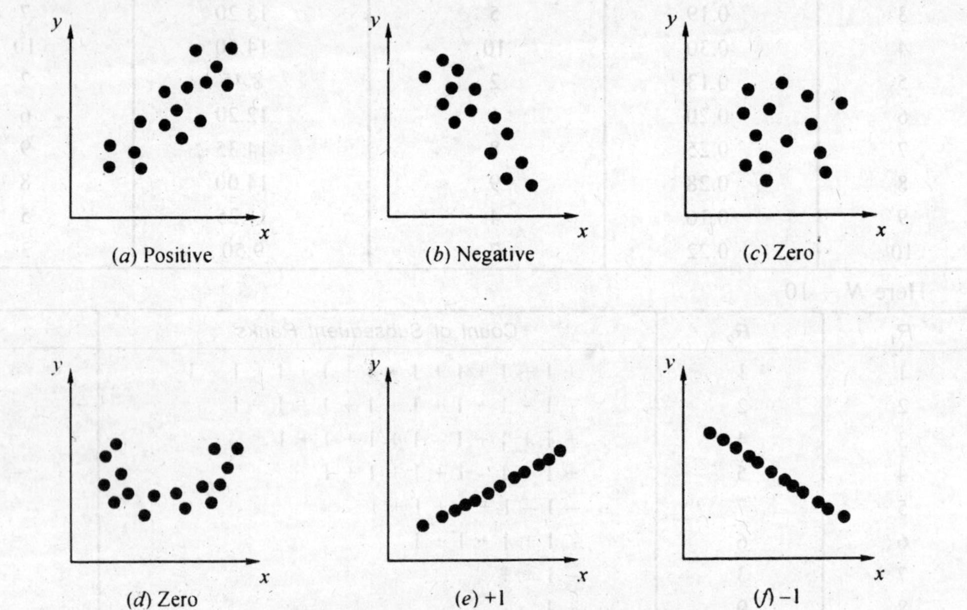

Fig. 13.3 Scatter diagrams showing different types and degress of correlation (*a, b, c, d, e,* & *f*).

Merits & Demerits of Scatter Diagram:

- **Merits:**

I. It is a simple and non mathematical method of studying correlation.

II. It is easy to understand.

III. It is not influenced by the size of the extreme items.

IV. It may be treated as the first step in investigating the relation between the two variables.

- **Demerits**

By applying this method we can get an idea about the direction of correlation & also whether it is high or low. But exact degree of correlation between the two variables can not be established by applying this method.

Example: Prove that correlation coefficient r lies between –1 and +1.

Solution: Let '*X*' and '*Y*' be the variables with '*N*' pairs of observation with means $\overline{X}, \overline{Y}$ and standard deviation σ_x and σ_y respectively.

If we denote

$$U = \frac{X - \overline{X}}{dx} \quad V = \frac{Y - \overline{Y}}{dy}$$

Then $$U^2 = \sum\left(\frac{(X - \overline{X})}{\sigma x}\right)^2 = \frac{\sum (x - \overline{x})^2}{\sigma x^2} = \frac{n\sigma x^2}{\sigma x^2} = n$$

$\because$ $$\sigma x^2 = \frac{1}{n}\sum (X - \overline{X})^2$$

Similarly

$$V^2 = \left(\frac{\sum (Y - \overline{Y})^2}{\sigma x}\right) = \left(\frac{\sum (Y - \overline{Y})^2}{\sigma y^2}\right) = \frac{n\sigma y^2}{\sigma y^2} = n$$

Again $$UV = \sum\left(\frac{X - \overline{X}}{\sigma x}\right) = \left(\frac{Y - \overline{Y}}{\sigma y^2}\right) = \frac{(X - \overline{X})(Y - \overline{Y})}{\sigma x \sigma y}$$

$$UV = nr$$

$\because$ $$r = \frac{1}{n}\frac{\sum (X - \overline{X})(Y - \overline{Y})}{\sigma x \sigma y}$$

$$nr = \frac{\sum (X - \overline{X})(Y - \overline{Y})}{\sigma x \sigma y}$$

$(U + V)^2$ can never be negative, because it is a perfect square. Similarly $(U - V)^2$ can never be negative.

So $\sum (U + V)^2 \geq 0$

or $\sum (U^2 + V^2 + 2UV) \geq 0$

or $\sum (n + n + 2nr) \geq 0$

or $\sum (2n + 2nr) \geq 0$

or $2n(1 + r) \geq 0$

or $1 + r \geq 0 (\because n > 0)$

$\therefore \quad r \geq -1$ (1)

or $\sum (U - V)^2 \geq 0$

or $\sum (U^2 - V^2 - 2UV) \geq 0$

or $(n + n - 2nr) \geq 0$

or $2n(1 - r) \geq 0$

or $1 - r \geq 0 \ (\because n > 0)$

$\therefore \quad r \leq 1$ (2)

From (1) & (2) we get

$-1 \leq r \leq 1 \text{ or} \ |r| \leq 1$

So we say that the numerical value of correlation coefficient (*r*) lies between –1 and +1.

Write down the correct answer: correlation coefficient can take (I) any positive value, (II) Any value between –1 and 2 (III) Any real value (IV) Any value between 0 and 2.

Ans.: III

Example: State in each case whether you would expect to find a positive correlation, a negative correlation or no correlation:

I. The ages of husband & wife.
II. Shoe size & intelligence.
III. Years of education & income.
IV. Amount of rainfall and yield of crop.

Solution:

I. Positive.
II. No correlation.
III. Positive.
IV. Positive.

Example: If two variants are independent, their correlation co-efficient is zero. Is the converse true?

Solution:

I. If two variants are independent, their correlation co-efficient is zero. But the converse is not true.
II. A zero correlation does not necessarily signify that the variables are independent. It only implies that there is no linear relationship between the variables. However non linear relationship may be present.

Q: Length (cm) and weight (g) of 10 fishes are as follows.

Length (X):	32	38	48	43	40	22	41	69	35	64
Weight (Y):	20	22	28	26	25	18	24	32	21	30

Calculate correlation coefficient. *C.U. 2003*

• Partial Correlation:

The study of relationship between two variables excluding the effect of some other variables is called **partial correlation.**

I. Partial correlation provides better relationship between two variables X_1 & X_2, partial out the effect of variables X_3.

$$r_{12.3} = \frac{r_{12} - r_{13}\, r_{23}}{\sqrt{\left(1 - r_{13}^2\right)\left(1 - r_{23}^2\right)}}$$

r_{12}, r_{13} and r_{23} are the simple correlation coefficient between the variables X_1 & X_2, X_1 & X_3 and X_2 & X_3 respectively.

II. The Partial correlation co efficient lies between –1 and +1.

• Features:

I. Partial correlation aims at eliminating or partialling out the effects of the other variables.
II. To examine the effect of one variable upon another offer eliminating the effects of all the other variables.
III. It is the part of product moment r between two given variables.
IV. There are different orders of partial correlation according to the number of variables to be eliminated or partialled out.

- **First order partial *r*:**

I. In this type, correlation exist between two variables (X_1 & X_2) partialling out another variables. (X_3) correlated with both of them.

II. **Example: $r_{12,3}$ partial out the variable X_3 to measure the correlation between X_1 & X_2 without the effect of X_3.**

$$r_{12,3} = \frac{r_{12} - r_{13}\,r_{23}}{\sqrt{\left(1-r_{13}^2\right)\left(1-r_{23}^2\right)}}$$

- **Second order partial *r*:**

I. It involves four inter correlated variables & measures the correlation between two of them. Partialling out the other two of them.

II. **Example: $r_{12,34}$ correlates X_1 and X_2 partialling out the effects of X_3 & X_4.**

$$r_{12,34} = \frac{r_{12.3} - r_{14.3}\,r_{24.3}}{\sqrt{\left(1-r_{14.3}^2\right)\left(1-r_{24.3}^2\right)}} \quad \text{or} \quad r_{12,34} = \frac{r_{12,4} - r_{13.4}\,r_{23.4}}{\sqrt{\left(1-r_{13.4}^2\right)\left(1-r_{23.4}^2\right)}}$$

- r_{12} is the correlation coefficient between X_1 and X_2 ignoring altogether any influences of X_3. $r_{12,3}$ is also the correlation coefficient between X_1 and X_2 after eliminating the influence of X_3 from both.
- r_{12} is total correlation coefficient & $r_{12,3}$ is partial correlation coefficient.
- r_{12} is same as r_{21}. Similarly $r_{12,3}$ is the same as $r_{21.3}$.

Example 1. Given r_{12} = +0.65, r_{13} = +0.60 and r_{23} = +0.90 calculate $r_{12,3}$.

Solution:

$$r_{12,3} = \frac{r_{12} - r_{13}\cdot r_{23}}{\sqrt{\left(1-r_{13}^2\right)\left(1-r_{23}^2\right)}} = \frac{0.65 - 0.60 \times 0.90}{\sqrt{\left[1-(.60)^2\right]\left[1-(.90)^2\right]}}$$

$$= \frac{0.65 - 0.54}{\sqrt{(1-0.36)(1-0.81)}} = \frac{0.11}{\sqrt{0.64 \times 019}}$$

$$= \frac{0.11}{\sqrt{0.1216}} = \frac{0.11}{0.3487} = 0.315 = 0.32$$

Example 2. Given r_{12} = +0.41, r_{13} = 0.71 and r_{23} = 0.5. Find $r_{12.3}$.

Solution:

$$r_{12.3} = \frac{0.41 - 0.71 \times 0.5}{\sqrt{\left[1-(0.71)^2\right]\left[1-(0.5)^2\right]}} = \frac{0.41 - 0.355}{\sqrt{(1-0.5041)(1-0.25)}}$$

$$= \frac{0.055}{\sqrt{0.4959 \times 0.75}} = \frac{0.055}{\sqrt{0.3719}} = \frac{0.055}{0.60985} = +0.09$$

Example 3. The product moment *r* scores (r_{12}) between O_2 consumption (X_1 ml/minute) and tracheal ventilation volume (X_2 ml/minute) we found to be +0.75 in a sample of 53 *Anopheles* mosquitoes; the *r* scores (r_{13}) between O_2 consumption (X_1 ml/minute) and atmospheric O_2 tension (X_3 mm/Hg) a mounted to +0.35; while *r* scores (r_{23}) between tracheal ventilation (X_2 ml/minute) and atmospheric O_2 tension (X_3 mlHg) was found to be +0.25. Find if there is a significant partial correlation between X_1 and X_2, eliminating X_3.

($\alpha = 0.05$)

Critical *t* scores:

$t_{0.01(52)} = 2.007$ $t_{0.01(51)} = 2.008$ $t_{0.05(50)} = 2.009$

Solution:

$r_{12} = +0.75$ $r_{13} = +0.35$ $r_{23} = +0.25$ $n = 53$

$$r_{123} = \frac{r_{12} - r_{13}.r_{23}}{\sqrt{\left[1 - r_{13}^2\right]\left[1 - r_{23}^2\right]}} = \frac{0.75 - 0.35 \times 0.25}{\sqrt{\left[1 - (0.35)^2\right]\left[1 - (0.25)^2\right]}}$$

$$= \frac{0.75 - 0.0875}{\sqrt{(1 - 0.1225)(1 - 0.0625)}} = \frac{0.6625}{\sqrt{0.8775 \times 0.9375}}$$

$$= \frac{0.6625}{\sqrt{0.8226562}} = \frac{0.6625}{0.9070} = 0.73$$

[Computed *r* is transformed into *t* score for interpretation. Using standard error (*Sr*) of *r*. The *df* of the computed *t* is taken here as $n - 3$.]

$$Sr_{123} = \sqrt{\frac{1 - r_{123}^2}{n - 3}} = \sqrt{\frac{1 - (0.73)^2}{53 - 3}} = \sqrt{\frac{1 - 0.5329}{50}}$$

$$= \sqrt{\frac{0.4671}{50}} = \sqrt{0.009342} = 0.09665 = 0.097$$

$$t = \frac{r_{123}}{Sr_{123}} = \frac{0.73}{00.97} = 7.526$$

$df = n - 3 = 53 - 3 = 50$

Critical $t_{0.05(50)} = 2.009$

The computed *t* exceeds the critical $t_{0.05}$. So *P* is too low & the *Ho* is rejected.

Hence there is significant partial correlation between X_1 & X_2 eliminating X_3 ($P < 0.05$).

Example 4. The product moment *r* scores (r_{12}) between gill weights (X_1 gms) and trunk lengths (X_2 cm) was found to be +0.55 in a sample of 43 carp fishers; the *r* scores (r_{13}) between their gill weights (X_1 gms) and body weights (X_3 gms) a mounted to 0 + .30. While *r* scores (r_{23}) between trunk lengths (X_2 gm) and body weights (X_3 gm) was found to be +0.28.

Find if there is a significant partial correlation between X_1 & X_2 eliminating X_3.

($\alpha = 0.01$)

Critical *t* scores: $t_{0.01(42)} = 2.698$ $t_{0.01(41)} = 2.701$ $t_{0.01(40)} = 2.704$.

Solution:

$r_{12} = +0.55$ $r_{13} = +0.30$ $r_{23} = +0.28$ $n = 43$

$$r_{123} = \frac{r_{12} - r_{13}.r_{23}}{\sqrt{\left(1 - r_{12}^2\right)\left(1 - r_{23}^2\right)}} = \frac{0.55 - 0.30 \times 0.28}{\sqrt{\left[1 - (0.30)^2\right]\left[1 - (0.28)^2\right]}}$$

$$= \frac{0.55 - 0.30 \times 0.28}{\sqrt{(1 - .09)(1 \quad 0.0784)}} = \frac{0.466}{\sqrt{0.91 \times 0.9216}}$$

$$= \frac{0.466}{0.91578} = \frac{0.466}{0.916} = 0.508$$

[Standard error of '*r*' *i.e., Sr* & *df* = *n* – 3]

$$Sr_{123} = \sqrt{\frac{1 - r_{123}^2}{n - 3}} = \sqrt{\frac{1 - (0.508)^2}{43 - 3}} = \sqrt{\frac{1 - .258}{40}}$$

$$= \sqrt{\frac{0.742}{40}} = \sqrt{0.01855} = \sqrt{0.019} = 0.136$$

$$t = \frac{r_{123}}{Sr_{123}} = \frac{0.508}{0.136} = 3.735$$

$$df = n - 3 = 43 - 3 = 40 \quad \alpha = 0.01 \; t_{0.01(40)} - 2.704$$

The computed t exceeds the critical $t_{0.01(40)} = 2.704$. So P is too low ($P < 50$). The Ho is rejected.

Hence there is significant partial correlation between X_1 and X_2 partial ling or eliminating X_3.

Example 5. In a sample of 53 foot ball players, the product moment *r* values between the stroke volume of heart (X_1), the venous return (X_2) and the vascular peripheral resistance (X_3) were found to be as follows. $r_{12} = +0.65$ $r_{13} = +0.12$ $r_{23} = +0.25$. Compute the partials *r* between stroke volume and the venous return, eliminating the effect of peripheral resistance and test its significance.

Solution: $r_{12} = +0.65$ $r_{13} = +0.12$ $r_{23} = +0.25$.

$$r_{12,3} = \frac{r_{12} - r_{13}.r_{23}}{\sqrt{\left(1 - r_{13}^2\right)\left(1 - r_{23}^2\right)}} = \frac{0.65 - 0.12 \times 0.25}{\sqrt{\left[1 - (0.12)^2\right]\left[1 - (0.25)^2\right]}}$$

$$= \frac{0.65 - 0.03}{\sqrt{(1 - 0.0144)(1 - 0.0625)}} = \frac{0.62}{\sqrt{(0.9856)(0.9375)}}$$

$$= \frac{0.62}{\sqrt{0.924}} = \frac{0.62}{0.961249} = 0.644994 = 0.645$$

$$t = \frac{r_{12,3}}{\sqrt{\left(\frac{1 - r_{12,3}^2}{n - 3}\right)}} = \frac{0.645}{\sqrt{\frac{1 - (0.645)^2}{53 - 3}}} = \frac{0.645}{\sqrt{\frac{1 - 0.416}{50}}}$$

$$= \frac{0.645}{\sqrt{\frac{0.58397}{50}}} = \frac{0.645}{\sqrt{0.0116794}} = \frac{0.645}{0.10807} = 5.968$$

$$df = n - 3 = 53 - 3 = 50$$

For a two tail test critical value of $t_{.01(50)} = 2.678$ and critical $t_{.001(50)} = 3.496$.

The computed t is higher than table value of t for 0.001 level. So the partial r is significant *i.e.*, ($P < 0.001$).

Example 6. In a group of 163 students of N.D College Howrah, the product moment r values between intelligence test scores (X_1), tension test scores (X_2) and age (X_3) were found to be as follows: $r_{12} = +0.46$ $r_{13} = +0.35$ $r_{23} = +0.17$. Compute $r_{12,3}$ and find whether it is significant or not.

Solution: $r_{12} = +0.46$ $r_{13} = +0.35$ $r_{23} = +0.17$.

Partial correlation *i.e.*, $$r_{12,3} = \frac{r_{12} - r_{13}.r_{23}}{\sqrt{\left(1-r_{13}^2\right)\left(1-r_{23}^2\right)}} = \frac{0.46 - 0.35 \times 0.17}{\sqrt{1-(0.35)^2 \times 1-(0.17)^2}}$$

$$= \frac{0.46 - 0.0595}{\sqrt{(1-0.1225)(1-0.0287)}} = \frac{0.4005}{\sqrt{0.8775 \times 0.9711}}$$

$$r_{12,3} = \frac{0.4005}{\sqrt{0.8521402}} = \frac{0.4005}{0.9231144} = 0.43385 = +0.434$$

$$t = \frac{r_{12,3}}{\sqrt{\frac{1-(r_{12,3}^2)}{n-3}}} = \frac{0.434}{\sqrt{\frac{1-(0.434)^2}{163-3}}} = \frac{0.434}{\sqrt{\frac{1-0.188356}{160}}}$$

$$= \frac{0.434}{\sqrt{\frac{0.811644}{160}}} = \frac{0.434}{\sqrt{0.0050727}}$$

$$= \frac{0.434}{0.0712228} = 6.935543 = 6.936$$

$$df = n - 3 = 163 - 3 = 160$$

For a two tail test critical $t_{.01} = 2.576$ and critical $t_{.001} = 3.291$.

Here the computed 't' is higher than the critical or table value of t for 0.001 level. So the partial r is significant *i.e.*, $P < 0.001$.

Example 7. The product moment r score (r_{12}) between glomerular filtration rate (X_1 ml/min) and glomerular blood pressure (X_2 ml/Hg) was found to be +0.68 in 53 patient in a health centre of Serampore the r scores (r_{13}) between glomerular filtration rate (X_1 ml/min) and plasma protein osmotic pressure (X_3 mm/Hg) amounted to +0.32; while r scores (r_{23}) between glomerular blood pressure (X_2 mm/Hg) and plasma protein osmotic pressure (X_3 mm/Hg) was found to be +0.18. Find if there is a significant partial correlation between X_1 and X_2 partialling X_3.

Critical t scores: $t_{0.01(52)} = 2.007$ $t_{0.01(51)} = 2.008$ $t_{0.01(50)} = 2.009$

Solution: $r_{12} = +0.68$ $r_{13} = +0.32$ $r_{23} = +0.18$

$$r_{12,3} = \frac{r_{12} - r_{13}.r_{23}}{\sqrt{\left(1-r_{12}^2\right)\left(1-r_{23}^2\right)}} = \frac{0.68 - (0.32 \times 0.18)}{\sqrt{\left[1-(.68)^2\right]\left[1-(.18)^2\right]}}$$

$$= \frac{0.68 - 0.0576}{\sqrt{(1-0.4624)(1-0.0324)}} = \frac{0.06224}{\sqrt{0.5376 \times 0.9676}}$$

$$= \frac{0.06224}{\sqrt{0.5201817}} = \frac{0.6224}{0.721236} = 0.8629$$

The SE of r *i.e.*, Sr

$$Sr_{123} = \sqrt{\frac{1-r_{23}^2}{n-3}} = \sqrt{\frac{1-(0.8629)^2}{53-3}} = \sqrt{\frac{1-0.74459}{50}}$$

$$= \sqrt{\frac{0.2554}{50}} = \sqrt{0.005108} = 0.07147$$

$$t = \frac{r_{123}}{Sr} = \frac{0.8629}{0.07147}$$

• Multiple Correlation:

The study of quantitative assessment of the magnitude and direction of correlation between a given variable and the joint influence of two or more variables is called *multiple correlation.*

$$R_{1.23} = \sqrt{\frac{r_{12}^2 + r_{13}^2 - 2r_{12}r_{13}r_{23}}{1-r_{23}^2}}$$

The $R_{1.23}$ is the *multiple correlation coefficient*. The squared value of multiple correlation coefficient $\left(R_{1.23}^2\right)$ is called the *coefficient* of *determination*. It is always considered as positive: $0 \le R_{1.23} \le 1$.

• **Features:**

I. Here the single variable being correlated is called *criterion* while the variable whose combination is being correlated are called *predictors*.

II. Both criterion and predictors are continuous measurement variables without gap in this scale of scores.

III. Scores of each variable should have normal *i.e.*, symmetrical distribution.

IV. The existence of linear association between the scores of each pair of variable.

• **Calculation of multiple linear correlation:**

I. It is basically calculated from the product moment r values of each pair of variables.

II. For calculating the multiple linear correlation coefficient ($R_{1.23}$) between variables (criterion) X_1 the combination of two other variable (predictors) *i.e.*, X_2 and X_3; the β coefficient (β_2 & β_3) are first calculated from the product moment r values (r_{12}, r_{13} and r_{23}) between the respective variables.

III. β_2 & β_3 act as measures of proportion of variance of the criterion (X_1); associated with the variances of respective predictors (X_2 & X_3).

IV. $\beta_2 = \frac{r_{12} - r_{13}r_{23}}{1-r_{23}^2}$ $\beta_3 = \frac{r_{13} - r_{12}r_{23}}{1-r_{23}^2}$

$$\therefore \quad R_{1.23} = \sqrt{\beta_2 r_{12} + \beta_3 r_{13}} \quad or \quad R_{1.23} = \sqrt{\frac{r_{12}^2 + r_{12}^2 - 2r_{12}r_{13}r_{23}}{1-r_{23}^2}}$$

• R =1 the multiple correlation is linear.

• R = 0 the linear multiple correlation does not exist.

• **Coefficient of multiple determination:**

It is a measure of that proportion of the total variation of the criterion which is determined by the variation of the combined contribution of all the predictors.

$R_{123}^2 = \beta_2 r_{12} + \beta_3 r_{13}$ Criterion = X_1

Predictors = X_2 & X_3.

• **Coefficient of multiple non determination:**

It is a measure of that proportion of the total variance of criterion which is determined by the independent combined contribution of the predictors.

$$K^2_{1.23} = 1 - R^2_{1.23}$$

Example 1. The product moment r scores (r_{12}) between gill weights (X_1) and body weight (X_2) was found to be +0.80 in a sample of 33 fishes; the *r* scores (r_{13}) between their gill weight (X_1) and body length (X_3) amounted to +0.20, while the *r* score (r_{23}) between their body weight (X_2) and body length (X_3) was found to be + 0.30. Find if there is significant multiple linear correlation between (X_1) and the combination of (X_2) and (X_3) (α = 0.01).

Critical *t* scores: $t_{0.01(32)}$ = 2.733, $t_{0.01(31)}$ = 2.744 & $t_{0.01(30)}$ = 2.750.

Solution: $t_{12} = 0.80$ $r_{13} = 0.20$ $r_{23} = 0.30$ $n = 33$

β coefficient are β_2 and β_3

$$\beta_2 = \frac{r_{12} - r_{13}r_{23}}{1 - r^2_{23}} = \frac{0.80 - 0.20 \times 0.30}{1 - (0.30)^2} = \frac{0.80 - 0.60}{1 - 0.09} = \frac{0.20}{0.91} = 0.219 = 0.22$$

$$\beta_3 = \frac{r_{13} - r_{12}r_{23}}{1 - r^2_{23}} = \frac{0.20 - 0.80 \times 0.30}{1 - (0.30)^2} = \frac{0.20 - 0.16}{1 - 0.09} = \frac{0.04}{0.91} = 0.044$$

$$R_{1,23} = \sqrt{\beta_2 r_{12} + \beta_3 r_{13}} = \sqrt{0.22 \times 0.80 + 0.04 \times 0.20}$$

$$= \sqrt{0.176 + 0.008} = \sqrt{0.184} = 0.428 = 0.43$$

[*SE* of $R_{1,23}$ *i.e.*, $SR_{1,23}$]

$$SR_{1,23} = \frac{1}{\sqrt{n-3}} = \frac{1}{\sqrt{33-3}} = \frac{1}{\sqrt{30}} = \frac{1}{5.477} = \frac{1}{5.5} = 0.181$$

$$t = \frac{R_{1,23}}{SR_{1,23}} = \frac{0.43}{0.181} = 2.375 = 2.4$$

$df = n\ 3 = 33 - 3 = 30$ $\alpha = 0.05$ Critical value $t_{0.05(30)} = 2.750.$

As the computed *t* does not exceeds the critical $t_{0.05}$, *P* is too high ($P > 0.05$). The *Ho* is accepted the computed $R_{1,23}$ is non significant.

Example 2. The product moment *r* scores (r_{12}) between gill weight (X_1 gm) and trunk length (X_2 cm) was found to be +0.55 in a sample of 43 *labeo rohita*; the *r* scores (r_{13}) between their gill weight (X_1 gms) and body weights (X_3 gms) amounted to +0.30; while r scores (r_{23}) between their trunk lengths (X_2 cm) body weight (X_3 gms) was found to be +0.28. Find whether or not there is a significant multiple linear correlation between the combination of X_1 and X_2. (α = 0.01)

Critical *t* scores: $t_{0.01(42)}$ = 2.698 $t_{0.01(41)}$ = 2.701 $t_{0.01(40)}$ = 2.704.

Solution: $r_{12} = 0.55$ $r_{13} = 0.30$ $r_{23} = 0.28$ $n = 43$

$$\beta_2 = \frac{r_{12} - r_{13}r_{23}}{1 - r^2_{23}} = \frac{0.55 - 0.30 \times 0.28}{1 - (0.28)^2} = \frac{0.55 - 0.084}{1 - 0.0784} = \frac{0.466}{0.926} = 0.503$$

$$\beta_3 = \frac{r_{13} - r_{12}r_{23}}{1 - r^2_{23}} = \frac{0.30 - 0.55 \times 0.28}{1 - (0.28)^2} = \frac{0.30 - 0.154}{1 - 0.0784} = \frac{0.146}{0.9216} = 0.1584$$

$$R_{1,23} = \sqrt{B_2 r_{12} + B_3 r_{13}} = \sqrt{0.503 \times 0.55 + 0.158 \times 0.30}$$

$$= \sqrt{0.277 + 0.0474} = \sqrt{0.3244} = 0.569$$

SE of $R_{1,23}$ *i.e.*,

$$SR_{1,23} = \frac{1}{\sqrt{n-3}} = \frac{1}{\sqrt{43-3}} = \frac{1}{\sqrt{40}} = \frac{1}{6.32} = 0.158$$

$$t = \frac{R_{1,23}}{SR_{1,23}} = \frac{0.569}{0.158} = 3.60$$

$df = n - 3 = 43 - 3 = 40$ $\alpha = 0.01$ Critical value $t_{0.05(40)} = 2.704$.

As the computed t exceeds the critical $t_{0.05}$. P is too low ($P < 0.05$). The *Ho* is rejected and the computed $R_{1,23}$ is significant.

Example 3. Find whether or not there is a significant multiple linear correlation between O_2 consumption (X_1 ml/minute) and tracheal ventilation volume (X_2 ml/minute) partialling out atmospheric O_2 tension (X_3 mm/Hg). Using the following data of a sample of 53 grasshoppers. ($\alpha = 0.05$)

$r_{12} = +0.75$, $r_{13} = +0.35$ & $r_{23} = +0.25$

$t_{0.05(52)} = 2.007$, $t_{0.05(51)} = 2.008$ & $t_{0.05(50)} = 2.009$,

Solution: $r_{12} = +0.75$, $r_{13} = +0.35$, $r_{23} = +0.25$ & $n = 53$

$$\beta_2 = \frac{r_{12} - r_{13}r_{23}}{1 - r_{23}^2} = \frac{0.75 - 0.35 \times 0.25}{1 - (0.25)^2} = \frac{0.75 - 0.0875}{1 - 0.0625} = \frac{0.6625}{0.9375}$$

$$= 0.7066 = 0.71$$

$$\beta_3 = \frac{r_{13} - r_{12}r_{23}}{1 - r_{23}^2} = \frac{0.35 - 0.75 \times 0.25}{1 - (0.25)^2} = \frac{0.35 - 0.1875}{1 - 0.0625} = \frac{0.1625}{0.9375}$$

$$= 0.1733 = 0.17$$

$$R_{1,23} = \sqrt{B_2 r_{12} + B_3 r_{13}} = \sqrt{0.71 \times 0.75 + 0.17 \times 0.35}$$

$$= \sqrt{0.5325 + 0.0595} = \sqrt{0.592} = 0.7694 = 0.77$$

SE of $R_{1,23}$ *i.e.*,

$$SR_{1,23} = \frac{1}{\sqrt{n-3}} = \frac{1}{\sqrt{53-3}} = \frac{1}{\sqrt{50}} = \frac{1}{7.07} = 0.141$$

$$t = \frac{R_{1,23}}{SR_{1,23}} = \frac{0.569}{0.158} = 3.60$$

$df = n - 3 = 43 - 3 = 40$ $\alpha = 0.01$ Critical value $t_{0.05(40)} = 2.704$.

As the computed e exceeds the critical $t_{0.05}$. P is too low ($P < 0.05$). The *Ho* is rejected & the computed $R_{1,23}$ is significant.

Example 4. From the following data calculate multiple correlation co efficient

(*a*) $r_{12} = 0.77$ $r_{13} = 0.72$ $r_{23} = 0.52$

(*b*) $r_{12} = 0.70$ $r_{13} = 0.06$ $r_{23} = 0.4$

(a) $$R_{1,23} = \sqrt{\frac{r_{12}^2 + r_{13}^2 - 2r_{12}r_{13}r_{23}}{1 - r_{23}^2}}$$

$$= \sqrt{\frac{(0.77)^2 + (0.72)^2 - 2 \times 0.77 \times 0.72 \times 0.52}{1 - (0.52)^2}}$$

$$= \sqrt{\frac{0.5929 + 0.5184 - 0.576576}{1 - 0.2704}}$$

$$= \sqrt{\frac{1.1113 - 0.576576}{0.7296}} = \sqrt{\frac{0.534724}{0.7296}} = \sqrt{0.73229} = 0.856 = 0.86$$

(b) $r_{12} = 0.70$ $r_{13} = 0.06$ $r_{23} = 0.4$

$$R_{12.3} = \sqrt{\frac{(0.7)^2 + (0.6)^2 - 2 \times 0.7 \times 0.6 \times 0.4}{1 - (0.4)^2}}$$

$$= \sqrt{\frac{49 + 0.36 - 0.336}{1 - 0.16}} = \sqrt{\frac{0.85 - 0.336}{0.84}}$$

$$= \sqrt{\frac{0.514}{0.84}} = \sqrt{0.6119047} = 0.7822 = 0.78$$

Example 5. From the following data, calculate multiple correlation coefficient ($R_{12.3}$) and partial correlation coefficient ($r_{1.23}$).

$r_{12} = +0.65$, $r_{13} = +0.60$ and $r_{23} = +0.90$

Solution: Multiple correlation coefficient *i.e.,* $R_{1,23}$.

$$R_{12.3} = \sqrt{\frac{r_{12}^2 + r_{13}^2 - 2r_{12}r_{23}r_{13}}{1 - r_{23}^2}} = \sqrt{\frac{(0.65)^2 + (0.60)^2 - 2 \times 0.65 \times 0.60 \times 0.90}{1 - (0.90)^2}}$$

$$= \sqrt{\frac{0.4225 + 0.36 - 0.702}{1 - 0.81}} = \sqrt{\frac{0.7825 - 0.702}{0.19}}$$

$$= \sqrt{\frac{0.0805}{0.19}} = \sqrt{0.42368} = 0.6509 = 0.651$$

Or by using β_2 & β_3 we can also get $R_{1,23}$.

$$\beta_2 = \frac{r_{12} - r_{13}r_{23}}{1 \ (r_{23})^2} = \frac{0.65 - 0.60 \times 0.90}{1 - (0.90)^2} = \frac{0.65 - 0.54}{1 - 0.81} = \frac{0.11}{0.19} = 0.5789 = 0.579$$

$$\beta_3 = \frac{r_{12} - r_{13}r_{23}}{1 - (r_{23})^2} = \frac{0.60 - 0.65 \times 0.90}{1 - (0.90)^2} = \frac{0.60 - 0.585}{1 - 0.81} = \frac{0.015}{0.19} = 0.0789$$

$$R_{1.2.3} = \sqrt{\beta_2 r_{12} + \beta_3 r_{12}} = \sqrt{0.579 \times 0.65 + 0.0789 \times 0.60} = \sqrt{0.376 + 0.473}$$

$$= \sqrt{0.4233} = 0.6506 = 0.651$$

The partial correlation coefficient.

$$r_{123} = \frac{r_{12} - r_{13}r_{23}}{\sqrt{(1-r_{13}^2)(1-r_{23}^2)}} = \frac{0.65 - 0.60 \times 0.90}{\sqrt{\left[1-(0.60)^2\right]\left[1-(0.90)^2\right]}}$$

$$= \frac{0.65 - 0.54}{\sqrt{(1-0.36)(1-0.81)}} = \frac{0.11}{\sqrt{0.64 \times 0.19}} = \frac{0.11}{\sqrt{0.1216}}$$

$$= \frac{0.11}{0.3487} = \frac{0.11}{0.349} = 0.315 = 0.32$$

Example 6. Given the following coefficients, calculate the multiple correlation coefficient.

$r_{12} = 0.41$, $r_{13} = 0.71$ and $r_{23} = 0.5$

Solution:

$$R_{1.23} = \sqrt{\frac{r_{12}^2 + r_{13}^2 - 2r_{12}r_{13}r_{23}}{1-r_{23}^2}} = \sqrt{\frac{(0.41)^2 + (0.71)^2 - 2 \times 0.41 \times 0.71 \times 0.5}{1-(0.5)^2}}$$

$$= \sqrt{\frac{0.1681 + 0.5041 - 0.2911}{1-0.25}} = \sqrt{\frac{0.6722 - 0.2911}{0.75}}$$

$$= \sqrt{\frac{0.3811}{0.75}} = \sqrt{0.508133} = 0.712 = +0.71$$

Example 7. Find whether or not there is a significant multiple linear correlation between glomerular filtration rate (X_1 ml/min) on glomerular blood pressure (X_2 mm/Hg) and capsular fluid pressure (X_3 mm/Hg) using the following data of a sample of 40 monkeys of Puri (Orissa) ($\alpha = 0.05$)

$r_{12} = +0.82$, $r_{13} = -0.21$ and $r_{23} = 0.18$

$t_{0.05(37)} = 2.026$, $t_{0.05(38)} = 2.024$ and $t_{0.05(40)} = 2.021$

Solution:

$$R_{1,23} = \sqrt{\frac{(0.82)^2 + (-0.21)^2 - 2 \times 0.82 \times (-0.21) \times 0.18}{1-(0.18)^2}}$$

$$= \sqrt{\frac{0.6724 + 0.0441 - (-0.061991)}{1-0.0324}} = \sqrt{\frac{0.6724 + 0.0441 + 0.06992}{0.9676}}$$

$$= \sqrt{\frac{0.778492}{0.9676}} = \sqrt{0.8045597} = \sqrt{0.8046} = 0.89699 = +0.8969$$

SE of $R_{1,23}$ *i.e.*,

$$SR_{1,23} = \frac{1}{\sqrt{n-3}} = \frac{1}{\sqrt{40-3}} = \frac{1}{\sqrt{37}} = \frac{1}{6.082} = 0.16441$$

$$t = \frac{R_{1,23}}{SR_{1,23}} = \frac{0.8969}{0.1644} = 5.455$$

The critical t value $t_{0.05(37)} = 2.026$. Computed t value is higher & significantly different.

14

CHAPTER

REGRESSION

Regression is used to denote estimation or prediction of the average value of one variable for a specified value of the other variable. One of the variables is called **independent** or the **explained variable** and the other is called **dependent** or the **explaining variable**.

> "Regression is the measure of the average relationship between two or more variables in terms of the original units of the data." *M.M. Blair*

The estimation or prediction is done by means of suitable equation derived on the basis of available bivariate data. Such an equation is known as *Regression equation* and its geometrical representation is called *Regression curve*.

I. Regression Equation of X on Y is

$$X - \bar{X} = b_{xy}(Y - \bar{Y})$$

$$= r\frac{\sigma x}{\sigma y}(Y - \bar{Y})$$ [It estimates X for given Value of Y]

X = Value of x

$\bar{X}$ = Mean of x

σx = Standard deviation of x series

r = Correlation coefficient

Y = Value of Y

$\bar{Y}$ = Mean of Y

σy = Standard deviation of y series

b = Slope or coefficient of regression

II. Regression Equation Y on X is

$$Y - \bar{Y} = b_{yx}(X - \bar{X})$$

$$= r\frac{\sigma y}{\sigma x}(X - \bar{X})$$ [It estimates Y for a given value of X]

- **Regression Lines:**

If a bivariate data are plotted as points on graph paper, it will be found that the concentration point follows a certain pattern showing the relationship between the variables. When the trend points are found to be linear, we determine the best fitting straight line by *Least Square Method. Such straight lines which are used to obtain best estimates of one variable for given values of the other, are called* ***regression lines***.

If two variables are linearly related, then that relation can be expressed as $Y = bx + a$.

Where 'b' is the slope of the line relating Y to X and 'a' is the 'Y' intercept of that line.

> A *line of regression* is the straight line which gives the best fit in the least square sense to given sets of data.

- **Regression Coefficient:**

I. The regression coefficient (b) is an expression of how much (*on the average*) one dependent variable (Y) may be expected to change per unit change in some other independent variable (X).

II. It is denoted by letter 'b'.

III. The regression coefficient of Y on X is

$$= byx = r\frac{\sigma y \ (S.D.of\ Y\ series)}{\sigma x \ (S.D.of\ X\ series)}.$$

IV. The regression coefficient of X on Y is

$$= bxy = r\frac{\sigma x \ (S.D.of\ X\ series)}{\sigma y \ (S.D.of\ Y\ series)}$$

• Types of Regression:

(*a*) Simple regression:

I. Here the dependent variable (criterion) is a function of a single independent variable (predictor)

II. The score of the dependent variable is predicted from the given scores of the single predictor.

Example: Height of person on his weight.

(*b*) Multiple regression:

I. Here the dependent variable (criterion) is a function of two or more predictors.

II. The scores are predicted from the scores of more than one predictor.

III. It may be linear or nonlinear.

Example: Thyroid calcitonin on combination of thyroxine secretion & serum calcium.

(*c*) Linear regression:

I. Here the dependent variable (criterion) is linearly correlated with the predictor (independent variable).

II. The scores of the dependent variables are predicted by working out an equation for a straight line, depending on the linear association between the two.

The statistical analysis employed to find out the exact position of the straight line is known as linear regression analysis.

(*d*) Nonlinear regressions:

If the criterion (dependent variable) has a nonlinear correlation with the predictor (independent variable), the scores of the criterion have to be predicted in terms of a curved line like a sigmoid or hyperbolic or exponential curve, according to their form of association.

• Properties of Regression:

1. It is an expression of the dependent variable (criterion) as a function of the independent variable (predictor).
2. A regression can be worked out only when there is a significant correlation between the dependent (criterion) and the independent (predictor) variable.
3. Regression predicts only a probable score of the criterion on a given score of the predictor.
4. When a pair of variables correlated with one another, regression can be worked in two ways. viz. (*i*) a regression variable X as criterion on variable Y as predictor and (*ii*) another regression of variable Y as criterion on variable X as predictor.
5. A regression equation is worked out using a statistic called the regression coefficient.

• Method of Studying Regression:

There are two methods: (*a*) Graphic method and (*b*) Algebraic method.

(*a*) Graphic method:

I. The points are plotted on a graph paper representing pairs of values of concerned variables.

II. In this diagram independent variable is taken on the horizontal axis and dependent variable on the vertical axis.

III. These points give a picture of a scatter diagram. A regression line may be drawn in between these points by free hand or by a scale rule.

(*b*) Algebraic method:

I. A regression line is a straight line fitted to the data by the method of least squares.

II. It indicates the best probable mean value of one variable corresponding to the mean value of the other.

III. There are always two regression lines constructed for the relationship between the two variables. *viz.*, *X* and *Y*.

Thus one regression line shows regression of *X* upon *Y* and the other shows regression *Y* upon *X*.

Linear regression:

Let the equation in general terms:

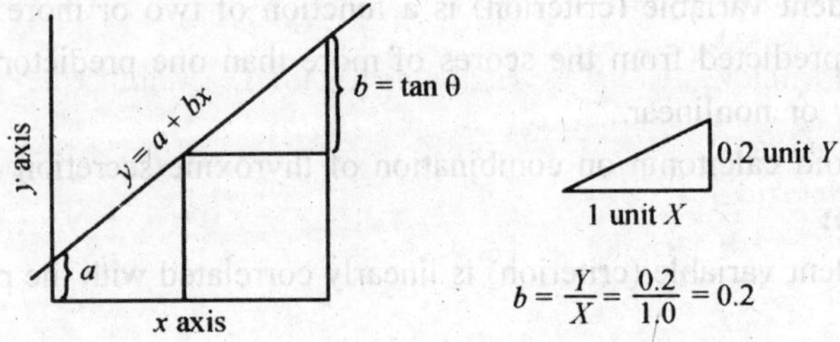

Fig. 14.1. Intercept & Slope for regression line

I. $Y = a + bx$ where *y* and *x* represent two variables, '*a*' is the *y* intercept or distance between the *x* axis and the point where the line crosses the *y* axis.

II. '*b*' is the slope or increase in the *y* value per unit change in *x* value.

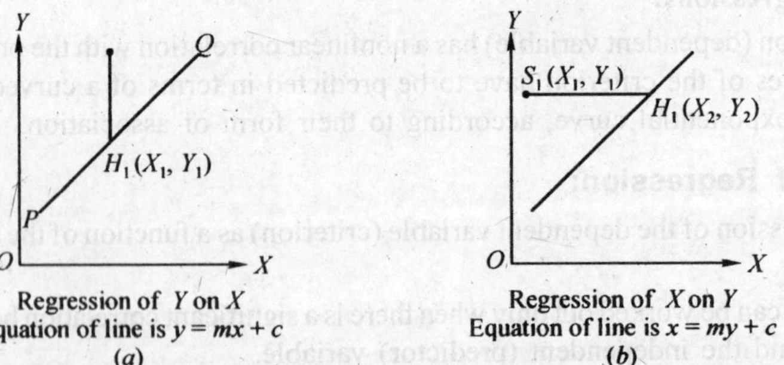

Fig. 14.2. (*a*) & (*b*) show regression *Y* on *X* and *X* on *Y* respectively.

(*A*) I. If the line of regression is so chosen that the sum of squares of deviation parallel to the axis of *y* is minimized [Fig. 14.2 (*a*)], it is called the line of regression of *Y* on *X* and it gives the best estimate of *Y* for any given value of *X*.

II. Its equation is $y = a + bx$.

III. The slope of the line b in the equation is known as the regression coefficient. It shows that y changes b times as fast as x.

IV. Symbolically the regression coefficient of y on x is b_{yx}.

(B) I. If the line of regression is so chosen that the sum of squares of deviations parallel to the axis of x is minimized [Fig. 14.2 (*b*)], it is called the line of regression of X on Y and it gives the best estimate of x for any value of y.

II. The regression equation in this case is $x = a + by$.

III. The regression coefficient of x on y is b_{xy}.

Regression Y on X	*Regression X on Y*
Line of regression $Y = mx + c$ The coefficient of x *i.e.*, m represent the regression coefficient of Y on X *i.e.*, myx.	Line of regression $X = my + c$ The coefficient of y *i.e.*, m represent the regression coeffcient of X on Y *i.e.*, mxy.

• Computation of Linear Regression:

(List of formulae)

1. Regression Y on X:

(a) Regression equation:

$$y = mx + c$$
$$y - \bar{y} = byx\,(x - \bar{x})$$

(b) Regression coefficient: $bxy = \dfrac{\sigma y}{\sigma x}$

(i) When the deviations are taken from the mean:

$$dx = x - \bar{x}$$
$$dy = y - \bar{y}$$
$$byx = \frac{\sum dxdy}{\sum d_x^2} \quad \text{or} \quad byx = \frac{\sum xy - n(\bar{x}\bar{y})}{\sum x^2 - n(\bar{x})^2}$$

(ii) When the deviations are taken from the assumed mean:

$$u = x - a \qquad v = y - b$$

(a and b assumed mean of X and Y series respectively)

$$byx = \frac{\sum uv - \dfrac{\sum u \sum v}{n}}{\sum u^2 - \dfrac{\left(\sum u\right)^2}{n}}$$

(iii) When the original values (raw scores) are used:

$$byx = \frac{n\sum xy - \left(\sum x\right)\left(\sum y\right)}{n\sum x^2 - \left(\sum x\right)^2} = \frac{\sum xy - \dfrac{\left(\sum x\right)\left(\sum y\right)}{n}}{\sum x^2 - \dfrac{\left(\sum x\right)^2}{n}}$$

2. Regression X on Y:

(a) Regression equation: $X = my + c$

$$X - \bar{X} = bxy\,(Y - \bar{Y}).$$

(*b*) **Regression coefficient:** $byx = \dfrac{\sigma x}{\sigma y}$

I. When the deviations are taken from the mean:

$$dx = X - \bar{X} \quad dy = Y - \bar{Y}$$

$$bxy = \frac{\sum dx \sum dy}{\sum d^2 y} \quad \text{or } bxy = \frac{\sum xy - x(\bar{x}, \bar{y})}{\sum y^2 - n(\bar{Y})^2}$$

II. When the deviations are taken from assumed mean:

$$u = x - a \quad v = y - b$$

$$bxy = \frac{\sum uv - \dfrac{\sum u \sum v}{n}}{\sum v^2 - \dfrac{\left(\sum v\right)^2}{n}}$$

III. When the original values are used (i.e., raw score):

$$bxy = \frac{n\,xy - \sum x \sum y}{n\sum y^2 - \sum y^2} \quad \text{or} \quad bxy = \frac{\sum xy - \left(\sum x \sum y\right)n}{\sum y^2 - \left(\sum y\right)^2 n}$$

- **Properties of Regression Coefficients:**

1. Correlation coefficient is the geometric mean between the regression coefficient:

The regression coefficients are

$$byx = r\frac{\sigma y}{\sigma x} \text{ and } bxy = r\frac{\sigma x}{\sigma y}$$

$$\therefore \text{ GM between them } = \sqrt{byx \cdot bxy} = \sqrt{\frac{r\sigma_y}{\sigma_x} \times \frac{r\sigma_x}{\sigma_y}} = \sqrt{r^2} = r$$

Thus the correlation coefficient is the geometric mean between the two regression coefficients.

$$r^2 = byx.bxy \qquad [\because r = \sqrt{byx.bxy}]$$

$$= \frac{\sum dx\,dy}{n\sigma_x^2} \cdot \frac{\sum dx\,dy}{n\sigma_y^2}$$

$$\therefore \quad r = \frac{\sum dx\,dy}{n\sigma_x \sigma_y}$$ It is the product moment formula due to Karl Pearson

$$r = \sqrt{dx\,\sigma y}$$

$$r = \frac{\sum \sigma x\,\sigma y}{n\,\sigma x\,\sigma y}.$$

2. If one of the regression coefficients is greater than unity numerically, the other must be less than unity numerically:

The regression coefficients are

$$byx = \frac{r\,\sigma_y}{\sigma_x} \text{ and } bxy = r\,\frac{\sigma_x}{\sigma_y}$$

Let $byx \geq 1$ ...(1)

Since $byx \cdot bxy = r^2 \le 1$ $(\because -1 \le r \le 1)$

$$bxy \le \frac{1}{byx} < 1$$

Similarly if $bxy > 1$, then $byx < 1$

3. The correlation coefficient and regression coefficients have the same sign:

Regression coefficient of Y on X

$$= byx = r\frac{\sigma y}{\sigma x}$$

Regression coefficient of X on Y

$$= bxy = r\frac{\sigma x}{\sigma y}$$

Since σ_x and σ_y are both positive, therefore byx and bxy have the same sign as that of r.

4. Arithmetic mean of regression coefficients is greater than the correlation coefficient:

$$\frac{byx + bxy}{2} > r$$

or $$\frac{r\dfrac{\sigma_y}{\sigma_x} + r\dfrac{\sigma_x}{\sigma_y}}{2} > r$$

or $$r\left(\frac{\sigma_y^2 + \sigma_x^2}{\sigma_x \sigma_y}\right) > 2r$$

or $$\sigma_x^2 + \sigma_y^2 > 2\sigma_x \sigma_y$$

or $\sigma_x^2 + \sigma_y^2 - 2\sigma_x \sigma_y > 0$ or $(\sigma_x - \sigma_y)^2 > 0$

or it is known that A.M. > G.M.

$\therefore$ $$\frac{bxy + byx}{2} > \sqrt{bxy \times byx}$$ $(\because r = \sqrt{bxy \times byx})$

$\therefore$ A.M. of regression coefficient is greater than correlation coefficient.

5. Regression coefficients are independent of the origin but not scale:

Let $u = \dfrac{X - A}{h}$ $v = \dfrac{Y - B}{k}$ [A, B, h and k are constant]

$$byx = r\frac{\sigma_y}{\sigma_x} = r\frac{k\sigma_y}{h\sigma_x} = \frac{k}{h}\left(r\frac{\sigma_y}{\sigma_x}\right) = \frac{k}{h}b_{vu}$$

Similarly, $$bxy = \frac{h}{k}\left(r\frac{\sigma_x}{\sigma_y}\right) = \frac{h}{k}b_{uv}$$

Thus both byx & bxy are independent of a and b but not of h and k.

Difference between Correlation and Regression

Correlation	*Regression*
1. Correlation is the relationship between two or more variables which vary in sympathy with the other in the same or the opposite direction.	1. Regression is a mathematical measure showing the average relationship between two variables.
2. Here both the variables *i.e.*, x and y are random variables.	2. Here x is a random variable and y is fixed. Sometimes both the variables may be random variables.
3. It finds out the degree of relationship between two variables [not cause and effect of the variable].	3. It indicates the cause and effect relationship between the variables.
4. It is used for testing and verifying the relationship between two variables.	4. It is used for prediction of one value in respect to the other given value.
5. The coefficient correlation is a relative measure. The range of relationship lies between ±1.	5. Regression coefficient is an absolute figure. If we know the value of independent variable, we can find the value of dependent variable.
6. It has limited application because it is confined only to linear relationship between the variables.	6. It has wide application as it studies linear and non-linear relationship between the variables.
7. If the coefficient correlation is positive, then the two variables are positively correlated and vice-versa.	7. The regression coefficient explains that the decrease in one variable is associated with the increase in the other variable.

Example 1: Compute *byx* for the following data:

x, y : (5, 2), (7, 4), (8, 3), (4, 2), (6, 4).

Solution:

x	y	xy	x^2
5	2	10	25
7	4	28	49
8	3	24	64
4	2	08	16
6	4	24	36
30	15	94	190

$$\sum x = 30 \quad \sum y = 15 \quad \sum xy = 94 \quad \sum x^2 = 190 \quad n = 5$$

$$byx = \frac{\sum xy - \frac{\sum x \sum y}{n}}{\sum x^2 - \frac{(\Sigma x)^2}{n}} = \frac{94 - \frac{30 \times 15}{5}}{190 - \frac{(30)^2}{5}} = \frac{94-90}{190-180} = \frac{4}{10} = \frac{2}{5} = 0.4$$

$\therefore \quad bxy = 0.4.$

Example 2: Find the two regression equations from the following pairs of observations on X and Y: (1, 2), (2, 3), (3, 5), (4, 6), (5, 4). Hence find the predicted value of Y when $X = 2.5$ and the predicted value of X when $Y = 4.5$.

Solution:

X	Y	x^2	y^2	XY
1	2	1	4	2
2	3	4	9	6
3	5	9	25	15
4	6	16	36	24
5	4	25	16	20
15	20	55	90	67

$$\sum X = 15 \sum Y = 20 \quad n = 5 \quad \sum XY = 67 \quad \bar{X} = \frac{15}{5} = 3 \quad \bar{Y} = \frac{20}{5} = 4$$

$$\sum X^2 = 55 \quad \sum Y^2 = 90$$

$$bxy = \frac{\sum XY - \frac{\sum X \sum Y}{n}}{\sum Y^2 - \frac{\left(\sum Y\right)^2}{n}} = \frac{67 - \frac{15 \times 20}{5}}{90 - \frac{(20)^2}{5}} = \frac{67 - 60}{90 - 80} = \frac{7}{10} = 0.7$$

$$byx = \frac{\sum XY - \frac{\sum X \sum Y}{n}}{\sum X^2 - \frac{\left(\sum X\right)^2}{n}} = \frac{67 - \frac{15 \times 20}{5}}{55 - \frac{(15)^2}{5}} = \frac{67 - 60}{55 - 45} = \frac{7}{10} = 0.7$$

The regression equation Y on X is

$$Y - \bar{Y} = byx\,(X - \bar{X})$$

or $\quad Y - 4 = 0.7\,(X - 3)$

or $\quad Y = 0.7X\ 2.1 + 4$

$\quad = 0.7X + 1.9 \qquad$ ...(1)

The regression equation X on Y is

$$X - \bar{X} = bxy\,(Y - \bar{Y})$$

$X - 3 = 0.7\,(Y - 4)$

or $\quad X = 0.7Y - 2.8 + 3$

$X = 0.7Y + 0.2 \qquad$ ...(2)

If $x = 2.5$, then from Equation (1), we get

$Y = 0.7 \times 2.5 + 1.9$

$Y = 1.75 + 1.9 = 3.65$

$\therefore \quad Y = 3.65$

If $y = 4.5$, then from Equation (2), we get

$X = 0.7 \times 4.5 + 0.2$

$X = 3.15 + 0.2 = 3.35$

$\therefore \quad X = 3.35.$

Example 3: Calculate regression coefficients *byx* and *bxy* for the following data:

$$\sum X = 55 \quad \sum Y = 88 \quad \sum XY = 586 \quad \sum X^2 = 385 \quad \sum Y^2 = 1114 \quad n = 10$$

Solution:

$$bxy = \frac{\sum XY - \frac{\sum X \sum Y}{n}}{\sum Y^2 - \frac{\sum Y^2}{n}} = \frac{586 - \frac{55 \times 88}{10}}{1114 - \frac{(88)^2}{10}}$$

$$= \frac{586 - 484}{1114 - 774.4} = \frac{102}{339.6} = \frac{1020}{3396} = 0.3$$

$$bxy = \frac{\sum XY - \frac{\sum X \sum Y}{n}}{\sum X^2 - \frac{\left(\sum X\right)^2}{n}} = \frac{586 - \frac{55 \times 88}{10}}{385 - \frac{(55)^2}{10}} = \frac{586 - 484}{385 - 302.5}$$

$$= \frac{102}{82.5} = \frac{1020}{825} = 1.236 = 1.2.$$

Example 4: Find the correlation coefficient in each of the following cases: (*i*) *byx* = 0.4 and *bxy* = 0.9, (*ii*) *byx* = 1.6 and *bxy* = 0.4, (*iii*) *byx* = –0.3 and *bxy* = –1.2.

Solution: (*i*) $r = \sqrt{byx \times bxy} = \sqrt{0.4 \times 0.9} = \sqrt{0.36} = \pm 0.6$

(*ii*) $r = \sqrt{1.6 \times 0.4} = \sqrt{0.64} = \pm 0.8$

(*iii*) $r = \sqrt{-0.3 \times -1.2} = \sqrt{0.36} = \pm 0.6$

Here *byx* & *bxy* are both negative. Hence *r* must be negative *i.e.*, –0.6.

Example 5: From the following data, calculate (*a*) correlation coefficient and (*b*) standard deviation of *Y* (σ_y). *X* = 0.85*Y* *Y* = 0.89*X*, $\sigma_x = 3$

Solution:

(*a*) $r = \sqrt{bxy \times byx}$ $\quad r = \sqrt{0.85 \times 0.89} = \sqrt{0.7565} = 0.869$

(*b*) $$bxy = r \times \frac{\sigma_x}{\sigma_y}$$

or $$0.85 = 0.869 \times \frac{3}{\sigma_y}$$

or $$\sigma_y = \frac{0.869 \times 3}{0.85} = \frac{2.607}{0.85} = 3.067.$$

Example 6: Find the regression coefficients *byx* and *bxy* of *Y* on *X* on *Y* respectively, if standard deviations of *X* and *Y* are 4 and 3 respectively and correlation coefficient between *X* and *Y* is 0.8.

Solution: $\sigma_x = 4 \quad \sigma_y = 3 \quad r = 0.8$

$$bxy = r\frac{\sigma_x}{\sigma_y} \quad byx = r\frac{\sigma_Y}{\sigma_X}$$

$$bxy = 0.8 \times \frac{4}{3} \quad byx = 0.8 \times \frac{3}{4}$$

$$bxy = 1.066 \quad byx = 0.6$$

$$bxy = 1.067.$$

Example 7: The correlation coefficient between *X* and *Y* is 0.60. If the variance of *x* = 225, the variance of *Y* = 400, mean of *X* = 10 and mean of *Y* = 20, find the equation of the regression lines of (*i*) *Y* on *X* and (*ii*) *X* on *Y*.

Solution:

Variance of X *i.e.*, $\sigma_x^2 = 225$ $\therefore$ $\sigma_x = \sqrt{225} = 15$

Variance of Y *i.e.*, $\sigma_y^2 = 400$ $\therefore$ $\sigma_y = \sqrt{400} = 20$

$r = 0.60$ $\bar{X} = 10$ $\bar{Y} = 20$

$$byx = r\frac{\sigma_y}{\sigma_x} = 0.6 \times \frac{20}{15} = 0.8 \quad bxy = 0.6 \times \frac{15}{20} = 0.45$$

(*i*) Regression equation Y on X *i.e.*,

$$Y - \bar{Y} = byx\,(X - \bar{X})$$
$$Y - 20 = 0.8\ (X - 10)$$
$$Y = 0.8X - 8 + 20$$
$$Y = 0.8X + 12$$

(*ii*) Regression equation of X on Y *i.e.*,

$$X - \bar{X} = bxy\,(Y - \bar{Y})$$
$$X - 10 = 0.45\ (Y - 20)$$
$$= 0.45Y - 9$$
$$X = 0.454 - 9 + 10 = 0.45Y + 1.$$

Example 8: You are given the following results of two variables X and Y:

$$\bar{X} = 36 \quad \bar{Y} = 85 \quad \sigma_x = 11 \quad \sigma_y = 8 \quad r(X, Y) = 0.66.$$

Find the two regression equations and estimate the value of X when $Y = 75$.

Solution:

$$bxy = r\frac{\sigma_x}{\sigma_y} = 0.66 \times \frac{11}{8} = \frac{7.26}{8} = 0.9075$$

$$byx = r\frac{\sigma_y}{\sigma_x} = 0.66 \times \frac{8}{11} = \frac{5.28}{11} = 0.48$$

The regression equation X on Y is

$$X - \bar{X} = bxy\,(Y - \bar{Y})$$
$$X - 36 = 0.9075\ (Y - 85)$$
$$= 0.9075Y - 77.1375$$
$$X = 0.9075Y + 36 - 77.1375$$
$$= 0.9075Y - 41.1375$$

The regression equation Y on X is

$$Y - \bar{Y} = byx\,(X - \bar{X})$$
$$Y - 85 = 0.48\ (X - 36)$$
$$Y = 0.48X - 17.28 + 85$$
$$Y = 0.48X + 67.72$$

When $Y = 75$ the value of X will be

$$X = 0.9075\ Y\quad 41.1375$$
$$= 0.9075 \times 75 - 41.1375$$
$$= 68.0625 - 41.1375$$
$$= 26.925$$

Example 9: You are given the following data:

$$\overline{X} = 36 \quad \overline{Y} = 85 \quad \sigma_x = 11 \quad \sigma_y = 8 \quad r_{xy} = 0.66.$$

Calculate value of Y when $X = 10$.

Solution:
$$byx = r\frac{\sigma_Y}{\sigma_X} = 0.66 \times \frac{8}{11} = \frac{5.28}{11} = 0.48$$

$$Y - \overline{Y} = byx\,(X - \overline{X})$$

$$Y - 85 = 0.48 \times 10 - 0.48 \times 36$$

$$Y = 4.8 - 17.28 + 85$$

$$= 89.8 - 17.28$$

$$= 72.52.$$

Example 10: The equation of two lines are $8X - 10Y + 66 = 0$ and $40X - 18Y = 214$. The variance of x is 9. Find: (*i*) The mean of X and Y (*ii*) regression coefficient *byx* and *bxy* (*iii*) correlation coefficient between X and Y (*iv*) standard deviation of Y (*v*) the value of Y for $x = 2$ (*vi*) the value of X for $Y = 3$.

Solution: The equation of the two lines of regression are

$$8X - 10Y + 66 = 0$$

or $\quad 8X - 10Y = -66 \quad ...(1) \quad$ Multiplying with 5

$$40X - 18Y = 214 \quad ...(2)$$

$$40X - 50Y = -330 \quad ...(3)$$

Subtracting $\quad 40X - 18Y = 214 \quad ...(2)$

$$- \quad + \quad -$$

$$-32Y = -544$$

or
$$Y = \frac{544}{32} = 17$$

Substituting the value of Y in Equation (1), we get

$$8X - 10.17 = -66$$

or
$$8X = -66 + 170$$

$$X = \frac{104}{8} = 13$$

(*i*) The mean value of $X = 13$ and $Y = 17$.

Rewriting the Equation (1),

$$8X - 10Y = -66$$

(*ii*)
$$10Y = +8X + 66$$

$$Y = 0.8X + 6.6 \text{ (regression equation)}$$

$$byx = r\frac{\sigma_y}{\sigma_x} = 0.8$$

Rewriting the Equation (2),

$$40X = 18Y + 214$$

$$X = \frac{18}{40}Y + \frac{214}{40}$$

$$X = 0.45Y + 5.35$$

$$bxy = r\frac{\sigma_x}{\sigma_y} = 0.45$$

(*iii*) $r^2 = bxy \times byx = .45 \times 0.8 = 0.36 \quad \therefore r = \pm 0.6$

(*iv*) Variance $\sigma_x^2 = 9 \quad \sigma_x = 3$

(*v*) $Y = 0.8 \times 2 + 6.6 \qquad X = 0.45X + 5.35$

$= 1.6 + 6.6 = 8.2 \qquad = 0.45X\ 3 + 5.35 = 1.35 + 5.35 = 6.7.$

Example 11: Equations of two lines of regression are 4*X* + 3*Y* + 7 = 0 and 3*X* + 4*Y* + 8 = 0. Find (*i*) mean of *X* and *Y* (*ii*) regression coefficient *bYX* and *bXY* and correlation coefficient between *X* and *Y*.

Solution: The equations of the two lines of regression are

$$4X + 3Y + 7 = 0$$

or $4X + 3Y = -7$ ...(1) Multiplying by (3)

$3X + 4Y + 8$ or $3X + 4Y = -8$ ...(2) Multiplying by (4)

$$12X + 9Y = -21$$

$$12X + 16Y = -32$$

substracting

$$-7Y = 11$$

or $7Y = -11$

$$Y = -\frac{11}{7}$$

Substituting the value of *Y* on (1) equation

$$4X + 3\left(-\frac{11}{7}\right) = -7$$

$$4X - \frac{33}{7} = -7$$

or $4X = \frac{33}{7} - 7 = \frac{33-49}{7} = -\frac{16}{7}$

$$4X = -\frac{16}{7}$$

$\therefore \quad X = -\frac{16}{7 \times 4} = -\frac{4}{7}$

1. The mean of *X* *i.e.*, $\overline{X} = -\frac{4}{7}$ and mean of *Y* *i.e.*, $\overline{Y} = -\frac{11}{7}$
2. Rewriting the Equation (1),

$$4X = -3Y - 7$$

or $X = -\frac{3}{4}Y - \frac{7}{4} \quad \therefore bxy = -\frac{3}{4}$

Similarly, the Equation (2)

$$3X + 4Y = -8$$

$$4Y = -3X - 8$$

$$Y = -\frac{3}{4} \times -2 \quad \therefore byx = -\frac{3}{4}$$

3. Coefficient correlation *i.e.*, r

$$r = \sqrt{bxy \times byx} \quad byx \text{ and } bxy \text{ are both negative}$$

$$= \sqrt{-\frac{3}{4} \times -\frac{3}{4}} = \sqrt{+\frac{9}{16}} = \pm\frac{3}{4}$$

So, $r = -\frac{3}{4}$

Example 12: *X* and *Y* are a pair of correlated variables. Ten observations of their values (*X*, *Y*) have the following results:

$$\sum X = 55 \quad \sum Y = 55 \quad \sum XY = 350 \quad \sum X^2 = 385$$

Predict the values of *Y* when the value of *X* is 6.

Solution: $n = 10 \quad \sum X = 55 \quad \sum Y = 55 \quad \sum XY = 350$

$$\overline{X} = \frac{55}{10} = 5.5 \quad \overline{Y} = \frac{55}{10} = 5.5$$

$$byx = \frac{\sum XY - \frac{\sum X \sum Y}{n}}{\sum X^2 - \frac{\left(\sum X\right)^2}{n}} = \frac{350 - \frac{55 \times 55}{10}}{385 - \frac{(55)^2}{10}} = \frac{350 - 302.5}{385 - 302.5}$$

$$= \frac{47.5}{82.5} = 0.57575 = 0.5758$$

The equation of regression line of *Y* on *X* is

$$Y - \overline{Y} = byx\,(X - \overline{X})$$

$$Y - 5.5 = 0.5758\,(X - 5.5)$$

$$Y - 5.5 = 0.5758X - 3.1669$$

$\therefore$ $Y = 0.5758X + 5.5 - 3.1669 = 0.5758X + 2.3331$

When X is 6 Y will be $Y = 0.5758X\ 6 + 2.3331$

$$= 3.4548 + 2.331$$

$$= 5.7879.$$

Example 13: *X* and *Y* are a pair of correlated variables *X* and *Y*. The following results are obtained:

$$\sum X = 15 \quad \sum Y = 25 \quad \sum X^2 = 55 \quad \sum XY = 83 \quad \sum Y^2 = 135 \quad n = 5.$$

Find the equation of the lines of regression & estimate the value of *X* & *Y* if *X* = 8 and *Y* = 12.

Solution: $\sum X = 15 \quad \overline{X} = \frac{15}{5} = 3 \quad \sum Y = 25 \quad \overline{Y} = \frac{25}{5} = 5 \quad \sum XY = 83$

$$\sum X^2 = 55 \quad \sum Y^2 = 135 \quad n = 5$$

$$bxy = \frac{\sum XY - \frac{\sum X \sum Y}{n}}{\sum Y^2 - \frac{\left(\sum Y\right)^2}{n}} = \frac{83 - \frac{15 \times 25}{5}}{135 - \frac{(25)^2}{5}} = \frac{83 - 75}{135 - 125} = \frac{8}{10} = \frac{4}{5} = 0.8$$

The regression line of X on Y is

$$X - \bar{X} = bxy\,(Y - \bar{Y})$$

$$X - 3 = 0.8\,(Y - 5)$$

$$x = 0.8Y - 4.0 + 3 = 0.8Y - 1$$

$$byx = \frac{83 - \frac{15 \times 25}{5}}{55 - \frac{(15)^2}{5}} = \frac{83 - 75}{55 - 45} = \frac{8}{10} = 0.8$$

The regression line of Y on X is

$$Y - \bar{Y} = byx\,(X - \bar{X})$$

$$Y - 5 = 0.8\,(X - 3)$$

$$Y = 0.8X - 2.4 + 5 = 0.8X - 2.6$$

If $y = 12$, then $x = 0.8Y - 1$

$$= 0.8 \times 12 - 1 = 9.6 - 1 = 8.6$$

If $X = 8$

Then $Y = 0.8X - 2.6$

$$= 0.8 \times 8 - 2.6 = 6.4 - 2.6 = 3.8.$$

Example 14: Find the regression coefficient of y on x for the following data:

$$\sum X = 24 \quad \sum Y = 44 \quad \sum X^2 = 164 \quad \sum Y^2 = 574 \quad \sum XY = 306 \quad n = 4.$$

Solution:

$$byx = \frac{\sum XY - \frac{\sum X \sum Y}{n}}{\sum X^2 - \frac{\left(\sum X\right)^2}{n}} = \frac{306 - \frac{24 \times 44}{4}}{164 - \frac{(24)^2}{4}} = \frac{306 - 264}{164 - 144} = \frac{42}{20} = 2.1.$$

Example 15: For observations of pairs (X, Y) of the variables X and Y. the following results are obtained:

$$\sum X = 125 \quad \sum Y = 100 \quad \sum X^2 = 1650 \quad \sum Y^2 = 1500 \quad \sum XY = 50 \quad n = 25.$$

Find the equation of line of regression of x on y. Estimate the value of x if $y = 5$.

Solution:

$$bxy = \frac{\sum XY - \frac{\sum X \sum Y}{n}}{\sum Y^2 - \frac{\left(\sum Y\right)^2}{n}}$$

$$= \frac{50 - \frac{125 \times 100}{25}}{1500 - \frac{(100)^2}{25}} = \frac{50 - 500}{1500 - 400} = -\frac{450}{1100} = -\frac{9}{22}$$

$$\bar{X} = \frac{125}{25} = 5 \quad \bar{Y} = \frac{100}{25} = 4$$

$$X - \bar{X} = bxy\,(Y - \bar{Y})$$

or $$X - 5 = -\frac{9}{22}(Y - 4)$$

$$22(X - 5) = -9Y + 36$$

$$22X - 110 = -9Y + 36$$

$$22X + 9Y - 146 = 0$$

$$22X = 146 - 9Y$$

$$22X = 146 - 9 \times 5 = 146 - 45$$

$$X = \frac{101}{22} = 4.5903 = 4.591.$$

Example 16: You are given the following data:

	X	Y
Mean	**20**	**25**
SD	**5**	**4**

Correlation coefficient between X and Y is 0.6. Find two regression equations and linear estimate of Y (*i*) when $X = 30$ and (*ii*) X when $Y = 28$.

Solution: $$bxy = r\frac{\sigma_x}{\sigma_y} = 0.6 \times \frac{5}{4} = 0.75$$

$$byx = r\frac{\sigma_y}{\sigma_x} = 0.6 \times \frac{4}{5} = 0.48$$

$$\bar{X} = 20 \quad \bar{Y} = 25$$

Regression equation of X on Y is

$$X - \bar{X} = bxy(Y - \bar{Y})$$

$$X - 20 = 0.75(Y - 25)$$

or $$X = 0.75Y - 18.75 + 20 = 0.75Y + 1.25$$

Therefore, $$X = 0.75Y + 1.25$$

Regression equation of Y on X is

$$Y - \bar{Y} = byx(X - \bar{X})$$

$$Y - 25 = 0.48(X - 20)$$

or $$Y = 0.48X - 9.6 + 25 = 0.48X + 15.4$$

$\therefore$ $$Y = 0.48X + 15.4$$

(*i*) When $$X = 30$$

$$Y = 0.48 \times 30 + 15.4$$

$\therefore$ $$Y = 14.4 + 15.4 = 29.8$$

(*ii*) When $$Y = 28$$

$$X = 0.75 \times 28 + 1.25$$

$$X = 21 + 1.25$$

$\therefore$ $$X = 22.25.$$

Example 17: The following results were worked out from the scores in Mathematics and Statistics in a test examination of Serampore College:

	Score in Mathematics (X)	Score in Statistics (Y)
Mean	**80**	**50**
Standard deviation	**15**	**10**

Coefficient of correlation is 0.4. Estimate the marks in Mathematics obtained by a student who has scored 60 marks in Statistics.

Solution: X denotes marks in Mathematics and Y denotes marks in Statistics. As we have to calculate marks in Mathematics for a student who scored 60 marks in Statistics, hence we have to fix regression equation X on Y.

$$X - \overline{X} = r \frac{\sigma_x}{\sigma_y} (Y - \overline{Y})$$

$$X - 80 = 0.4 \frac{15}{10} (Y - 50)$$

$$X - 80 = 0.6Y - 30$$

or $$X = 0.6Y - 30 + 80 = 0.6Y + 50$$

He scored 60 marks in Statistics *i.e.*, $X = 60$

$$X = 0.6 \times 60 + 50 = 36 + 50 = 86$$

Hence he scored 60 marks in Statistics, he likely scored 86 marks in Mathematics.

Example 18: The following results were worked out from the scores in Biophysics and Biostatistics in the 3rd semester M.Sc. Zoology examination of Serampore College.

	Score in Biophysics (X)	Score in Biostatistics (Y)
Mean	**39.5**	**47.5**
Standard deviation	**10.8**	**17.8**

Correlation coefficient is 0.42. Find both regression lines. Estimate the marks in Biostatistics obtained by a student who scored 50 in Biophysics.

Solution: X denotes marks in Biophysics and Y denotes marks in Biostatistics. As we have to calculate marks in Biostatistics for a student who scored 50 marks in Biophysics, hence we have to fix regression of Y on X.

Regression line Y on X *i.e.*

$$Y - \overline{Y} = r \frac{\sigma_y}{\sigma_x} (X - \overline{X})$$

$$Y - 47.5 = 0.42 \frac{17.8}{10.8} (X - 39.5)$$

$$Y - 47.5 = \frac{7.476}{10.8} (X - 39.5)$$

$$Y - 47.5 = 0.692X - 27.334$$

$$Y = 0.692X + 47.5 - 27.33Y$$

$$= 0.692X + 20.166$$

He scored 50 marks in Biophysics *i.e.*, $X = 50$, therefore, marks in Biostatistics will be

$$Y = 0.692 \times 50 + 20.166$$
$$= 34.6 + 20.166$$
$$= 54.766.$$

Example 19: Give reasons to justify that following statement can not be true:
(*i*) *byx* = 0.8 and *bxy* = –0.3, (*ii*) *byx* = 0.8 and *bxy* = 2.4.

Solution: (*i*) We have $r^2 = bxy \times byx = 0.8 \times -0.3 = -0.24 < 0$ which is impossible. Hence the statement $byx = 0.8$ and $bxy = -0.3$ cannot be true.

(*ii*) We have $r^2 = byx \times bxy = 0.8 \times 2.4 = 1.92 > 1$ which is observed since $-1 \leq r \leq 1$. Hence statement $byx = 0.8$ and $bxy = 2.4$ cannot be true.

Example 20: Find the mean of *X*, the mean of *Y*, the regression coefficients from the two regression equations: 3*X* = *Y* and 4*Y* = 3*X*. Find also correlation coefficient.

Solution:
$$3X = Y \quad \text{or} \quad 3X - Y = 0 \quad ...(i)$$
$$4Y = 3X \quad \text{or} \quad -3X + 4Y = 0 \quad ...(ii)$$

By adding
$$+3Y = 0$$
$$Y = 0$$

$$-3X + 4Y = 0 \quad ...(2)$$

By substituting the value of *Y i.e.*, $Y = 0$ in $4Y = 3X$

$$3X = 0 \quad \therefore X = 0$$

Therefore, mean of *X i.e.*, $\bar{X} = 0$ and mean of *Y i.e.*, $\bar{Y} = 0$.

Regression equation of *X* on *Y i.e.*,

$$3X - Y = 0$$

or
$$3X = Y + 0$$
$$X = \frac{1}{3} Y + 0 \quad \therefore \text{Regression coefficient is } \frac{1}{3}$$
$$bxy = \frac{1}{3} = 0.333$$

Regression equation of *Y* on *X i.e.*,

$$4Y - 3X = 0$$
$$4Y = 3Y + 0$$
$$Y = \frac{3}{4} Y + 0 \quad \text{Regression coefficient} = \frac{3}{4}$$
$$\therefore \quad byx = \frac{3}{4} = 0.75$$
$$r^2 = bxy \times byx$$
$$r^2 = \frac{1}{3} \times \frac{3}{4} = \frac{1}{4} = 0.25$$
$$\therefore \quad r = \sqrt{.25} = 0.5 = \frac{1}{2}$$

Mean $\quad \bar{X} = 0 \quad \bar{Y} = 0$

Regression coefficient $bxy = \frac{1}{3} = 0.333$

$$byx = \frac{3}{4} = 0.75$$

Correlation coefficient $= 0.5 = \frac{1}{2}$.

Example 21: Find mean of the variables *X* and *Y* and the correlation coefficient from the following regression equations: $2Y - X = 50$ $3Y - 2X = 10$.

Solution: $2Y - X = 50$ ...(1) Multiplying by 2

$3Y - 2X = 10$ ...(2)

$$4Y - 2X = 100$$

By substracting $3Y - 2X = 10$

$+$

$$Y = 90 \qquad \therefore \quad \overline{Y} = 90$$

By substituting *Y* value to Equation (2)

$$3 \times 90 - 2X = 10$$

or

$$-2X = 10 - 270$$

$$+2X = +260$$

$$X = 130$$

$$\therefore \quad \overline{X} = 130$$

$$\overline{X} = 130 \qquad \overline{Y} = 90$$

Regression coefficient *Y* on *X* *i.e.*,

$$2Y - X = 50$$

$$2Y = X + 50$$

$$Y = \frac{1}{2}X + 25$$

$$\therefore \quad Y = 0.5X + 25$$

$$\therefore \quad byx = 0.5$$

Regression equation *X* on *Y* *i.e.*,

$$3Y - 2X = 10$$

$$2X = 3Y - 10$$

$$X = \frac{3}{2}Y - 5$$

$$X = 1.5Y - 5$$

$$\therefore \quad bxy = 1.5$$

Correlation coefficient *i.e.*,

$$r^2 = byx \times bxy$$

$$= 0.5 \times 1.5$$

$$= 0.75$$

or

$$= \sqrt{0.75} = \pm 0.8660 = \pm 0.87.$$

Example 22: The correlation coefficient between the two variables X & Y is $r = 0.60$. If $\sigma_x = 1.50$, $\sigma_y = 2.00$ $\bar{X} = 10$, $\bar{Y} = 20$, find the equation of the regression (*i*) X on Y (*ii*) Y on X.

Solution: Regression equation X on Y *i.e.*,

$$X - \bar{X} = r\frac{\sigma_x}{\sigma_y}(Y - \bar{Y}) = 0.60 \times \frac{1.5}{2.00}(Y - 20)$$

$$X - 10 = 0.45\,(Y - 20)$$

$$= 0.45Y - 9$$

$$X = 0.45Y - 9 + 10$$

$$\therefore \quad X = 0.45Y + 1$$

Regression equation Y on X *i.e.*,

$$Y - \bar{Y} = r\frac{\sigma_y}{\sigma_x}(X - \bar{Y})$$

$$Y - 20 = 0.60 \times \frac{2.00}{1.5}(X - 10)$$

$$Y - 20 = 0.8\,(X - 10)$$

$$Y = 0.8X - 8 + 20$$

$$Y = 0.8X + 12$$

Regression equation X on Y = $X = 0.45Y + 1$

Regression equation Y on X = $Y = 0.8X + 12$.

Example 23: The following results were obtained from measurement of body length (Y) and body weight (X) of 25 *Anabas sp.*

$$\sum X = 1165 \quad \sum X^2 = 56947 \quad \sum XY = 9024.4 \quad \sum Y = 185.2 \quad \sum Y^2 = 1437.24$$

Find an appropriate regression line of body length on body weight.

[C.U. (M.Sc zoology) = 2007]

Solution: The regression equation of body length (Y) on body weight (X) *i.e.*,

$$byx = \frac{n\sum XY - \sum X \sum Y}{n\sum X^2 - \left(\sum X\right)^2}$$

$$= \frac{25 \times 9024.4 - 1165 \times 185.2}{25 \times 56947 - (1165)^2}$$

$$= \frac{225610 - 215758}{1423675 - 1357225}$$

$$= \frac{9852}{66450}$$

$$= 0.14826 = 0.148$$

The regression equation Y on X is

$$Y - \bar{Y} = byx\,(X - \bar{X})$$

$$\bar{Y} - \frac{185.2}{25} = 7.408$$

$$\overline{X} - \frac{1165}{25} = 46.6$$

$\therefore$ $Y - 7.408 = 0.148\ (X - 46.6)$

$Y - 7.408 = 0.148X - 46.6 \times 0.148$

or $Y - 7.408 = 0.148X - 6.8968$

or $Y - 7.408 = 0.148X - 6.897$

$Y = 0.148X - 6.897 + 7.408$

$= 0.148X + 0.511$

or $Y = 0.511 + 148X.$

Example 24: The following results were obtained from records of age (x) and systolic blood pressure (y) of a group of 10 teachers of Serampore College.

Variable	*Age (X)*	*Blood pressure (Y)*
Mean	**5.3**	**142**
Variance	**130**	**165**

$$= \sum (X - \overline{X})\ (Y - \overline{Y}) = 1220$$

Find the appropriate regression equation and estimate the blood pressure of a teacher whose age is 50 years.

Solution: The appropriate regression equation is regression of Y on X *viz.*,

$$Y - \overline{Y} = byx\ (X - \overline{Y})$$

where $byx = \dfrac{\text{cov}\ (x, y)}{\sigma_x^2\ (\text{variance of } x)}$

$$\text{cov}\ (x, y) = \frac{\left(\sum X - \overline{X}\right)(Y - \overline{Y})}{n} \qquad \left[\begin{array}{c} n = 10 \\ \overline{X} = 53 \quad \overline{Y} = 142 \end{array}\right]$$

$$= \frac{1220}{10} = 122$$

so that $byx = \dfrac{122}{130} = 0.938 = 0.94$

The regression equation is therefore

$Y - 142 = 0.94\ (X - 53)$

or $Y = 0.94X - 53 \times 0.94 + 142$

$\therefore$ $Y = 0.94X - 49.82 + 142$

$= 0.94X + 92.18$

When age is 50 years *i.e.*, $X = 50$, the blood pressure will be *i.e.*,

$Y = 0.94 \times 50 + 92.18$

$= 47 + 92.18$

$= 139.18$

The estimated blood pressure is 139.18.

Example 25: From the following results, obtain the two regression equations and estimate the yield of crops when the rainfall is 22 cms and the amount of 9 cms rainfall when the yield is 600 kg.

Variable	*Yield in kg (Y)*	*Rainfall in cm (X)*
Mean	**508.4**	**26.7**
S.D.	**36.8**	**4.6**

Coefficient correlation between yield and rainfall = 0.52. *[C.A. 1976]*

Solution: (*i*) To estimate yield of crops (Y) we have to use regression equation Y on X and the rainfall (X), the regression equation of X on Y.

(*ii*) $\bar{X} = 26.7$ S.D. $(\sigma_x) = 4.6$

$\bar{Y} = 508.4$ S.D. $(\sigma_y) = 36.8$ $r = 0.52$

Therefore $$bxy = r \cdot \frac{\sigma_y}{\sigma_x} = 0.52 \times \frac{36.8}{4.6} = 0.52 \times 8 = 4.16$$

$$bxy = r \cdot \frac{\sigma_x}{\sigma_y} = 0.52 \times \frac{4.6}{36.8} = 0.52 \times 0.125 = 0.065$$

(*iii*) So that regression equation of Y on X is

$$Y - \bar{Y} = bxy\ (X - \bar{X})$$

$$Y - 508.4 = 4.16\ (X - 26.7)$$

$$Y = 4.16X - 4.16 \times 26.7 + 508.4$$

$$Y = 4.16X - 111.072 + 508.4$$

$$Y = 4.16X + 397.328$$

Regression equation of X on Y is

$$= X - \bar{X} = bxy\ (Y - \bar{Y})$$

$$X - 26.7 = 0.065\ (Y - 508.4)$$

$$X - 26.7 = 0.065Y - 0.065 \times 508.4$$

$$X - 26.7 = 0.065Y - 33.046$$

$$X = 0.065Y + 26.7 - 33.046$$

$$X = 0.065Y - 6.346$$

(*iv*) When rainfall (X) is 22 cms, the yield of crops will be

$$Y = 4.16 \times 22 + 397.328$$

or $$Y = 91.52 + 397.328 = 488.848 \text{ kg}$$

When yield of crops is 600 kg, the rainfall will be

$$X = 0.065 \times 600 - 6.346$$

$$X = 39 - 6.346 = 32.654 \text{ cms.}$$

Example 26: From the following information of an agricultural area of West Bengal, estimate the probable crop yield when rainfall is 30 inches:

Variables	*Rainfall in inches (X)*	*Crop yield in unit per acre (Y)*
Mean	**25**	**40**
S.D.	**3**	**6**

Coefficient correlation between the variables 0.65.

Solution: To estimate crop yield (Y) we have to use regression equation Y on X *i.e.*,

$$Y - \bar{Y} = byx\,(X - \bar{X})$$

$$\bar{X} = 25 \qquad \sigma_x = 3$$

$$\bar{Y} = 40 \qquad \sigma_y = 6 \qquad r = 0.65$$

Therefore

$$byx = r \cdot \frac{\sigma_y}{\sigma_x} = 0.65 \times \frac{6}{3} = 0.65 \times 2 = 1.30$$

$$Y - \bar{Y} = 1.3\,(X - 25)$$

$$Y - 40 = 1.3X - 25 \times 1.3$$

$$Y = 1.3X - 32.5 + 40$$

$$Y = 1.3X + 7.5$$

When rainfall is 30 inches *i.e.*, $X = 30$ inches

$$Y = 1.3 \times 30 + 7.5$$

$$= 39 + 7.5 = 46.5 \text{ units}$$

Therefore, the crop yield will be 46.5 unit per acre.

Example 27: Work out the linear regression equation of trunk lengths (Y mm) on wing lengths (X mm), using the following data from a sample of 10 grasshoppers:

Animal	Trunk length (Y)	Wing length (X)
1	45	20
2	55	35
3	45	25
4	50	30
5	60	40
6	70	45
7	35	22
8	55	40
9	50	30
10	35	23

[Vidyasagar Univ. (M. Sc Zool) 2005]

Solution:

Sl. No.	Y	$Y - \bar{Y} = dy$	$(Y - \bar{Y})^2 = dy^2$	X	$X - \bar{X} = dx$	$(X - \bar{X})^2 = dx^2$	$dx \times dx$
1.	45	45 − 50 = −5	25	20	20 − 31 = −11	121	−5 × − 11 = 55
2.	55	55 − 50 = +5	25	35	35 − 31 = +4	16	+5 × 4 = 20
3.	45	45 − 50 = −5	25	25	25 − 31 = −6	36	−5 × − 6 = 30
4.	50	50 − 50 = 0	0	30	30 − 31 = −1	01	0 × − 2 = 00
5.	60	60 − 50 = +10	100	40	40 − 31 = +9	81	+10 × 9 = 90
6.	70	70 − 50 = +20	400	45	45 − 31 = +14	196	+20 × 14 = 280
7.	35	35 − 50 = −15	225	22	22 − 31 = −9	81	−15 × −9 = 135
8.	55	55 − 50 = +5	25	40	40 − 31 = +9	81	+5 × 9 = 45
9.	50	50 − 50 = 0	0	30	30 − 31 = −1	01	0 × −1 = 00
10.	35	35 − 50 = −15	225	23	23 − 31 = −8	04	−15 × −8 = 120
	500	$\sum d_y = 0$	1050	310	$\sum d_x = 0$	678	775

$\sum Y = 500$ $\sum X = 310$ $\sum dydx = 775$

$\bar{Y} = \frac{500}{10} = 50$ $\bar{X} = \frac{310}{10} = 31$

$\sum(Y - \bar{Y})^2 = dy^2 = 1050$ $\sum(X - \bar{X})^2 = dx^2 = 678$

$$byx = \frac{\sum dx\, dy}{\sum dx^2} = \frac{775}{678} = 1.14$$

Regression equation Y on X is $Y - \bar{Y} = byx\,(X - \bar{X})$

$$Y - 50 = byx\,(X - \bar{X})$$
$$Y - 50 = 1.14\,(X - 31)$$
$$Y = 1.14 \times -1.14 \times 31 + 50$$
$$Y = 1.14 \times -35.34 + 50$$
$$Y = 1.14 \times +14.66.$$

Example 28: The following results were obtained from the measurement of body weight (X) in gm and brain AChE in μ moles (Y) of 25 *channa* sp fishes:

$$\sum X = 97.1 \quad \sum X^2 = 459.8 \quad \sum XY = 444.68 \quad \sum Y = 124.5 \quad \sum Y^2 = 643.81$$

Find the appropriate regression equation and estimate brain AChE activities for a fish having body weight 5 gm. *[C.U. (M.Sc. Zoology) 2007]*

Solution: $n = 25$

The regression equation of brain AChE activity (Y) on body weight (X). *i.e.,*

$$Y - \bar{Y} = byx\,(X - \bar{X})$$

$$byx = \frac{n\sum XY - \sum X \sum Y}{n\sum X^2 - \left(\sum X\right)^2} = \frac{25 \times 444.68 - 97.1 \times 124.5}{25 \times 459.8 - (97.1)^2}$$

$$= \frac{11117 - 12088.95}{11495 - 9428.41} = \frac{-971.95}{+2066.59} = -0.47$$

$$\bar{Y} = \frac{124.5}{25} = 4.98 \quad \bar{X} = \frac{97.1}{25} = 3.884$$

$$Y - 4.98 = -0.47\,(X - 3.884)$$
$$Y - 4.98 = -0.47X + 1.82548$$
$$Y = -0.47X + 1.82548 + 4.98$$
$$Y = -0.47X + 6.805$$

Body weight 5 gm (X) AChE activity (Y) will be

$$Y = -0.47 \times 5 + 6.805$$
$$= -2.35 + 6.805 = +4.455$$

AChE activity is 4.455μ moles.

Example 29: Work out the linear regression equation of wing length (Y in mm) on body weight (X in mg) of the mosquito *Anopheles subpictus* collected from a rice field area at night. *[Vidyasagar Univ. M.Sc (2008)]*

Sl. No.	*1*	*2*	*3*	*4*	*5*	*6*	*7*	*8*	*9*	*10*
Wing (Y) length	2.3	2.5	2.9	3.1	3.1	3.2	3.3	3.4	3.2	4.0
Body (X) weight	4.1	4.2	4.4	5.1	5.2	5.2	5.3	5.4	5.0	5.6

From your computed equation, find out the wing length (in mm) of a mosquito of 4.5 mg body weight.

Solution: (*i*) Let the assumed mean $a = 3.1$ and $b = 5.0$ for X and Y series respectively.

(*ii*) Let the deviation from assumed mean be $u = X - a$ and $v = Y - b$.

Sl. No.	*Wing weight Y*	*u = Y – 3.1*	u^2	*Body weight X*	*v = X – 5.0*	v^2	*u × v = uv*
1.	2.3	2.3 – 3.1 = –0.8	.64	4.1	4.1 – 5.0 = –0.9	0.81	–0.8 × –0.9 = 0.72
2.	2.5	2.5 – 3.1 = –0.6	.36	4.2	4.2 – 5.0 = –0.8	0.64	–0.6 × –0.8 = 0.48
3.	2.9	2.9 – 3.1 = –0.2	.04	4.4	4.4 – 5.0 = –0.6	0.36	–0.2 × –0.6 = 0.12
4.	3.1	3.1 – 3.1 = 0	00	5.1	5.1 – 5.0 = 0.1	0.01	0 × 0.1 = 0.00
5.	3.1	3.1 – 3.1 = 0	00	5.2	5.2 – 5.0 = 0.2	0.04	0 × 0.2 = 0.00
6.	3.2	3.2 – 3.1 = +0.1	0.01	5.2	5.2 – 5.0 = 0.2	0.04	+0.1 × 0.2 = 0.02
7.	3.3	3.3 – 3.1 = 0.2	0.04	5.3	5.3 – 5.0 = 0.3	0.09	+0.2 × 0.3 = 0.06
8.	3.4	3.4 – 3.1 = 0.3	0.09	5.4	5.4 – 5.0 = 0.4	0.16	+ 0.3 × 0.4 = 0.12
9.	3.2	3.2 – 3.1 = 0.1	0.01	5.0	5.0 – 5.0 = 00	00	+0.1 × 0 = 0.00
10.	4.0	4.0 – 3.1 = 0.9	0.81	5.6	5.6 – 5.0 = 0.6	0.36	+0.9 × 0.6 = 0.54
		+0.16 – 0.16 = 0	2.00		+0.18 – 2.30 = – 2.12	2.51	$\sum uv = 2.06$

$$\sum u = 0 \quad \sum u^2 = 2 \qquad\qquad \sum v = -2.12 \quad \sum v^2 = 2.51$$

$$\overline{Y} = a + \frac{\sum u}{n} = 3.1 + \frac{0}{10} = 3.1 \qquad \overline{X} = 5.0 + \frac{2.12}{10} = 5.0 - 0.212$$

$$= 4.788 = 4.79$$

The line of regression Y on X *i.e.*,

$$Y - \overline{Y} = byx\ (X - \overline{X})$$

$$byx = \frac{\sum uv - \frac{\sum u \sum v}{n}}{\sum v^2 - \frac{\left(\sum v\right)^2}{n}} = \frac{2.08 - \frac{0 \times -0.5}{10}}{2.51 - \frac{(0.5)^2}{10}} = \frac{2.08}{2.51 - .025} = \frac{2.08}{2.485} = 0.837 = 0.84$$

$$Y - 3.1 = 0.84\ (X - 4.79)$$

$$Y - 3.1 = 0.84X - 4.02$$

$$Y = 0.84X - 4.02 + 3.1$$

$$= 0.84X - 0.92$$

When body weight (X) is 4.5 mg, the wing length will be

$$Y = 0.84 \times 4.5 - 0.92 = 3.78 - 0.92 = 2.86 \text{ mm}$$

Example 30: Work out linear regression of O_2 consumption (Y ml/min) on tracheal ventilation (X ml/min) for the following sample of beetles:

Beetle nos.	1	2	3	4	5	6	7	8
X ml	**75**	**86.2**	**77**	**74**	**65**	**84.3**	**82**	**70.5**
Y ml	**2.5**	**3.3**	**3.0**	**2.4**	**2.0**	**3.2**	**3.1**	**2.0**

[Vidyasagar University Msc. (2001) 2003]

Solution: (*i*) Let the assumed mean $a = 75$ and 2.5 for X and Y respectively.

(*ii*) Let the deviation from assumed mean be $u = X - a$ and $v = Y - b$

Sl. No.	X	u = X – 75	u^2	Y	v = Y – 2.5	v^2	u × v = uv
1.	75	75 – 75 = 0	0	2.5	2.5 – 2.5 = 0	0	0 × 0 = 0
2.	86.2	86.2 – 75 = 11.2	125.44	3.3	3.3 – 2.5 = 0.8	0.64	11.2 × 0.8 = 8.96
3.	77	77 – 75 = 2	4	3.0	3.0 – 2.5 = 0.5	0.25	2 × 0.5 = 1.00
4.	74	74 – 75 = –1	1	2.4	2.4 – 2.5 = –0.1	0.01	–1 × –0.1 = 0.10
5.	65	65 – 75 = –10	100	2.0	2.0 – 2.5 = –0.5	0.25	–10 × –0.5 = 5.00
6.	84.3	84.3 – 75 = 9.3	86.49	3.2	3.2 – 2.5 = 0.7	0.49	9.3 × 0.7 = 6.51
7.	82	82 – 75 = 7	49	3.1	3.1 – 2.5 = 0.6	0.36	7 × 0.6 = 4.20
8.	70.5	70.5 – 75 = –4.5	20.25	2.0	2.0 – 2.5 = –0.5	0.25	–4.5 × –0.5 = 2.25
		$\sum u = 14.0$	$\sum u^2 = 386.18$		$\sum v = +2.6\ (-1.1) = 1.5$	2.25	$\sum uv = 28.02$

$$\overline{X} = a + \frac{14.0}{8} = 75 + 1.75 = 76.75 \quad \overline{Y} - 2.5 + \frac{1.5}{8} = 1.875 = 2687 = 2.69$$

$$byx = \frac{\sum uv - \frac{\sum u \sum v}{n}}{\sum v^2 - \frac{\left(\sum u\right)^2}{n}} = \frac{28.02 - \frac{14 \times 1.5}{8}}{386.18 - \frac{(14)^2}{8}} = \frac{28.02 - 2.625}{386.18 - \frac{196}{8}}$$

$$= \frac{25.395}{386.18 - 24.8} = \frac{25.395}{361.68} = 0.07$$

$$Y - \overline{Y} = byx\,(X - \overline{X})$$

$$Y - 2.69 = 0.07\,(X - 76.75)$$

$$Y = 0.07X \times 5.3725 + 2.69$$

$$= 0.07X - 2.6825.$$

Example 31: Obtain the equation of the line of regression of yield of rice (*Y*) and water (*X*)

Water in inches	**12**	**18**	**24**	**30**	**36**	**42**	**48**
Yield in tonnes	**5.27**	**5.68**	**6.25**	**7.21**	**8.02**	**8.71**	**8.42**

Estimate the most probable yield of rice for 40 inches of water.

Solution:

Si. No.	X	dx = (X – 30)	dx^2 $(X - 30)^2$	Y	dy = Y – 7.21	dy^2 $(Y - 7.21)^2$	dx dy
1.	12	12 – 30 = –18	324	5.27	5.27 – 7.21 = –1.94	= 3.7636	34.92
2.	18	18 – 30 = –12	144	5.68	5.68 – 7.21 = –1.53	= 2.34	18.36
3.	24	24 – 30 = –6	36	6.25	6.25 – 7.21 = –0.96	= 0.9216	5.76
4.	30	30 – 30 = 0	0	7.21	7.21 – 7.21 = 0	= 0	0
5.	36	36 – 30 = +6	36	8.02	8.02 – 7.21 = +.81	= .6561	4.86
6.	42	42 – 30 = +12	144	8.71	8.71 – 7.21 = +1.50	= 2.25	18.00
7.	48	48 – 30 = +18	324	8.42	8.42 – 7.21 = +1.21	= 1.4641	21.78
		$\sum dx = 0$	1008		$\sum dy = -91$	11.394	103.68

Let assume mean of X is $A_x = 30$ and $Y = A_y = 7.21$

$$\bar{X} = Ax + \frac{\sum dx}{N} = 30 + \frac{0}{7} = 30$$

$$\bar{Y} = Ay + \frac{\sum dy}{N} = 7.21 - \frac{.91}{7} = 7.21 = .13 = 7.08$$

$$bxy = \frac{\sum dx\, dy - \frac{\sum dx \sum dy}{N}}{\sum dy^2 - \frac{\left(\sum dy\right)^2}{N}} \qquad byx = \frac{\sum dy\, dx - \frac{\sum dy \sum ds}{N}}{\sum dx^2 - \frac{\left(\sum dx\right)^2}{N}}$$

$$= \frac{103.68 - \frac{0 \times (-.91)}{7}}{11.394 - \frac{(-.91)^2}{7}} \qquad = \frac{103.68 - \frac{(-.91) \times 0}{7}}{1008 - \frac{(0)^2}{7}}$$

$$= \frac{103.68}{11.394 - 0.1183} \qquad = \frac{103.68}{1008} = 0.103$$

$$= \frac{103.68}{11.2757} = 9.195$$

Regression equation of Y on X is

$$(Y - \bar{Y}) = byx\,(X - \bar{X})$$

$$Y - 7.08 = .103\,(X - 30)$$

$$Y = .103X - 3.09 + 7.08$$

$$= .103X + 3.99$$

Putting $X = 40''$

$$\therefore \quad Y = 0.103 \times 40 + 3.99$$

$$= 4.12 + 3.99 = 8.11 \text{ tons.}$$

Example 32: Find the regression of X or Y from the following data:

$\sum X = 24$, $\sum Y = 44$, $\sum XY = 306$, $\sum X^2 = 164$, $\sum Y^2 = 574$, $N = 4$

Find the value of X when $Y = 6$.

Solution: Here $\bar{X} = \frac{\sum X}{N} = \frac{24}{4} = 6$

$$\bar{Y} = \frac{\sum Y}{N} = \frac{44}{4} = 11$$

Regression coefficient $bxy = r\,\frac{\sigma x}{\sigma y}$

(as x on y)

$$= \frac{N \sum XY - \sum X \sum Y}{N \sum Y^2 - \left(\sum Y\right)^2} = \frac{4 \times 306 \quad 24 \times 44}{4 \times 574 \quad (44)^2}$$

$$= \frac{1224 - 1056}{2296 - 1936} = \frac{168}{360} = 0.46666 = 0.47$$

The regression equation of X on Y is

$$X - \bar{X} = r\frac{\sigma_x}{\sigma_y}(Y - \bar{Y})$$

$$X - 6 = 0.47\ (Y - 11)$$

or $$X - 6 = 0.47Y - 0.47 \times 11$$

or $$X = 0.47Y - 5.17 + 6$$

As $Y = 6$, So $$X = 0.47 \times 6 + 0.83$$

$$= 2.82 + 0.83 = 3.65.$$

Example 33: For 10 observations on price (X) and supply (Y), the following data were obtained:

$$\sum X = 130, \quad \sum Y = 220, \quad \sum X^2 = 2288, \quad \sum Y^2 = 5506, \quad \sum XY = 346.7$$

Obtain the line of regression of Y on X and estimate the supply when the price is 16 units.

Solution: $\bar{X} = \frac{130}{10} = 13$ $\quad \bar{Y} = \frac{220}{10} = 22$ $\quad \sum XY = 3467$

$N = 10$

Regression coefficient $(byx) = r\frac{\sigma y}{\sigma x}$

$$= \frac{N\sum XY - \sum X \sum Y}{N\sum X^2 - \left(\sum X\right)^2}$$

$$= \frac{10 \times 3467 - 130.220}{10 \times 2288 - (130)^2}$$

$$= \frac{34670 - 28600}{22880 - 16900}$$

$$= \frac{6070}{5980} = 1.0$$

The regression equation Y on X is

$$Y - \bar{Y} = r\frac{\sigma y}{\sigma x}(X - \bar{X})$$

$$Y - 22 = 1\ (X - 13)$$

$$Y = X - 13 + 22 \quad \text{or} \quad Y = X + 9$$

When the price is 16 units *i.e.*, $X = 16$, then supply $y = 16 + 9 = 25$.

Example 34: You are given the following data:

Variable	X	Y
Mean	47	96
Variance	64	81

Correlation coefficient (r) between X and Y = 0.36. Determine the equations of regression lines. Calculate Y when X = 50 and X when Y = 88.

Solution: $\bar{X} = 47$ $\quad \bar{Y} = 96$ $\quad r_{xy} = 0.36$

$\sigma x^2 = 64$ $\quad \sigma y^2 = 81$

$\sigma x = 8$ $\quad \sigma y = 9$

Regression line of Y on X $= Y - \bar{Y} = r \frac{\sigma y}{\sigma x} (X - \bar{X})$

$$Y - 96 = .36 \frac{9}{8} (X - 47)$$

$$Y - 96 = \frac{324}{800} (X - 47)$$

$$Y - 96 = 0.405X - 19.035$$

$$Y = 0.405X + 96 - 19.04$$

$$= 0.405X + 76.96$$

Regression equation X on Y $= X - \bar{X} = r \frac{\sigma x}{\sigma y} (Y - \bar{Y})$

$$X - 47 = .36 \times \frac{8}{9} (Y - 96)$$

$$= .04 \times 8 (Y - 96)$$

$$X - 47 = .32Y - 30.72$$

$$X = .32Y + 47 \quad 30.72$$

$$= .32Y + 16.28$$

When $X = 50$, we have $Y = .405X + 76.96$

$$= .405 \times 50 + 76.96$$

$$Y = 20.25 + 76.96 = 97.21$$

When $Y = 88$, we have $X = .32Y + 16.28$

$$= .32 \times 88 + 16.28$$

$$= 28.16 + 16.28 = 44.44.$$

Example 35. The following results were obtained from measurements of body wt.(x) in gm & brain AchE activity (y) of 25 Tilapia fish.

$$\sum x = 974 \sum x^2 = 459.80 \sum y = 124.5 \sum y^2 = 643.81 \sum xy = 444.68$$

Find out appropriate regression equation & estimate the brain AchE activity of a fish having body wt. 5gm. *[M. Sc (Zoology) C.U, 2002]*

Solution:

Now, we have to find out y value when $x = 5$ gm

$n = 25$

So, $\bar{x} = \frac{\sum x}{n} = \frac{97.4}{25} = 3.9$

$$\bar{y} = \frac{\sum y}{n} = \frac{124.5}{25} = 4.98$$

Regression coefficient (by x) $= r \frac{\sigma y}{\sigma x}$

(y on x)

$$by\ x = \frac{N\sum xy - \sum x.\sum y}{N\sum x^2 - \left(\sum x\right)^2}$$

$$= \frac{(25 \times 444.68) - (97.4 \times 124.5)}{(25 \times 459.8) - (97.4)^2}$$

$$= \frac{11117 - 12126.3}{11495 - 9486.76} = \frac{1009.3}{2008.24} = .5$$

The regression equation y on x is

$$y - \bar{y} = r\frac{\sigma y}{\sigma x}(x - \bar{x})$$

$$y - 4.98 = .5 \times (5 - 3.9)$$

$$y = .55 + 4.98$$

$$y = 5.53$$

Example 36. The following results were obtained from the measurement of body length (y) in cm & body wt (x) in gm of 25 *Anabas scandans* fishes.

$$\sum x = 1165 \sum x^2 = 56947 \sum xy = 9024.40 \sum y = 185.20 \sum y^2 = 1434.24$$

Find the appropriate regression equation and estimate the body length of a fish having weight 50 gms.

Solution:

Now we have to find the value of y when x = 50 gm, n = 25

Now $\bar{x} = \frac{\sum x}{n} = \frac{1165}{25} = 46.6$

$$\bar{y} = \frac{\sum y}{n} = \frac{185.20}{25} = 7.408$$

Regression coefficient (by x) $= r\frac{\sigma y}{\sigma x}$

(y on x)

$$= \frac{N\sum xy - \sum x \sum y}{N\sum x^2 - \left(\sum x\right)^2}$$

$$= \frac{(25 \times 9024.4) - (1165 \times 185.2)}{(25 \times 56947) - (1165)^2}$$

$$= \frac{225610 - 215758}{1423675 - 1357225} = \frac{9852}{66450} = .148$$

The regression equation y on x is

$$y - \bar{y} = r\frac{\sigma y}{\sigma x}(x - \bar{x})$$

$$y - 7.408 = 148(50 - 46.5)$$

$$y = .5032 + 7.408$$

$$y = 7.9112$$

- **Multiple Regression:**

It is basically the method of expressing the criterion by using two or more predictors.

- **Types:**

(*a*) Multiple linear regression.
(*b*) Multiple non linear regression.

- **Assumptions:**

I. All the variables (**predictors and criterions**) are continuous measurement variables.
II. Scores of variables have **unimodal** and **fairly** symmetrical distribution.
III. Linear association persists between the criterion and predictors.

- **Computation of Multiple Linear Regression:**

I. The multiple linear regression of criterion (X_1) on the combination of two predictors (X_2 and X_3) can be worked out by using partial regression coefficients ($b_{1,23}$ and $b_{1,32}$). Beta coefficients (β_2 & β_2) and standard deviations (s_1, s_2 & s_3) of all three variables (X_1, X_2 & X_3).
II. Beta coefficients (β_2 & β_2) are computed from the product moment r values.
III. Partial regression coefficients ($b_{1,23}$ and $b_{1,32}$) are computed by using beta coefficients and standard deviations (s_1, s_2 & s_3) which are the measures of slopes of regression lines of the criterion (X_1) on the predictors (X_2 and X_3).
IV. The Y intercept of the regression line *i.e.*, $a_{1,23}$ is computed by using partial regression coefficient and means of the variables.
V. $\widehat{X}$ is the predicted score of criterion.

I. $\beta_2 = \dfrac{r_{12} - r_{13}\, r_{23}}{1 - r_{23}^2} \qquad \beta_3 = \dfrac{r_{13} - r_{12}\, r_{23}}{1 - r_{23}^2}$

II. $b_{1,23} = \beta_2 \times \dfrac{S_1}{S_2} \qquad b_{1,32} = \beta_3 \times \dfrac{S_1}{S_3}$

III. $a_{1,23} = \overline{X}_1 - b_{1,23}\, \overline{X}_2 - b_{1,32}\, \overline{X}_3$

IV. $\hat{X}_1 = a_{1,23} + b_{1,23}\, X_2 + b_{1,32}\, X_3$

Example 1. Work out the multiple linear regression equation of X_1 on combination of X_2 and X_3 from the following data.

$\overline{X}_1 = 12.0 \quad S_1 = 2.82 \quad r_{12} = +0.72$
$\overline{X}_2 = 8.5 \quad S_2 = 2.10 \quad r_{13} = +0.21$
$\overline{X}_3 = 6.4 \quad S_3 = 1.21 \quad r_{23} = +0.23$

Solution: (*a*) The beta coefficients are computed from the product moment r values.

$$\beta_2 = \frac{r_{12} - r_{13}\, r_{23}}{1 - r_{23}^2} = \frac{0.72 - 0.21 \times 0.23}{1 - (0.23)^2} = \frac{0.72 - 0.483}{1 - 0.0529} = \frac{0.24}{0.95} = 0.25$$

$$\beta_3 = \frac{r_{13} - r_{12}\, r_{23}}{1 - r_{23}^2} = \frac{0.21 - 0.72 \times 0.23}{1 - (0.23)^2} = \frac{0.21 - 0.1656}{1 - 0.0529} = \frac{0.44}{0.95} = 0.46$$

(*b*) The partial regression coefficients (b) are computed using the beta (β) coefficients and standard deviations (σ).

$$b_{1,23} = \beta_2 \times \frac{S_1}{S_2} = 0.25 \times \frac{2.82}{2.10} = 0.25 \times 1.34 = 0.335 = 0.34$$

$$b_{1,32} = \beta_3 \times \frac{S_1}{S_3} = 0.46 \times \frac{2.82}{1.21} = 0.46 \times 2.33 = 1.071 = 1.07$$

(c) The partial regression coefficients and the means are used in computing $a_{1,23}$ *i.e.*, *Y* intercept the regression line.

$$a_{1,23} = \bar{X}_1 - b_{1,23}\,\bar{X}_2 - b_{1,32}\,\bar{X}_3 = 12.0 - 0.34 \times 8.5 - 1.07 \times 6.4$$
$$= 12.0 - 2.89 - 6.84 = 12.0 - 9.73 = 2.27$$

(d) The multiple regression equation of X_1 on X_2 and X_3 may be written as follows.

$$\hat{X}_1 = a_{1,23} + b_{1,23}\,X_2 + b_{1,32}\,X_3$$
$$= 2.27 + 0.34\,X_2 + 1.07\,X_3.$$

Example 2. Compute the multiple linear regression equation of X_1 on combinations of X_2 & X_3 from the following data.

$\bar{X}_1 = 60$ $\quad S_1 = 3$ $\quad r_{12} = 0.7$
$\bar{X}_2 = 70$ $\quad S_2 = 4$ $\quad r_{13} = 0.6$
$\bar{X}_3 = 100$ $\quad S_3 = 5$ $\quad r_{23} = 0.4$

Solution: (a) The beta coefficients are computed from the product moment *r* values.

$$\beta_2 = \frac{r_{12} - r_{13}\,r_{23}}{1 - r_{23}^2} = \frac{0.7 - 0.6 \times 0.4}{1 - (0.4)^2} = \frac{0.7 - 0.24}{1 - 0.16} = \frac{0.46}{0.84} = 0.5476 = 0.55$$

$$\beta_3 = \frac{r_{13} - r_{12}\,r_{23}}{1 - r_{23}^2} = \frac{0.6 - 0.7 \times 0.4}{1 - (0.4)^2} = \frac{0.6 - 0.28}{1 - 0.16} = \frac{0.32}{0.84} = 0.38$$

(b) The partial regression coefficients are computed using the beta (β) coefficients and standard deviations.

$$b_{12,3} = \beta_2 \times \frac{S_1}{S_2} = 0.55 \times \frac{3}{4} = 0.55 \times 0.75 = 0.4125 = 0.41$$

$$b_{13,2} = \beta_3 \times \frac{S_1}{S_3} = 0.38 \times \frac{3}{5} = 0.38 \times 0.6 = 0.228 = 0.23$$

(c) The partial regression coefficients and the means are used in computing $a_{1,23}$ as follows.

$$a_{1,23} = \bar{X}_1 - b_{12,3}\,\bar{X}_2 - b_{13,2}\,\bar{X}_3 = 60 - 0.41 \times 70 - 0.23 \times 100$$
$$= 60 - 28.7 - 23 = 60 - 51.7 = 8.3$$

(d) The multiple regression equation of X_1 on X_2 and X_3 may be written as follows.

$$\hat{X}_1 = a_{1,23} + b_{12,3}\,X_2 + b_{13,2}\,X_3$$
$$= 8.3 + 0.41\,X_2 + 0.23\,X_3.$$

Example 3. Compute the multiple linear regression of glomerular filtration rate (Y_1 ml/min) on glomerular blood pressure (X_2 mm Hg) and capsular fluid pressure (X_3 mm Hg) using the following data.

$\bar{Y}_1 = 120$ **ml/min** $\quad S_1 = 21.5$ **ml** $\quad r_{12} = +0.82$
$\bar{X}_2 = 58$ **mm Hg** $\quad S_2 = 14.2$ **mm Hg** $\quad r_{13} = -0.21$
$\bar{X}_3 = 18$ **mm Hg** $\quad S_3 = 3.5$ **mm Hg** $\quad r_{23} = +0.18$

Solution: (a) Computation of β coefficients.

$$\beta_2 = \frac{r_{12} - r_{13}\,r_{23}}{1 - r_{23}^2} = \frac{0.82 - (-0.21 \times 0.18)}{1 - (0.18)^2}$$

$$= \frac{0.82 - 0.038}{1 - 0.0324} = \frac{0.858}{0.9676} = 0.886 = 0.89$$

$$\beta_3 = \frac{r_{13} - r_{12}\, r_{23}}{1 - r_{23}^2} = \frac{-0.21 - 0.82 \times 0.18}{1 - (0.18)^2}$$

$$= \frac{-0.21 - 0.1476}{1 - 0.0324} = \frac{-0.3576}{0.9676} = -0.3695 = -0.37$$

(*b*) Calculation of partial regression coefficients.

$$b_{1,23} = \beta_2 \times \frac{S_1}{S_2} = 0.89 \times \frac{21.5}{14.2} = 0.89 \times 1.514 = 1.347 = 1.35$$

$$b_{1,32} = \beta_3 \times \frac{S_1}{S_3} = 0.37 \times \frac{21.5}{3.5} = 0.37 \times 6.143 = -2.27$$

(*c*) Computation of $a_{1,23}$ by using partial regression coefficients and means.

$$a_{1,23} = \overline{Y}_1 - b_{1,23}\, \overline{X}_2 - b_{1,32}\, \overline{X}_3 = 120 - 1.35 \times 58 - (-2.27) \times 18$$

$$= 120 - 78.3 - 40.86 = 160.86 - 78.3 = 82.56$$

(*d*) The multiple regression equation of Y_1 on X_2 and X_3 is written as follows.

$$\hat{Y}_1 = a_{1,23} + b_{123}\, X_2 + b_{132}\, X_3$$

$$= 82.56 + 1.35\, X_2 + (-2.27)\, X_3$$

$$= 82.56 + 1.35\, X_2 - 2.27\, X_3.$$

CHAPTER

ANALYSIS OF VARIANCES [ANOVA]

- **Anova:** It is a powerful statistical procedure for determining if differences in means are significant and for dividing the variance into components.
- **Variance** (σ^2) It is an absolute measure of dispersion of raw scores around the sample (group) mean, the dispersion of the scores resulting from their varying differences (*error terms*) from the means.

  The square of standard deviation is called variance and is denoted by σ^2.
- **Mean Square:** The measure of variability used in the analysis of variance is called a "Mean square".

  Sum of squared deviation from mean divided by degrees of freedom.

$$\text{Mean square} = \frac{\text{Sum of squared deviation from mean}}{\text{Degrees of freedom}}$$

- **Assumptions in the analysis of variance:**
  (*i*) The samples are independently drawn.
  (*ii*) The population are normally distributed, with common variance.
  (*iii*) They occur at random and independent of each other in the groups.
  (*iv*) The effects of various components are additive.
- **Technique for analysis of variance:**
  (*a*) *One-way Anova:* Here a single independent variable is involved.

  Example: Effect of pesticide (*independent variable*) on the oxygen consumption (*dependent variable*) in a sample of insect.

  (*b*) *Two-way Anova:* Here two independent variables are involved.

  Example: Effects of different levels of combination of a pesticide (*independent variable*) and an insect hormone (*independent variable*) on the oxygen consumption of a sample of insect.
- **Working Procedure:**
  (*i*) The procedure of calculation in direct method are lengthy as well as time consuming and this is not popular in practice for all purposes.
  (*ii*) Therefore a short-cut method based on the sum of the squares of the individual values are usually used.
  (*iii*) This method is more convenient.
- **Steps of calculation:**
  (*i*) Set up the null hypothesis and alternative hypothesis.
  (*ii*) Calculate total T of all the observations in all samples.

$$T = \sum X_1 + \sum X_2 + \sum X_3 ..... + \sum X_K$$

(*iii*) The correction factor is $\frac{T^2}{N}$.

(*iv*) Calculate sum of square deviation (total) *i.e.*,

$$SST = \sum X_1^2 + \sum X_2^2 + ... + \sum X_K^2 - \frac{T^2}{N}$$

(*v*) Now calculate sum of squares of deviation between the samples *i.e.*, *SSB*.

$$SSB = \left[\frac{\sum(X_1)^2}{n_1} + \frac{\sum(X_2)^2}{n_2} + \frac{\sum(X_3)^2}{n_3} + ... + \frac{\sum(X_K)^2}{n_K}\right] - \frac{T^2}{N}$$

(*vi*) Now calculate sum of squares of deviation within the samples (*SSW*)

$SSW - SST - SSB.$

(*vii*) Now calculate mean square deviation between the samples (*MSB*): Dividing *SSB* by the degrees of freedom $v_1 = k - 1$

$$MSB = \frac{SSB}{v_1}$$

In the same way, calculate mean square within the samples *MSW*. Dividing *SSW* by the degrees of freedom $V_2 = N - K$

$$MSW = \frac{SSW}{V_2}$$

(*viii*) Calculation of *F* statistic:

Both *MSB* and *MSW* are independent unbiased estimates of the same population variances.

Therefore $F = \frac{MSB}{MSW}$, degrees of freedom $V_1 = K - 1$ & $V_2 = N - K$,

In general $MSB > MSW$. If $MSB < MSW$, then $F = \frac{MSW}{MSB}$.

(*x*) Compare calculated value with the tabulated value and if greater then reject null hypothesis or lesser then accept the hypothesis.

- **ANOVA** = Statistical technique for analysis of variables.
- **ANCOVA** = Analysis of co-variance is used when groups are to be compared.
- **MANOVA** = Multivariate analysis of variance is used when there are more than two dependent variables.

Example 1. The following data give the yields on 12 plots of land in three samples under three varieties of fertilizers.

A	B	C
25	20	24
22	17	26
24	16	30
21	19	20

Is there any significant difference in the average yields of land under the three varieties of fertilizers?

Given that *F* at *df* (2, 9) at 5% level = 4.26.

Solution:

- *Null Hypothesis* (*Ho*): There is no significant difference in the average yields under the three varieties.
- *Alternative Hypothesis* (H_1): The difference in average yields is significant.
- *Calculation:*

Sample A		*Sample B*		*Sample C*	
X_1	X_1^2	X_2	X_2^2	X_3	X_3^2
25	625	20	400	24	576
22	484	17	289	26	676
24	576	16	256	30	900
21	441	19	361	20	400
$\sum X_1 = 92$	$\sum X_1^2 = 2126$	$\sum X_2 = 72$	$\sum X_2^2 = 1306$	$\sum X_3 = 100$	$\sum X_3^2 = 2552$

Now $$T = \sum X_1 + \sum X_2 + \sum X_3 = 92 + 72 + 100 = 264$$

$$\text{Correction factor} = \frac{T^2}{N} = \frac{(264)^2}{4+4+4} = \frac{264 \times 264}{12} = 5808$$

$$\text{The total sum squares } (SST) = \sum X_1^2 + \sum X_2^2 + \sum X_3^2 - \frac{T^2}{N}$$

$$= 2126 + 1306 + 2552 - 5808$$

$$= 5984 - 5808 = 176$$

Sum squares between the samples (*SSB*)

$$= \frac{\left(\sum X_1\right)^2}{n_1} + \frac{\left(\sum X_2\right)^2}{n_2} + \frac{\left(\sum X_3\right)^2}{n_3} - \frac{T^2}{N}$$

$$= \frac{92 \times 92}{4} + \frac{72 \times 72}{4} + \frac{100 \times 100}{4} - 5808$$

$$= 2116 + 1296 + 2500 - 5808$$

$$= 5912 - 5808 = 104$$

Degrees of freedom $V_1 = (df_t) = \text{K} - 1 = 3 - 1 = 2$

$$\text{Mean square between the sample } (MSB) = \frac{SSB}{V_1} = \frac{104}{2} = 52$$

Sum square within the samples (SSW) = $SST - SSB$

$$SSW = 176 - 104 = 72$$

Degrees of freedom $V_2 = (df_w) = N - K = (n_1 + n_2 + n_3) - K = 12 - 3 = 9$

$$\text{Mean square within the sample } (MSW) = \frac{SSB}{V_2} = \frac{72}{9} = 8$$

Anova Table for Fertilizers

Source of variation	*Sum of squares (SS)*	*Degreess of freedom (df)*	*Mean squares (MS)*	*Test statistic*
Between the samples	$SSB = 104$	$df_t = 2$	$MSB = \frac{SSB}{V_1} = 52$	$F = \frac{MSB}{MSW}$
Within the samples	$SSW = 72$	$df_w = 9$	$MSW = \frac{SSW}{V_2} = 8$	$= \frac{52}{8} = 6.5$
Total	$SST = 176$	$df = 11$		

- *Decision:* The calculated value of F at 0.05 at degrees of freedom $V_1 = 2$ & $V_2 = 9$ is 6.5. It is much greater than the given value of $F = 4.26$. Hence we reject the null hypothesis (Ho) at 0.05 level and conclude that the difference in average yields under the three varieties is significant.

Example 2. The varieties of *A*, *B*, *C* wheat were sown in 4 plots each and the following yields in quintal per acre were obtained:

A	B	C
8	7	2
4	5	5
6	5	4
7	3	4

Test the significance of difference between the yield of the varieties.

Solution:

- *Null Hypothesis* (Ho): Varieties are not significantly different from each other in their yielding capacities.
- *Alternative Hypothesis* (H_1): Varieties are significantly different from each other.
- *Calculation:*

Sample A		*Sample B*		*Sample C*	
X_1	X_1^2	X_2	X_2^2	X_3	X_3^2
8	64	7	49	2	04
4	16	5	25	5	25
6	36	5	25	4	16
7	49	3	09	4	16
$\Sigma X_1 = 25$	$\Sigma X_1^2 = 165$	$\Sigma X_2 = 20$	$\Sigma X_2^2 = 108$	$\Sigma X_3 = 15$	$\Sigma X_3^2 = 61$

Row = 4 Column = 3 $N = 4 + 4 + 4 = 12$

Now
$$T = \sum X_1 + \sum X_2 + \sum X_3 = 25 + 20 + 15 = 60$$

$$\text{Correction factor} = \frac{T^2}{N} = \frac{(60)^2}{12} = \frac{60 \times \cancel{60}^{5}}{\cancel{12}} = 300$$

$$\text{Total sum squares } (SST) = \sum X_1^2 + \sum X_2^2 + \sum X_3^2 - \frac{T^2}{N}$$

$$= (165 + 108 + 61) - 300$$

$$= 334 - 300 = 34$$

Degrees of freedom $V_1 = (df_t) = K - 1 = 3 - 1 = 2$

Degrees of freedom $V_2 = (df_w) = N - K = 12 - 3 = 9$

Sum square between the varieties (SSB)

$$SSB = \frac{\left(\sum X_1\right)^2}{n_1} + \frac{\left(\sum X_2\right)^2}{n_2} + \frac{\left(\sum X_3\right)^2}{n_3} - \frac{T^2}{N}$$

$$= \frac{(25)^2}{4} + \frac{(20)^2}{4} + \frac{(15)^2}{4} - 300$$

$$= \frac{25 \times 25}{4} + \frac{400}{4} + \frac{225}{4} - 300$$

$$= \frac{625 + 400 + 225}{4} - 300$$

$$= \frac{1250}{4} - 300 = 312.5 - 300 = 12.5$$

Sum square within the varieties (SSW)

$$SSW = SST - SSB = 34 - 12.5 = 21.5$$

$$\text{Mean square between varieties} = \frac{SSB}{df_t} = \frac{SSB}{V_1}$$

$$MSB = \frac{12.5}{2} = 6.25$$

$$\text{Mean square within varieties} = \frac{SSW}{dfw} = \frac{SSW}{V_2}$$

$$MSW = \frac{21.5}{9} = 2.3888 = 2.39$$

Anova table for varieties of wheat

Source of variation	*Sum of squares (SS)*	*Degreess of freedom (df)*	*Mean squares (MS)*	*Test statistic*
Between varieties	$SSB = 12.5$	$df_t = 2$	$MSB = 6.25$	$F = \frac{MSB}{MSW}$
Within varieties	$SSW = 21.5$	$df_w = 9$	$MSW = 2.39$	$= \frac{6.25}{2.39} = 2.615$
Total	34.0	$df = 11$		

- *Decision:* 5% tabulated value of 5 at 2 & 9 *df* is 4.26 and calculated F is 2.615, it is less than tabulated value. Thus our null hypothesis is accepted *i.e.*, varieties are not significantly different from each other.

Example 3. The following data showed the tracheal ventilation of two groups of *Anopheles subpictus* from two different habitats. ($\alpha = 0.05$)

Group A (X_1):	**86**	**75**	**78**	**80**	**75**	**81**	**85**	**78**	**80**	**82**
Group B (X_2):	**82**	**71**	**72**	**70**	**70**	**73**	**74**	**65**	**70**	**73**

Is there any significant difference in the mean tracheal ventilation scores (ml/minute) ?

F scores: $F_{0.05(2,19)}$: 352 $F_{0.05(1,18)}$: 4.41

$F_{0.05(1,19)}$: 4.38 $F_{0.05(2,18)}$: 3.55

Solution:

- *Null Hypothesis:* There is no significant difference between the two variables (***An subpictus***).
- *Alternative Hypothesis:* There is significant difference between the two variables.
- *Calculation:*

Table for calculating sum squares directly from raw data

Gr A	X_1	X_1^2	*Gr B*	X_2	X_2^2
1	86	7396	1	82	6724
2	75	5625	2	71	5041
3	78	6084	3	72	5184
4	80	6400	4	70	4900
5	75	5625	5	70	4900
6	81	6551	6	73	5329
7	85	7225	7	74	5476
8	78	6084	8	65	4225
9	80	6400	9	70	4900
10	82	6724	10	73	5329
	$\sum X_1 = 800$	$\sum X_1^2 = 64124$		$\sum X_2 = 720$	$\sum X_2^2 = 52008$

$N = n_1 + n_2 = 10 + 10 = 20$

Now $T = \sum X_1 + \sum X_2 = 800 + 720 = 1520$

Correction factor $\dfrac{T^2}{N} = \dfrac{(1520)^2}{20} = \dfrac{1520 \times 1520}{20} = 76 \times 1520 = 115520$

The total sum squares $(SST) = \sum X_1^2 + \sum X_2^2 - \dfrac{T^2}{N}$

$= 64124 + 52008 - 115520$

$= 116132 - 115520 = 612$

Sum squares between the samples (*SSB*)

$$SSB = \frac{\left(\sum X_1\right)^2}{n_1} + \frac{\left(\sum X_2\right)^2}{n_2} - \frac{T^2}{N}$$

$$= \frac{(800)^2}{10} + \frac{(720)^2}{10} - 115520$$

$= (64000 + 51840) - 115520$

$= 115840 - 115520 = 320$

Degrees of freedom $V_1 = (df_t) = K - 1 = 2 - 1 = 1$

Mean square between the samples (*MSB*) $= \dfrac{SSB}{V_1}$

$$= \frac{320}{1} = 320$$

Sum square within the samples (SSW) = $SST - SSB$

$$= 612 - 320 = 292$$

Degrees of freedom $V_2 = (df_w) = N - K = (n_1 + n_2) - K = 20 - 2 = 18$

Mean square within the samples (MSW) $= \frac{SSW}{V_2} = \frac{292}{18} = 16.22$

Anova Table

Source of variation	*Sum of squares*	*Degrees of freedom (df)*	*Mean square*	*Test statistic*
Between the samples	$SSB = 320$	$V_1 = df_t = 1$	$MSB = \frac{SSB}{V_1} = 320$	$F = \frac{MSB}{MSW}$
Within the sample	$SSW = 292$	$V_2 = df_w = 18$	$MSW = \frac{SSW}{V_2} = 16.22$	$= \frac{320}{16.22} = 19.73$
Total	612	$df = 19$		

- *Decision:* The calculated F is 19.73 at degrees of freedom (1 and 18). The tabulated value of F at 0.5% level in the df 1, 18 is 4.41. The computed F is higher than tabulated value. So the null hypothesis is rejected.

Example 4. Work out one-way anova to find whether or not there is a significant difference between the wing lengths (mm) of the following two groups of grasshoppers from two different habitats of Dankuni area of West Bengal.

Group A:	**5.0**	**4.0**	**3.8**	**4.3**	**4.5**	**5.3**	**2.8**	**4.6**	**5.1**	**4.6**
Group B:	**4.7**	**2.4**	**3.0**	**4.6**	**3.0**	**4.5**	**2.2**	**3.1**	**3.7**	**2.8**

***F* scores:** $F_{0.05(1,18)} = 4.41$ $F_{0.01(1,18)} = 8.28$

Solution:

- *Null Hypothesis:* There is no significant difference between the wing lengths of two groups of grasshoppers.
- *Alternative Hypothesis:* There is significant difference between the wing lengths of grasshoppers of different habitats.
- *Calculation:*

Gr A	X_1	X_1^2	Gr B	X_2	X_2^2
1	5.0	25.00	1	4.7	22.09
2	4.0	16.00	2	2.4	5.76
3	3.8	14.44	3	3.0	9.00
4	4.3	18.49	4	4.6	21.16
5	4.5	20.25	5	3.0	9.00
6	5.3	28.09	6	4.5	20.25
7	2.8	7.84	7	2.2	4.84
8	4.6	21.16	8	3.1	9.61
9	5.1	26.01	9	3.7	13.69
10	4.6	21.16	10	2.8	7.84
	$\Sigma X_1 = 44.0$	$\Sigma X_1^2 = 198.44$		$\Sigma X_2 = 34.0$	$\Sigma X_2^2 = 123.24$

$N = n_1 + n_2 = 10 + 10 = 20$

Now $T = \sum X_1 + \sum X_2 = 44.0 + 34.0 = 78.0$

Correction factor $\frac{T^2}{N} = \frac{(780)^2}{20} = \frac{780 \times 780}{20} = 304.2$

The total sum squares $(SST) = \sum X_1^2 + \sum X_2^2 - \frac{T^2}{N}$

$= 198.44 + 123.24 - 304.2 = 321.68 - 304.2 = 17.48$

Sum squares between the samples (SSB)

$\frac{\sum X_1^2}{n_1} + \frac{\sum X_2^2}{n_2} - \frac{T^2}{N} = \frac{(44)^2}{10} + \frac{(34)^2}{10} - 304.2$

$= 193.6 + 115.6 - 304.2 = 309.2 - 304.2 = 5$

Degrees of freedom $V_1 = (df_t) = K - 1 = 2 - 1 = 1$

Mean square between the samples $(MSB) = \frac{SSB}{V_1} = \frac{5}{1} = 5$

Sum squares within the samples $(SSW) = SST - SSB$

$= 17.48 - 5 = 12.48$

Degrees of freedom $V_2 = (df_w) = N - K = 20 - 2 = 18$

Mean square within the samples $(MSW) = \frac{SSW}{V_2} = \frac{12.48}{18} = 0.69$

Anova Table

Source of variation	*Sum of squares*	*Degrees of freedom (df)*	*Mean square*	*Test statistic*
Between the samples	$SSB = 5$	1	$MSB = 5.00$	$F = \frac{5.00}{0.69}$
Within the samples	$SSW = 12.48$	18	$MSW = 0.69$	$= 7.246 = 7.25$

• *Decision:*

I. The calculated value of F is 7.25 at degrees of freedom (1, 18) at 0.05 level. The tabulated value of F at 0.05 level in degrees of freedom (1, 18) is 4.4.

II. Therefore it is more than the tabulated value. So the null hypothesis is rejected ($P < 0.05$). The groups differ significantly.

Added variance component

Size of each group $(n) = 10$

Added Variance $= \frac{MSB - MSW}{n} = \frac{5.00 - 0.69}{0} = \frac{4.31}{10} = 0.43$

$\frac{\text{Added variance}}{MSW + \text{added variance}} = \frac{0.43}{0.69 + 0.43} = \frac{0.43}{1.12} = 0.38$

Example 5. A study was conducted in Arambagh subdivision to determine if the physical fitness prior to surgery of persons under going a knee surgery has any effect on time required in post-surgery physical therapy until successful rehabilitation. The survey record of 24 male

patients (between 18 and 30 years) and the number of days required for physical therapy and prior to fitness condition were noted and shown in the following table.

Prior physical condition	*Number of days of physical therapy*									
Poor	29	42	38	40	43	40	30	42		
Average	30	35	39	28	31	29	31	35	33	29
Good	32	26	20	21	22	23				

Test the claim that prior physical condition has an influence on the length of physical therapy.

Solution:

Null Hypothesis: Prior physical condition has no influence on the length of physical therapy.

- *Calculation:*

Poor		*Average*		*Good*	
X_1	X_1^2	X_2	X_2^2	X_3	X_3^2
29	841	30	900	32	1024
42	1764	35	1225	26	676
38	1444	39	1521	20	400
40	1600	28	784	21	441
43	1849	31	961	22	484
40	1600	29	841	23	529
30	900	31	961		
42	1764	35	1225		
		33	1089		
		29	841		
$\sum X_1 = 304$	$\sum X_1^2 = 11762$	$\sum X_2 = 320$	$\sum X_2^2 = 10348$	$\sum X_3 = 144$	$\sum X_3^2 = 3554$

$N = n_1 + n_2 + n_3 = 8 + 10 + 6 = 24$

Now $T = 304 + 320 + 144 = 768$

Correction factor $\dfrac{T^2}{N} = \dfrac{768 \times 768}{24} = 768 \times 32 = 24576$

$$\text{The total sum squares } (SST) = \sum X_1^2 + \sum X_2^2 + \sum X_3^2 - \frac{T^2}{N}$$

$$= 11762 + 10348 + 3554 - 24576$$

$$= 25664 - 24576 = 1088$$

$$\text{Sum square between the samples } (SSB) = \frac{\left(\sum X_1\right)^2}{n_1} + \frac{\left(\sum X_2\right)^2}{n_2} + \frac{\left(\sum X_3\right)^2}{n_3} - \frac{T^2}{N}$$

$$= \frac{(304)^2}{8} + \frac{(320)^2}{10} + \frac{(144)^2}{6} - 24576$$

$$= (304 \times 38 + 320 \times 32 + 144 \times 24) - 24576$$

$$= 11552 + 10240 + 3456 - 24576$$

$$= 25248 - 24576 = 672$$

Degrees of freedom $V_1 = (df_t) = K - 1 = 3 - 1 = 2$

Mean square between the samples $(MSB) = \frac{SSB}{V_1} = \frac{672}{2} = 336$

Sum square within the samples $(SSW) = SST - SSB$

$= 1088 - 672 = 416$

Degrees of freedom $V_2 = (df_w) = N - K = 24 - 3 = 21$

Mean square within the samples $(MSW) = \frac{SSB}{V_2} = \frac{416}{21} = 19.8$

Anova Table

Source of variation	*Sum of squares*	*Degrees of freedom*	*Mean square*	*Test statistic*
Between the samples	$SSB = 336$	$V_1 = df_t = 2$	$MSB = \frac{SSB}{V_1} = 336$	$F = \frac{336}{19.8} = 16.96$
Within the samples	$SSW = 416$	$V_2 = df_w = 21$	$MSW = \frac{SSB}{V_2} = 19.8$	

• *Decision:*

I. The calculated F is 16.96, on degrees of freedom (2, 21) at 0.05 level.

II. The tabulated F on 2, 21 at 0.05 level is 3.47. The calculated F is higher than tabulated F. So the null hypothesis is rejected *i.e.,* significant variance is present.

III. Therefore it appears that prior physical condition has an influence on the length of post-surgery physical therapy.

Example 6. Find out the P value of the following groups. Given [(2, 9) df at 0.05 level is 4.26].

Group A:	**2**	**4**	**6**	**8**
Group B:	**12**	**14**	**16**	**18**
Group C:	**25**	**28**	**30**	**35**

Solution:

• *Null Hypothesis:* No significant difference present among the groups.

• *Calculation:*

Group A		*Group B*		*Group C*	
X_1	X_1^2	X_2	X_2^2	X_3	X_3^2
2	4	12	144	25	625
4	16	14	196	28	784
6	36	16	256	30	900
8	64	18	324	35	1225
$\Sigma X_1 = 20$	$\Sigma X_1^2 = 120$	$\Sigma X_2 = 60$	$\Sigma X_2^2 = 920$	$\Sigma X_3 = 118$	$\Sigma X_3^2 = 3534$

$N = n_1 + n_2 + n_3 = 4 + 4 + 4 = 12$

Now $T = 20 + 60 + 118 = 198$

Correction factor $\frac{T^2}{N} = \frac{198 \times 198}{12} = 16.5 \times 198 = 3267$

The total sum squares $(SST) = 120 + 920 + 3534 - 3267$

$= 4574 - 3267 = 1307$

Sum square between the samples (SSB)

$$= \frac{(20)^2}{4} + \frac{(60)^2}{4} + \frac{(118)^2}{4} - 3267$$

$= (5 \times 20 + 15 \times 60 + 29.5 \times 118) - 3267$

$= 100 + 900 + 3481.0 - 3267$

$= 4481 - 3267 = 1214$

Degrees of freedom $V_1 = (df_i) = K - 1 = 3 - 1 = 2$

Mean square between the samples $(MSB) = \frac{SSB}{V_1} = \frac{1214}{2} = 607$

Sum square within the samples $(SSW) = SST - SSB$

$= 1307 - 1214 = 93$

Degrees of freedom $V_2 = (df_w) = N - K = 12 - 3 = 9$

Mean square within the samples $(MSW) = \frac{SSB}{V_2} = \frac{93}{9} = 10.33$

Anova Table

Source of variation	*Sum of squares*	*Degrees of freedom*	*Mean square*	*Test of statistic*
Between the samples	$SSB = 1214$	$V_1 = 2$	$MSB = \frac{SSB}{V_1} = 607$	$F = \frac{607}{10.33} = 58.76$
Within the samples	$SSW = 93$	$V_2 = 9$	$MSW = \frac{SSB}{V_2} = 10.33$	

- *Decision:*

I. Here calculated value of F in (2, 9) degrees of freedom at 0.05 level is 58.76. The tabulated value of F in (2, 9) df at 0.05 level is 4.26.

II. Therefore calculated value is much greater than table value. Hence the result is significant and null hypothesis is rejected.

Example 7. The following observed data are presented in the table. Apply anova to interpret the observed data at ($\alpha = 0.05$).

Group A:	**35**	**34**	**34**	**33**	**34**
Group B:	**32**	**32**	**31**	**28**	**29**
Group C:	**34**	**33**	**32**	**32**	**33**

Solution:

- *Null Hypothesis:* There is no significant difference between observed samples.

• *Calculations:* Each observation is reduced by 30.

A (X_1)	B (X_2)	C (X_3)	X_1^2	X_2^2	X_3^2
5	2	4	25	4	16
4	2	3	16	4	09
4	1	2	16	1	04
3	−2	2	09	4	04
4	−1	3	16	1	09
$\sum X_1 = 20$	$\sum X_2 = 2$	$\sum X_3 = 14$	$\sum X_1^2 = 82$	$\sum X_2^2 = 14$	$\sum X_3^2 = 42$

$n_1 = 5,\ n_2 = 5$ and $n_3 = 5$

$N = n_1 + n_2 + n_3 = 15$

Now $T = \sum X_1 + \sum X_2 + \sum X_3 = 20 + 2 + 14 = 36$

Correction factor $\dfrac{T^2}{N} = \dfrac{36 \times 36}{15} = 86.4$

The total sum squares $(SST) = \sum X_1^2 + \sum X_2^2 + \sum X_3^2 - \dfrac{T^2}{N}$

$$SST = 82 + 14 + 42 - 86.4 = 138 - 86.4 = 51.6$$

Sum squares between the samples $(SSB) = \dfrac{\left(\sum X_1\right)^2}{n_1} + \dfrac{\left(\sum X_2\right)^2}{n_2} + \dfrac{\left(\sum X_3\right)^2}{n_3} - \dfrac{T^2}{N}$

$$= \frac{20 \times 20}{5} + \frac{2 \times 2}{5} + \frac{14 \times 14}{5} - 86.4$$

$$= \frac{400 + 4 + 196}{5} - 86.4 = \frac{600}{5} - 86.4 = 120 - 86.4 = 33.6$$

Degrees of freedom $V_1 = (df_t) = 3 - 1 = 2$

Mean square between the samples $(MSB) = \dfrac{SSB}{V_1} = \dfrac{33.6}{2} = 16.8$

Sum square within the samples $(SSW) = SST - SSB = 51.6 - 33.6 = 18$

Degrees of freedom $V_2 = (df_w) = N - K = 15 - 3 = 12$

Mean square within the samples $(MSW) = \dfrac{SSB}{V_2} = \dfrac{18}{12} = 1.5$

Test statistic $\quad F - \dfrac{MSB}{MSW} = \dfrac{16.8}{1.5} = 11.2$

Analysis of Variance Table

Source of variation	*Sum of squares*	*Degrees of freedom*	*Mean square*	*Test of statistic*	*Tabulated F*
Between the samples	$SSB = 33.6$	2	$MSB = 16.8$		$F_{0.05(2,12)} = 3.89$
Within the samples	$SSW = 18$	12	$MSW = 1.5$	$F = \frac{16.8}{1.5} = 11.2$	$F_{0.01(2,12)} = 6.93$

• *Decision:*

I. Since the observed F value is 11.2 at 0.05 level at 2, 12 *df* is higher than the tabulated value of F at 0.05, & 2, 12 *df* Therefore the null hypothesis is rejected.

II. So there is a significant difference between the observed samples.

Example 8. The three wheat varieties are grown on 4 plots and allocated completely at random which are given in the table. Find out if differences are significant. [Tabulated value of *F* for *df* (2, 9) is 8.02 (at 1% level)].

Varieties	***Plots***			
	1	**2**	**3**	**4**
Variety A:	**26**	**28**	**27**	**31**
Variety B:	**18**	**21**	**19**	**22**
Variety C:	**16**	**19**	**18**	**19**

Solution:

• *Null Hypothesis:* The differences in varieties are significant.

• *Calculation:* Each observation is reduced by 20.

$A\ (X_1)$	$B\ (X_2)$	$C\ (X_3)$	$A\ (X_1^2)$	$B\ (X_2^2)$	$C\ (X_3^2)$
26 − 20 = 6	−2	−4	36	4	16
28 − 20 = 8	+1	−1	64	1	1
27 − 20 = 7	−1	−2	49	1	4
31 − 20 = 11	+2	−1	121	4	1
$\sum X_1 = 32$	$\sum X_2 = 0$	$\sum X_3 = -8$	$\sum X_1^2 = 270$	$\sum X_2^2 = 10$	$\sum X_3^2 = 22$

$n_1 = 4,\ n_2 = 4$ and $n_3 = 4$

$N = 4 + 4 + 4 = 12$

Now $T = \sum X_1 + \sum X_2 + \sum X_3 = 32 + 0 - 8 = 24$

$$\frac{T^2}{N} = \frac{24 \times 24}{12} = 48$$

The total sum squares $(SST) = (270 + 10 + 22) - 48$

$$= 302 - 48 = 254$$

The sum square between the samples $(SSB) = \frac{(32)^2}{4} + \frac{(0)^2}{4} + \frac{(-8)^2}{4} - \frac{T^2}{N}$

$$= \frac{1024 + 0 + 64}{4} - 48 = \frac{1088}{4} - 48 = 272 - 48 = 224$$

Degrees of freedom df_t or $(V_1) = K - 1 = 3 - 1 = 2$

Mean square between the samples $(MSB) = \frac{SSB}{V_1} = \frac{224}{2} = 112$

Sum square within the samples $(SSW) = SST - SSB = 254 - 224 = 30$

Degrees of freedom df_w or $(V_2) = N - K = 12 - 3 = 9$

Mean square within the samples $(MSW) = \frac{SSW}{V_2} = \frac{30}{9} = 3.33$

Anova Table

Source of variation	*Sum of squares*	*Degrees of freedom*	*Mean square*	*Test of statistic*	*Tabulated F*
Between the samples	$SSB = 224$	$V_1 = 2$	$MSB = \frac{SSB}{V_1} = \frac{224}{2} = 112$	$F = \frac{112}{3.33} = 33.63$	$F_{0.01} = 8.02$
Within the samples	$SSW = 30$	$V_2 = 9$	$MSW = \frac{SSW}{V_2} = \frac{30}{9} = 3.33$		

• *Decision:*

I. Here calculated value of F is 33.6 at (2, 9) degrees of freedom.

II. Tabulated F at 0.01 level in 2, 9 degrees of freedom is 8.02.

III. Therefore computed value of F is greater than table value, so the null hypothesis, *i.e.*, differences are significant, is proved.

Example 9. The three (X, Y, Z) rice varieties are grown on 4 plots and allocated completely at random which are given in the table.

Varieties	I	II	III	IV
X	8	4	6	7
Y	7	5	5	3
Z	2	5	4	4

Test the significance of difference between yield of varieties.

Solution:

• *Null Hypothesis:* The varieties are not significantly different from each other.

• *Calculation:*

X	X^2	Y	Y^2	Z	Z^2
8	64	7	49	2	04
4	16	5	25	5	25
6	36	5	25	4	16
7	49	3	09	4	16
25	165	20	108	15	61

$\sum X = 25 \sum X^2 = 165 \sum Y = 20 \sum Y^2 = 108 \sum Z = 15 \sum Z^2 = 61$

$N = 4 + 4 + 4 = 12$

Now $T = 25 + 20 + 15 = 60$

Correction factor $\dfrac{T^2}{N} = \dfrac{60 \times 60}{12} = 300$

The total sum squares $(SST) = (165 + 108 + 61) - 300$

$= 334 - 300 = 34$

Sum square between the samples $(SSB) = \dfrac{(25)^2}{4} + \dfrac{(20)^2}{4} + \dfrac{(15)^2}{4} - 300$

$= \dfrac{625 + 400 + 225}{4} - 300$

$= \dfrac{1250}{4} - 300 = 312.5 - 300 = 12.5$

Degrees of freedom $df_t = (V_1) = K - 1 = 3 - 1 = 2$

Mean square between the samples $(MSB) = \dfrac{SSB}{V_1} = \dfrac{12.5}{2} = 6.25$

Sum square within the samples $(SSW) = SST - SSB = 34 - 12.5 = 21.5$

Degrees of freedom $df_w = (V_2) = N - K = 12 - 3 = 9$

Mean square within the sample $(MSW) = \dfrac{SSB}{V_2} = \dfrac{21.5}{9} = 2.388 = 2.39$

Anova Table

Source of variation	Sum of squares	Degrees of freedom	Mean square	Test of statistic	Tabulated F
Between the samples	$SSB = 12.5$	$V_1 = 3$	$MSB = \dfrac{SSB}{V_1} = \dfrac{12.5}{3} = 6.25$	$F = \dfrac{6.25}{2.39} = 2.61$	$F_{0.05(2,9)} = 4.26$
Within the samples	$SSW = 21.5$	$V_2 = 9$	$MSW = \dfrac{SSW}{V_2} = \dfrac{21.5}{9} = 2.39$		

- *Decision:*

I. The calculated value of F is 2.61 and the tabulated value of F at 0.05 and (2, 9) df is 4.26.

II. The calculated value is less than the tabulated value.

III. Therefore our null hypothesis is correct *i.e.,* varieties are not significantly different from each other.

Example 10. The three different pesticide solutions are being compared to study their effectiveness in controlling pest. The data are given below.

Solutions	*Number of days*			
	1	**2**	**3**	**4**
X_1	**13**	**22**	**18**	**39**
X_2	**16**	**24**	**17**	**44**
X_3	**05**	**04**	**01**	**22**

Is there any significant difference in their effectiveness at 0.05 level in (2, 9) degrees of freedom (P = 5.14).

Solution:

- *Null Hypothesis:* There are no significant differences among the pesticide solutions.
- *Calculation:* Each observation is reduced by 10

X_1	X_2	X_3	X_1^2	X_2^2	X_3^2
03	06	–05	09	36	25
12	14	–06	144	196	36
08	07	–09	64	49	81
29	34	+12	841	1156	144
$\Sigma X_1 = 52$	$\Sigma X_2 = 61$	$\Sigma X_3 = -8$	$\Sigma X_1^2 = 1058$	$\Sigma X_2^2 = 1437$	$\Sigma X_3^2 = 286$

$N = 4 + 4 + 4 = 12$

Now $\quad T = 52 + 61 - 8 = 105$

Correction factor $\dfrac{T^2}{N} = \dfrac{105 \times 105}{12} = 918.75$

The total sum squares $(SST) = (1058 + 1437 + 286) - 918.75$

$= 2781 - 918.75 = 1862.25$

$$\text{Sum squares between the samples } (SSB) = \frac{(52)^2}{4} + \frac{(61)^2}{4} + \frac{(-8)^2}{4} - \frac{T^2}{N}$$

$$= \frac{2704 + 3721 + 64}{4} - 918.75$$

$$= 6489 - 918.75 = 5570.25$$

Degrees of freedom $df_t = (V_1) = K - 1 = 3 - 1 = 2$

$$\text{Mean squares between the samples } (MSB) = \frac{SSB}{V_1} = \frac{5570.25}{2} = 2785.125$$

Sum squares within the samples $(SSW) = SST - SSB = 5570.25 - 2785.125 = 2785.125$

Degrees of freedom $df_w - (V_2) - N - K = 12 - 3 = 9$

$$\text{Mean squares within the samples } (MSW) = \frac{SSW}{V_2} = \frac{2785}{9} = 309.444 = 309.45$$

Anova Table

Source of variation	Sum of squares	Degrees of freedom	Mean square	Test of statistic	Tabulated F
Between the samples	$SSB = 5570.25$	$V_1 = 2$	$MSB = 2785.125$	$F = \frac{2785.125}{309.45} = 9.0$	$F = 5.14$
Within the samples	$SSW = 2785.125$	$V_2 = 9$	$MSW = 309.45$		

- *Decision:* The calculated F is .9.0 at 0.01 and the tabulated F is 5.14. But our data reveals that the calculated F value is much more than table value, so the null hypothesis is rejected.

Example 11. In an experiment different concentration of plant cytokinin were applied on the leaves and the emergence of roots were tested. The following results were obtained.

	Number of roots emerged under different concentration of cytokinin (ppm)			
Replicates	**0**	**5**	**10**	**15**
1	**1**	**2**	**3**	**5**
2	**2**	**3**	**3**	**4**
3	**1**	**3**	**4**	**5**
4	**0**	**4**	**3**	**4**
5	**1**	**3**	**4**	**4**

Apply anova to test whether the different concentrations have any effect on root emergence [Given $F_{0.01}$ (3,16) = 5.3].

Solution:

- *Null Hypothesis:* There is no significant effect of cytokinin concentration on root emergence.
- *Alternative Hypothesis:* There is significant effect of concentration of cytokinin on roof emergence.
- *Calculation:* Table for calculating sums squares directly from raw data.

	X_1 (0)	X_1^2	Y_2 (5)	X_2^2	X_3 (10)	X_3^2	X_4 (15)	X_4^2
1	1	1	2	4	3	9	5	25
2	2	4	3	9	3	9	4	16
3	1	1	3	9	4	16	5	25
4	0	0	4	16	3	9	4	16
5	1	1	3	9	4	16	4	16
	$\Sigma X_1 = 5$	$\Sigma X_1^2 = 7$	$\Sigma X_2 = 15$	$\Sigma X_2^2 = 47$	$\Sigma X_3 = 17$	$\Sigma X_3^2 = 59$	$\Sigma X_4 = 22$	$\Sigma X_4^2 = 98$

$N = 5 + 5 + 5 + 5 = 20$

$$T = \sum X_1 + \sum X_2 + \sum X_3 + \sum X_4 = 5 + 15 + 17 + 22 = 59$$

$$\text{Correction factor} = \frac{T^2}{N} = \frac{59 \times 59}{20} = \frac{3481}{20} = 174.05$$

The total sum squares (*SST*) = (7 + 47 + 59 + 98) – 174.05

= 211 – 174.05 = 36.95

Sum squares between the samples (*SSB*)

$$= \frac{\left(\sum X_1\right)^2}{N} + \frac{\left(\sum X_2\right)^2}{N} + \frac{\left(\sum X_3\right)^2}{N} + \frac{\left(\sum X_4\right)^2}{N} - \frac{T^2}{N}$$

$$SSB = \frac{(5)^2}{5} + \frac{(15)^2}{5} + \frac{(17)^2}{5} + \frac{(22)^2}{5} - 174.05$$

$$= \frac{25}{5} + \frac{225}{5} + \frac{289}{5} + \frac{484}{5} - 174.05$$

= (5 + 45 + 57.8 + 96.8) – 174.05 = 204.6 – 174.05 = 30.55

Degrees of freedom = $V_1 = df_t = k - 1 = 4 - 1 = 3$

$$\text{Mean square between the sample } (MSB) = \frac{SSB}{V_1} = \frac{30.55}{3} = 10.18$$

Sum square within the sample (*SSW*) = *SST* – *SSB* = 36.95 – 30.55 = 6.4

Degrees of freedom = $V_2 = df_w = N - k = 20 - 4 = 16$

$$\text{Mean square within the sample } (MSW) = \frac{SSW}{V_2} = \frac{6.4}{16} = 0.4$$

Anova Table

Source of veriation	*Sum of squares*	*Degrees of freedom*	*Mean square*	*Test of statistic*	*Tabulated F*
Between the samples	$SSB = 30.5$	$V_1 = df_t = 3$	$MSB = \frac{SSB}{V_1} = \frac{30.55}{3} = 10.18$	$F = \frac{MSB}{MSW} = \frac{10.18}{0.4} = 25.45$	$F_{0.01}(3.16) = 5.3$
Within the samples	$SSW = 6.4$	$V_2 = df_w = 16$	$MSW = \frac{SSW}{V_2} = \frac{6.4}{16} = 0.4$		

Decision: The calculated *F* is 25.45 and the tabulated *F* is 5.3. But our data reveals that the calculated *F* is much more than tabulated *F*. So the null hypothesis is rejected. We can conclude that the cytokinin concentrations have some effect.

- **Bonferroni Adjustment: *t* test**

I. If the anova has yielded significant '*F*' ratio, one multiple comparison '*t*' test has to be undertaken for each chosen pair of group means.

II. In this test the difference between the group means of the chosen pair is converted into '*t*' score by using *SE* of their differences *i.e.*, within group variances or error variances (S^2W) and the group sizes.

(*a*) The difference between the group means *i.e.*, $\overline{X}_1 - \overline{X}_2$.

(*b*) *SE* of their difference *i.e.*, $S_{\overline{X}_1 - \overline{X}_2} = \sqrt{\dfrac{S^2W}{n_1} + \dfrac{S^2W}{n_2}}$

(c) Thus $t = \dfrac{\overline{X}_1 - \overline{X}_2}{S_{\overline{X}_1} - S_{\overline{X}_2}} = \dfrac{\overline{X}_1 - \overline{X}_2}{\sqrt{\dfrac{S^2W}{n_1} + \dfrac{S^2W}{n_2}}}$

n_1 & n_2 = size of the group
$df = N - k$
$N = n_1 + n_2$

III. By computing '*t*' the probability level (*P*) for such group is obtained. It is then compared with the critical '*t*' scores with same *df*.

IV. The adjusted probability (*P*′) is obtained by multiplying the '*P*' with total number (*K*′).

V. If this *P*′ is lower than or equal to either a chosen α or α of 0.05, relevant group means are considered to differ significantly from each other at or beyond that α ($P' \leq \alpha$).

Example: Workout one way anova from the following data. Is there any significant different between the means of X_1 and X_2 and between X_2 and X_3?

	1	2	3	4	5	6	7	8	9	10
X_1 =	90	120	90	120	80	97	100	97	130	86
X_2 =	55	54	68	70	60	35	60	58	40	80
X_3 =	30	28	35	25	15	28	38	30	21	50

Critical *F* values: $F_{0.01(2,26)}$ = 5.53, $F_{0.01(2,27)}$ = 5.49, $F_{0.01(2,28)}$ = 5.45, $F_{0.01(1,29)}$ = 4.18 & F0.01(1,26) = 4.22.

Solution:

- *Null Hypothesis:* There is no significant difference between the members.
- *Alternative Hypothesis:* There is significant difference between the members.
- *Calculation:*

X_1	X_1^2	X_2	X_2^2	X_3	X_3^2
90	8100	55	3025	30	900
120	14400	54	2916	28	784
90	8100	68	4624	35	1225
120	14400	70	4900	25	625
80	6400	60	3600	15	225
97	9409	35	1225	28	784
100	10000	60	3600	38	1444
97	9409	58	3364	30	900
130	16900	40	1600	21	441
86	7396	80	6400	50	2500
1010	104514	580	35254	300	9828

$$N = n_1 + n_2 + n_3 = 10 + 10 + 10 = 30$$

$$\overline{X}_1 = \frac{1010}{10} = 101 \quad \overline{X}_2 = \frac{580}{10} = 58 \quad \overline{X}_3 = \frac{300}{10} = 30$$

Now $T = \sum X_1 + \sum X_2 + \sum X_3 = 1010 + 580 + 300 = 1890$

$$\frac{T^2}{N} = \frac{1890 \times 1890}{30} = 63 \times 1890 = 119070$$

Total sum square (*SST*) = (104514 + 3524 +9828) − 119070
= 149596 − 119070 = 30526

The sum square between the samples (*SSB*) $= \frac{(1010)^2}{10} + \frac{(580)^2}{10} + \frac{(300)^2}{10} - 119070$

$= (101 \times 1010 + 580 \times 58 + 300 \times 30) - 119070$

$= (102010 + 33640 + 9000) - 119070$

$= 144650 - 119070 = 25580$

Degrees of freedom $= V_1 = df_t = k - 1 = 3 - 1 = 2$

Mean square between the sample $(MSB) = \frac{SSB}{V_1} = \frac{25580}{2} = 12790$

Sum square within the sample (*SSW*) = *SST* − *SSB* = 30526 − 25580 = 4946

Degrees of freedom $= V_2 = df_w = N - k = 30 - 3 = 27$

Mean square within the sample $(MSW) = \frac{SSW}{V_2} = \frac{4946}{27} = 183.19$

Anova Table

Source of veriation	*Sum of squares*	*Degrees of freedom*	*Mean square*	*Test of statistic*	*Tabulated F*
Between the samples	$SSB = 25580$	$V_1 = k - 1$ $= 3 - 1 = 2$	$MSB = \frac{SSB}{V_1}$ $= \frac{25580}{3} = 12790$	$F = \frac{12790}{183.19}$ $= 69.82$	$F_{0.01}(1,27) = 5.49$
Within the samples	$SSW = 4946$	$V_2 = N - k$ $= 30 - 3 = 27$	$MSW = \frac{SSW}{V_2}$ $= \frac{4946}{27} = 18379$		

• *Decision:*

I. Here calculated value of $F = 69.82$ at (2,27) degrees of freedom.

II. Tabulated F at 0.01 level in 2,27 degrees of freedom is 5.49.

III. Therefore computed value is higher than tabulated value. *Ho* is rejected *i.e.,* significant differences are present between the groups.

As the F test yielded significant score so the difference $\overline{X}_1 - \overline{X}_2$ & $\overline{X}_2 - \overline{X}_3$ are subjected to multiple comparison t test *i.e.,* with **Bonferroni modification**.

$$t = \frac{\bar{X}_1 - \bar{X}_2}{\sqrt{\frac{S^2W}{n_1} + \frac{S^2W}{n_2}}} \qquad S^2W = 183.19 \therefore \sqrt{\frac{S^2W}{n_1} + \frac{S^2W}{n_2}} = \sqrt{\frac{183.19}{10} + \frac{183.19}{10}}$$

$$= \sqrt{\frac{366.38}{10}} = \sqrt{36.64} = 6.053$$

$$t = \frac{\bar{X}_1 - \bar{X}_2}{S_{\bar{X}_1 - \bar{X}_2}} = \frac{101.0 - 58.0}{6.053} = \frac{43}{6.053} = 7.10 \qquad \left[\because S_{\bar{X}_1} - \bar{X}_2 = \sqrt{\frac{S^2 w}{n_1} + \frac{S^2 w}{n_2}}\right]$$

$$t = \frac{\bar{X}_2 - \bar{X}_3}{S_{\bar{X}_1 - \bar{X}_2}} = \frac{58 - 30}{6.053} = \frac{28}{6.053} = 4.625 = 4.63$$

$$df = N - k = 30 - 3 = 27$$

Critical t score: $t_{.02(27)} = 2.473$, $t_{.01(27)} = 2.771$ & $t_{.001(27)} = 3.690$

Comparing the calculated t scores with critical t score it is found that $(\bar{X}_1 - \bar{X}_2)$ is significant beyond 0.001 level ($P < 0.001$) & $\bar{X}_2 - \bar{X}_3$ is also beyond 0.001 *i.e.*, ($P < 0.001$).

To apply Boniferroni modification each P is obtained by the t test is multiplied by the number k' of paired comparison to get corrected probabilities P' of the Ho being correct.

$$k' = 2 \quad P' = k'P$$

$\therefore$ *For* $\bar{X}_1 - \bar{X}_2 = P' = k'P = 2 \times (< 0.001) = < 0.002$

For $\bar{X}_2 - \bar{X}_3 = P' = k'P = 2 \times (< 0.001) = < 0.002.$

Two-way Anova:

I. It is used to investigate the simultaneous effects of two independent variables on a dependent variable.

II. Here one variable is represented along the rows and others framed along the columns.

(*a*) Two-way Anova with replications

I. It is used when every combination of the two factors–one level of each-has been applied on more than one individual.

II. The effect of each combination of independent variable is given by the replicated observations in a group of individuals.

(*b*) Two-way Anova without replication

I. It is worked out when every combination of the two factors-one level of each-has been applied on only one individual.

II. The effect of each such combination of independent variables is given by a single observation only.

16 CHAPTER

NON-PARAMETRIC STATISTICS

- **Parameter:** It is a summary value or numerical index like mean, median, standard deviation or variance of a variable for the entire population.
- **Parametric test:** Most commonly used statistical methods are called parametric because they are involved in testing the values of parameter (mean, median or standard deviation).
- **Non-parametric test:** Non-parametric test or methods are mathematical procedures concerned with the treatment of standard statistical problems when the assumption of normality is replaced by general assumption concerning the distribution function. It is also called distribution free test.

• Features:

I. It entails very few assumptions.

II. It works out without using any pre-computed statistic as an estimate of parameter.

III. It can be used for very small sample.

IV. It does not require normal distribution of the variables.

V. It can be computed by very simple method.

• Merits & Demerits of Non-Parametric Test:

• Merits:

I. It can be applied in all types of data.

II. It is generally simple to understand and very easy to compute and apply.

III. It has greater range of applicability.

IV. It does not require lengthy and laborious calculations.

V. It does not need pre-computed statistics.

• Demerits:

I. It is often wasteful of information and less efficient.

II. It sometimes pays for freedom from assumption.

III. This procedure throws away information.

IV. This procedure has lack of power.

• Types of Non-parametric test:

1. Mann-Whitney U-test or Rank sum test.

I. This method is used to determine the significant difference between the two independent sample groups.

II. This test is also known as U-test.

2. Kruskal-Wallis test or H-test.

I. It is a kind of rank sum test.

II. It is a rank dependent one-way anova and interpreted using critical chi square values.

3. The sign test for paired data.

I. It is based on the direction (+ or – sign) of a pair of observations and not on their numerical magnitudes.

II. It is discarded if the paired observation difference is zero.

4. One Sample Run test.

I. This test deals with the randomness with which the sample items have been selected.

II. It is based on the order in which the sample observations are obtained.

5. Kolmo Gorove-Smirnov test.

I. It is worked out to determine whether there is a significant difference between the observed frequency distribution and a theoretical distribution.

II. It is also known as **K-test** *i.e.*, another method of measuring goodness of fit of a frequency distribution.

III. It is more powerful and easier to apply.

6. Wilcoxon Signed Rank test.

I. It accounts for the magnitude of differences between paired values and not only their signs.

II. It is useful in comparing two population.

7. Median test for independent sample.

This test is used to find out the significance of differences between means of two or more independent groups using a common median of those groups.

8. Kenoal Test for Concordance.

I. It is applicable where two sets of ranking individuals are available.

II. It is used to test the significance of more than two sets of ranking individuals.

17 CHAPTER

STATISTICAL TABLES

TABLE – 1

Table A: Distribution of t values ("Student's" distribution)

Probabilities

df	*.90*	*.70*	*.50*	*.30*	*.10*	*.05*	*.01*	*.001*
1	.16	.51	1.00	2.00	6.31	12.70	63.66	636.62
2	.14	.44	82	1.39	2.92	4.30	9.92	31.60
3	.14	.42	.76	1.25	2.35	3.18	5.84	12.92
4	.13	.41	.74	1.19	2.13	2.78	4.60	8.61
5	.13	.41	.73	1.16	2.02	2.57	4.03	6.87
6	.13	.40	.72	1.13	1.94	2.45	3.71	5.96
7	.13	.40	.71	1.12	1.90	2.36	3.50	5.41
8	.13	.40	.71	1.11	1.86	2.30	3.36	5.04
9	.13	.40	.70	1.10	1.83	2.26	3.25	4.78
10	.13	.40	.70	1.09	1.81	2.23	3.17	4.59
11	.13	.40	.70	1.09	1.80	2.20	3.10	4.44
12	.13	.40	.70	1.08	1.78	2.18	3.05	4.32
13	.13	.39	.69	1.08	1.77	2.16	3.01	4.22
14	.13	.39	69	1.08	1.76	2.14	2.98	4.14
15	.13	.39	.69	1.07	1.75	2.13	2.95	4.07
16	.13	.39	.69	1.07	1.75	2.12	2.92	4.02
17	.13	.39	.69	1.07	1.74	2.11	2.90	3.96
18	.13	.39	.69	1.07	1.73	2.10	2.88	3.92
19	.13	.39	.69	1.07	1.73	2.09	2.86	3.88
20	.13	.39	.69	1.06	1.72	2.09	2.84	3.85
21	.13	.39	.69	1.06	1.72	2.08	2.83	3.82
22	.13	.39	.69	1.06	1.72	2.07	2.82	3.79
23	.13	.39	.68	1.06	1.71	2.07	2.81	3.77
24	.13	.39	.68	1.06	1.71	2.06	2.80	3.74
25	.13	.39	.68	1.06	1.71	2.06	2.79	3.72
26	.13	.39	.68	1.06	1.71	2.06	2.78	3.71
27	.13	.39	.68	1.06	1.70	2.05	2.77	3.69
28	.13	.39	.68	1.06	1.70	20.5	2.76	3.67
29	.13	.39	.68	1.06	1.70	20.4	2.76	3.66
30	.13	.39	.68	1.06	1.70	2.04	2.75	3.65
40	.13	.39	.68	1.05	1.68	2.02	2.70	3.55
60	.13	.39	.68	1.05	1.67	2.00	2.66	3.46
120	.13	.39	.68	1.04	1.66	1.98	2.62	3.37
∞	.13	.38	.67	04	1.64	1.96	2.58	3.29

Abridged from Table III of Fisher and Yates, *Statistical Tables for Biological, Agricultural and Medical Research*, published by Oliver & Boyd Ltd., Edinburgh, and by permission of the authors and publishers.

Table B. Critical values of *t*.
Levels of significance for two-tail test

df	*10*	*.05*	*.02*	*.01*	*.001*	*df*	*.10*	*.05*	*.02*	*.01*	*.001*
01	6.314	12.706	31.821	63.657	636.619	34	1.6991	2.032	2.441	2.728	3.601
02	2.920	4.303	6.965	9.925	31.598	35	1.6990	2.030	2.438	2.724	3.591
03	2.353	3.182	4.541	5.841	12.941	36	1.6888	2.028	2.434	2.719	3.582
04	2.132	2.776	3.747	4.604	8.610	37	1.6887	2.026	2.431	2.715	3.574
05	2.015	2.571	3.365	4.032	6.859	38	1.6886	2.024	2.429	2.712	3.566
06	1.943	2.447	3.143	3.707	5.959	39	1.6885	2.023	2.426	2.708	3.558
07	1.895	2.365	2.998	3.499	5.405	40	1.684	2.021	2.423	2.704	3.551
08	1.860	2.306	2.896	3.355	5.041	41	1.683	2.020	2.421	2.701	3.544
09	1.833	2.262	2.821	3.250	4.781	42	1.682	2.018	2.418	2.698	3.538
10	1.812	2.228	2.764	3.169	3.169	43	1.681	2.017	2.416	2.695	3.532
11	1.796	2.201	2.718	3.106	4.437	44	1.680	2.015	2.414	2.692	3.526
12	1.782	2.179	2.681	3.055	4.318	45	1.679	2.014	2.412	2.690	3.520
13	1.771	2.160	2.650	3.012	4.221	46	1.679	2.013	2.410	2.687	3.515
14	1.761	2.145	2.624	2.977	4.140	47	1.678	2.012	2.408	2.685	3.510
15	1.753	2.131	2.602	2.947	4.073	48	1.677	2.011	2.407	2.682	3.505
16	1.746	2.120	2.583	2.921	4.015	49	1.677	2.010	2.405	2.680	3.500
17	1.740	2.110	2.567	2.898	3.965	50	1.676	2.009	2.403	2.678	3.496
18	1.734	2.101	2.552	2.878	3.922	51	1.675	2.008	2.402	2.676	3.492
19	1.729	2.093	2.539	2.861	3.883	52	1.675	2.007	2.400	2.674	3.488
20	1.725	2.086	2.528	2.845	3.850	53	1.674	2.006	2.399	2.672	3.484
21	1.721	2.080	2.518	2.831	3.819	54	1.674	2.005	2.397	2.670	3.480
22	1.717	2.074	2.508	2.819	3.792	55	1.673	2.004	2.396	2.668	3.476
23	1.714	2.069	2.500	2.807	3.767	56	1.673	2.003	2.395	2.667	3.473
24	1.711	2.064	2.492	2.797	3.745	57	1.672	2.002	2.394	2.665	3.470
25	1.708	2.060	2.485	2.787	3.725	58	1.672	2.002	2.392	2.663	3.466
26	1.706	2.056	2.479	2.779	3.707	59	1.671	2.001	2.391	2.662	3.463
27	1.703	2.052	2.473	2.771	3.690	60	1.671	2.000	2.390	2.660	3.460
28	1.701	2.048	2.467	2.763	3.674	70	1.667	1.994	2.381	2.648	3.435
29	1.699	2.045	2.462	2.756	3.659	80	1.664	1.990	2.374	2.639	3.416
30	1.697	2.042	2.457	2.750	3.646	90	1.662	1.987	2.368	2.632	3.402
31	1.696	2.040	2.453	2.744	3.633	100	1.660	1.984	2.364	2.626	3.390
32	1.6994	2.037	2.449	2.738	3.622	120	1.658	1.980	2.358	2.617	3.373
33	1.6692	2.035	2.445	2.733	3.611	∞	1.645	1.960	2.326	2.576	3.291
	.05	.025	.01	.005	.0005		.05	025	01	.005	.0005

Levels of significance for one-tail test

Table taken from Table III of R. A. Fisher and F. Yates, *Statistical Tables for Biological, Agricultural and Medical Research*, published by Longman Group Ltd., London (previously published by Oliver & Boyd, Edinburgh), by permission of the authors and publishers.

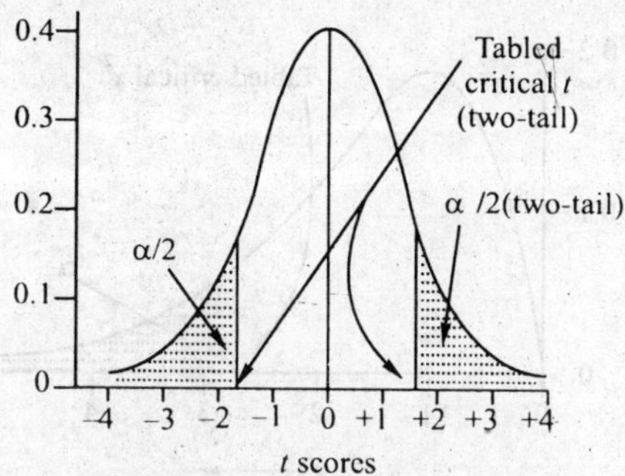

TABLE – 2

Table A. Critical values of chi square. Figures at the top of the table indicate levels of significance (α).

df	*.50*	*.30*	*.20*	*.10*	*.05*	*.02*	*.01*	*.001*
01	0.46	1.07	1.64	2.71	3.84	5.41	6.64	10.83
02	1.39	2.41	3.22	4.60	5.99	7.82	9.21	13.82
03	2.37	3.66	4.64	6.25	7.82	9.84	11.34	16.27
04	3.36	4.88	5.99	7.78	9.49	11.67	13.28	18.46
05	4.35	6.06	7.29	9.24	11.07	13.39	15.09	20.52
06	5.35	7.23	8.56	10.64	12.59	15.03	16.81	22.46
07	6.35	8.38	9.80	12.02	14.07	16.62	18.48	24.32
08	7.34	9.52	11.03	13.36	15.51	18.17	20.09	26.12
09	8.34	10.66	12.24	14.68	16.92	19.68	21.67	27.88
10	9.34	11.78	13.44	15.99	18.31	21.16	23.21	29.59
11	10.34	12.90	14.63	17.28	19.68	22.62	24.72	31.26
12	11.34	14.01	15.81	18.55	21.03	24.05	26.22	32.91
13	12.34	15.12	16.98	19.81	22.36	25.47	27.69	34.53
14	13.34	16.22	18.15	21.06	23.68	26.87	29.14	36 12
15	14.34	17.32	19.31	22.81	25.00	28.26	30.58	37.70
16	15.34	18.42	20.46	23.54	26.30	29.63	32.00	39.29
17	16.34	19.51	21.62	24.77	27.59	31.00	33.41	40.75
18	17.34	20.60	22.76	25.99	28.87	32.35	34.80	42.31
19	18.34	21.69	23.90	27.20	30.14	33.69	36.19	43.82
20	19.34	22.78	25.04	28.41	31.41	35.02	37.57	45.82
21	20.34	23.86	26.17	29.62	32.67	36.34	38.93	46.80
22	21.34	24.94	27.30	30.81	33.92	37.66	40.29	48.27
23	22.34	26.02	28.43	32.01	35.17	38.97	41.64	49.73
24	23.34	27.10	29.55	33.20	36.42	40.27	42.98	51.18
25	24.34	28.17	30.68	34.38	37.65	41.57	44.31	52.62
26	25.34	29.25	31.80	35.56	38.88	42.86	45.64	54.05
27	26.34	30.32	32.91	36.74	40.11	44.14	46.96	55.48
28	27.34	31.39	34.03	37.92	41.34	45.42	48.28	56.89
29	28.34	32.46	35.14	39.09	42.56	46.69	49.59	58.30
30	29.34	33.53	36.25	40.26	43.77	47.96	50.89	59.70

Table taken from Table IV of R. A. Fisher and F. Yates, *Statistical Table for Biological, Agricultural and Medical Research*, published by Longman Group Ltd., London (previously published by Oliver & Boyd, Edinburgh), by permission of the authors and publishers.

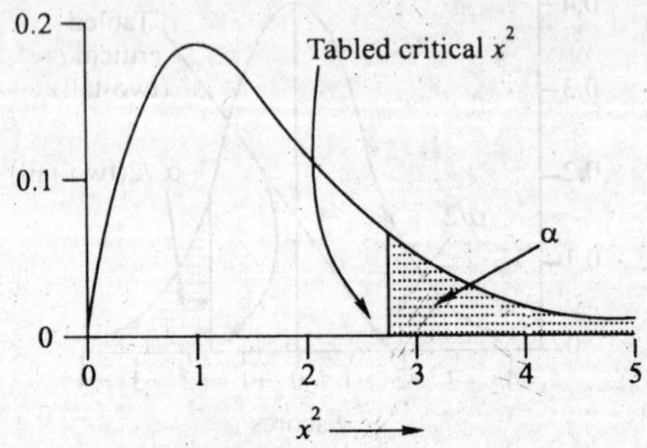

Table B: The probabilities of exceeding different chi-square values for degrees of freedom from 1 to 50 when the expected hypothesis is true.

	Probabilities									
df	.95	.90	.70	.50	.30	.20	.10	.05	.01	.001
1	004	.016	.15	.46	1.07	1.64	2.71	3.84	6.64	10.83
2	.10	.21	.71	1.39	2.41	3.22	4.61	5.99	9.21	13.82
3	.35	.58	1.42	2.37	3.67	4.64	6.25	7.82	11.35	16.27
4	.71	1.06	2.20	3.36	4.88	5.99	7.78	9.49	13.28	18.47
5	1.15	1.61	3.00	4.35	6.06	7.29	9.24	11.07	15.09	20.52
6	1.64	2.20	3.83	5.35	7.23	8.56	10.65	12.59	16.81	22.46
7	2.17	2.83	4.67	6.35	8.38	9.80	12.02	14.07	18.48	24.32
8	2.73	3.49	5.53	7.34	9.52	11.03	13.36	15.51	20.09	26.13
9	3.33	4.17	6.39	8.34	10.66	12.24	14.68	16.92	21.67	27.88
10	3.94	4.87	7.27	9.34	11.78	13.44	15.99	18.31	23.21	29.59
11	4.58	5.58	8.15	10.34	12.90	14.63	17.28	19.68	24.73	31.26
12	5.23	6.30	9.03	11.34	14.01	15.81	18.55	21.03	26.22	32.91
13	5.89	7.04	9.93	12.34	15.12	16.99	19.81	22.36	27.69	34.53
14	6.57	7.79	10.82	13.34	16.22	18.15	21.06	23.69	29.14	36.12
15	7.26	8.55	11.72	14.34	17.32	19.31	22.31	25.00	30.58	37.70
20	10.85	12.44	16.27	19.34	22.78	25.04	28.41	31.41	37.57	45.32
25	14.61	16.47	20.87	24.34	28.17	30.68	34.38	37.65	44.31	52.62
30	18.49	20.60	25.51	29.34	33.53	36.25	40.26	43.77	50.89	59.70
50	34.76	37.69	44.31	49.34	54.72	58.16	63.17	67.51	76.15	86.66

Abridged from Table IV of Fisher and Yates, *Statistical Tables for Biological, Agricultural and Medical Research,* Oliver & Boyd Ltd., Edinburgh, by permission of the authors and publishers.

TABLE – 3

Significance points of F

(1% values)

Table – A

Degree of freedom (n_2)	Degree of freedom (n_1) 1	2	3	4	5	6	7	8	9	α
1	4052	4999	5403	5625	5764	5859	5928	5981	6022	6366
2	98.50	99.00	99.17	99.25	99.30	99.33	99.36	99.37	99.39	99.50
3	34.12	30.82	29.46	28.71	28.24	27.91	27.67	27.49	27.35	26.13
4	21.20	18.00	16.69	15.98	15.52	15.21	14.98	14.80	14.66	13.46
5	16.26	13.27	12.06	11.39	10.97	10.67	10.46	10.29	10.16	9.02
6	13.75	10.92	9.78	9.15	8.75	8.47	8.26	8.10	7.98	6.88
7	12.25	9.55	8.45	7.85	7.46	7.19	6.99	6.84	6.72	5.65
8	11.26	8.65	7.59	7.01	6.63	6.37	6.18	6.03	5.91	4.86
9	10.56	8.02	6.99	6.42	6.06	5.80	5.61	5.47	5.35	4.31
10	10.04	7.56	6.55	5.99	5.64	5.39	5.20	5.06	4.94	3.91
11	9.65	7.21	6.22	5.67	5.32	5.07	4.89	4.74	4.63	3.60
12	9.33	6.93	5.95	5.41	5.06	4.82	4.64	4.50	4.39	3.36
13	9.07	6.70	5.74	5.21	4.86	4.62	4.44	4.30	4.19	3.17
14	8.86	6.51	5.56	5.04	4.69	4.46	4.28	4.14	4.03	3.00
15	8.68	6.36	5.42	4.89	4.56	4.32	4.14	4.00	3.89	2.87
16	8.53	6.23	5.29	4.77	4.44	4.20	4.03	3.89	3.78	2.75
17	8.40	6.11	5.18	4.67	4.34	4.10	3.93	3.79	3.68	2.65
18	8.29	6.01	5.09	4.58	4.25	4.01	3.84	3.71	3.60	2.57
19	8.18	5.93	5.01	4.50	4.17	3.94	3.77	3.63	3.52	2.49
20	8.10	5.85	4.94	4.43	4.10	3.87	3.70	3.56	3.46	2.42
30	7.56	5.39	4.51	4.02	3.70	3.47	3.30	3.17	3.07	2.01
∝	6.63	4.61	3.78	3.32	3.02	2.80	2.64	2.51	2.41	1.00

Significance points of F
(5% values)

TABLE – B

Degrees of freedom (n_2)	Degrees of freedom (n_1) 1	2	3	4	5	6	7	8	9	∝
1	161.4	199.5	215.7	224.6	230.2	234.0	236.8	238.9	240.5	245.3
2	18.15	19.00	19.16	19.35	19.30	19.33	19.35	19.37	19.38	19.50
3	10.13	9.55	9.28	9.12	9.01	8.94	8.89	8.85	8.81	8.53
4	7.71	6.94	6.59	6.39	6.26	6.16	6.09	6.04	6.00	5.63
5	6.61	5.79	5.41	5.19	5.05	4.95	4.88	4.82	4.77	4.36
6	5.99	5.14	4.76	4.53	4.39	4.28	4.21	4.15	4.10	3.67
7	5.59	4.74	4.35	4.12	3.97	3.87	3.79	3.73	3.68	3.23
8	5.32	4.46	4.07	3.84	3.69	3.58	3.50	3.44	3.39	2.93
9	5.12	4.26	3.86	3.63	3.48	3.37	3.29	3.23	3.18	2.71
10	4.96	4.10	3.71	3.48	3.33	3.22	3.14	3.07	3.02	2.54
11	4.84	3.98	3.59	3.36	3.20	3.09	3.01	2.95	2.90	2.40
12	4.75	3.88	3.49	3.26	3.11	3.00	2.91	2.85	2.80	2.30
13	4.67	3.80	3.41	3.18	3.02	2.92	2.83	2.77	2.71	2.21
14	4.60	3.74	3.34	3.11	2.96	2.85	2.76	2.70	2.65	2.13
15	4.54	3.68	3.29	3.06	2.90	2.79	2.71	2.64	2.59	2.07
16	4.49	3.63	3.24	3.01	2.85	2.74	2.66	2.59	2.54	2.01
17	4.45	3.59	3.20	2.96	2.81	2.70	2.61	2.55	2.49	1.96
18	4.41	3.55	3.16	2.93	2.77	2.66	2.58	2.51	2.46	1.92
19	4.38	3.52	3.13	2.90	2.74	2.63	2.54	2.48	2.42	1.88
20	4.35	3.49	3.10	2.87	2.71	2.60	2.51	2.45	2.39	1.84
30	4.17	3.32	2.92	2.69	2.53	2.42	2.33	2.27	2.21	1.62
∝	3.84	3.00	2.60	2.37	2.21	2.10	2.01	1.94	1.88	1.00

TABLE – 4

Ordinates at specific x/σ or z scores and areas from the mean to the z scores of the unit normal curve.

$\frac{x}{\sigma}$	Area	Ordinate	$\frac{x}{\sigma}$	Area	Ordinate	$\frac{x}{\sigma}$	Area	Ordinate	$\frac{x}{\sigma}$	Area	Ordinate
.00	.0000	.3989	.55	.2088	.3429	1.10	.3643	.2179	1.65	.4505	.1023
.01	.0040	.3989	.56	.2173	.3410	1.11	.3665	.2155	1.66	.4515	.1006
.02	.0080	.3989	.57	.2157	.3391	1.12	.3686	.2131	1.67	.4525	.0989
.03	.0120	.3988	.58	.2190	.3372	1.13	.3708	.2107	1.68	.4535	.0973
.04	.0160	.3986	.59	.2224	.3352	1.14	.3729	.2083	1.69	.4545	.0957
.05	.0199	.3984	.60	.2257	.3332	1.15	.3749	.2059	1.70	.4554	.0940
.06	.0239	.3982	.61	.2291	.3312	1.16	.3770	.2036	1.71	.4564	.0925
.07	.0279	.3980	.62	.2324	.3292	1.17	.3790	.2012	1.72	.4573	.0909
.08	.0319	.3977	.63	.2357	.3271	1.18	.3810	.1989	1.73	.4582	.0893
.09	.0359	.3973	.64	.2389	.3251	1.19	.3830	.1965	1.74	.4591	.0878
.10	.0398	.3970	.65	.2422	.3230	1.20	.3849	.1942	1.75	.4599	.0863
.11	.0438	.3965	.66	.2454	.3209	1.21	.3869	.1919	1.76	.4608	.0848
.12	.0478	.3961	.67	.2486	.3187	1.22	.3888	.1895	1.77	.4616	.0833
.13	.0517	.3956	.68	.2517	.3166	1.23	.3907	.1872	1.78	.4625	.0818
.14	.0557	.3951	.69	.2549	.3144	1.24	.3925	.1849	1.79	.4633	.0804
.15	.0596	.3945	.70	.2580	.3123	1.25	.3944	.1826	1.80	.4641	.0790
.16	.0636	.3939	.71	.2611	.3101	1.26	.3962	.1804	1.81	.4649	.0775
.17	.0675	.3932	.72	.2642	.3070	1.27	.3980	.1781	1.82	.4656	.0761
.18	.0714	.3925	.73	.2673	.3056	1.28	.3997	.1758	1.83	.4664	.0748
.19	.0753	.3918	.74	.2703	.3034	1.29	.4015	.1736	1.84	.4671	.0734
.20	.0793	.3910	.75	.2734	.3011	1.30	.4032	.1714	1.85	.4678	.0721
.21	.0832	.3902	.76	.2764	.2989	1.31	.4049	.1691	1.86	.4686	.0707
.22	.0871	.3894	.77	.2794	.2966	1.32	.4066	.1669	1.87	.4693	.0694
.23	.0910	.3885	.78	.2823	.2943	1.33	.4082	.1647	1.88	.4699	.0681
.24	.0948	.3876	.79	.2852	.2920	1.34	.4099	.1626	1.89	.4706	.0669
.25	.0987	.3867	.80	.2881	.2897	1.35	.4115	.1604	1.90	.4713	.0656
.26	.1026	.3857	.81	.2910	.2874	1.36	.4131	.1582	1.91	.4719	.0644
.27	.1064	.3847	.82	.2939	.2850	1.37	.4147	.1561	1.92	.4726	.0632
.28	.1103	.3836	.83	.2967	.2827	1.38	.4162	.1539	1.93	.4732	.0620
.29	.1141	.3825	.84	.2995	.2803	1.39	.4177	.1518	1.94	.4738	.0608
.30	.1179	.3814	.85	.3023	.2780	1.40	.4192	.1497	1.95	.4744	.0596
.31	.1217	.3802	.86	.3051	.2756	1.41	.4207	.1476	1.96	.4750	.0584
.32	.1255	.3790	.87	.3078	.2732	1.42	.4222	.1456	1.97	.4756	.0573
.33	.1293	.3778	.88	.3106	.2709	1.43	.4236	.1435	1.98	.4761	.0562
.34	.1331	.3765	.89	.3133	.2685	1.44	.4251	.1415	1.99	.4767	.0551
.35	.1368	.3752	.90	.3159	.2661	1.45	.4265	.1394	2.00	.4772	.0540
.36	.1406	.3739	.91	.3186	.2637	1.46	.4279	.1374	2.01	.4778	.0529
.37	.1443	.3725	.92	.3212	.2613	1.47	.4292	.1354	2.02	.4783	.0519
.38	.1480	.3712	.93	.3238	.2589	1.48	.4306	.1334	2.03	.4788	.0508
.39	.1517	.3697	.94	.3264	.2565	1.49	.4319	.1315	2.04	.4793	.0498
.40	.1554	.3683	.95	.3289	.2541	1.50	.4332	.1295	2.05	.4798	.0488

x/σ	Area	Ordinate	x/σ	Area	Ordinate	x/σ	Area	Ordinate	x/σ	Area	Ordinate
.41	.1591	.3668	.96	.3315	.2516	1.51	.4345	.1276	2.06	.4803	.0478
.42	.1628	.3653	.97	.3340	.2492	1.52	.4357	.1257	2.07	.4808	.0468
.43	.1664	.3637	.98	.3365	.2468	1.53	.4370	.1238	2.08	.4812	.0459
.44	.1700	.3621	.99	.3389	.2444	1.54	.4382	.1219	2.09	.4817	.0449
.45	.1736	.3605	1.00	.3413	.2420	1.55	.4394	.1200	2.10	.4821	.0440
.46	.1772	.3589	1.01	.3438	.2396	1.56	.4406	.1182	2.11	.4826	.0431
.47	.1808	.3572	1.02	.3461	.2371	1.57	.4418	.1163	2.12	.4830	.0422
.48	.1844	.3555	1.03	.3485	.2347	1.58	.4429	.1145	2.13	.4834	.0413
.49	.1879	.3538	1.04	.3508	.2323	1.59	.4441	.1127	2.14	.4838	.0404
.50	.1915	.3521	1.05	.3531	.2299	1.60	.4452	.1109	2.15	.4842	.0395
.51	.1950	.3503	1.06	.3554	.2275	1.61	.4463	.1092	2.16	.4846	.0387
.52	.1985	.3485	1.07	.3577	.2251	1.62	.4474	.1074	2.17	.4850	.0379
.53	.2019	.3467	1.08	.3599	.2227	1.63	.4484	.1057	2.18	.4854	.0371
.54	.2054	.3448	1.09	.3621	.2203	1.64	.4495	.1040	2.19	.4857	.0363

Ordinates and areas of the unit normal curve (continued).

x/σ	Area	Ordinate	x/σ	Area	Ordinate	x/σ	Area	Ordinate	x/σ	Area	Ordinate
2.20	.4861	.0355	2.50	.4938	.0175	2.80	.4974	.0079	3.10	.4990	.0033
2.21	.4864	.0347	2.51	.4940	.0171	2.81	.4975	.0077	3.11	.4991	.0032
2.22	.4868	.0339	2.52	.4941	.0167	2.82	.4976	.0075	3.12	.4991	.0031
2.23	.4871	.0332	2.53	.4943	.0163	2.83	.4977	.0073	3.13	.4991	.0030
2.24	.4875	.0325	2.54	.4945	.0158	2.84	.4977	.0071	3.14	.4992	.0029
2.25	.4878	.0317	2.55	.4946	.0154	2.85	.4978	.0069	3.15	.4992	.0028
2.26	.4881	.0310	2.56	.4948	.0151	2.86	.4979	.0067	3.16	.4992	.0027
2.27	.4884	.0303	2.57	.4949	.0147	2.87	.4979	.0065	3.17	.4992	.0026
2.28	.4887	.0297	2.58	.4951	.0143	2.88	.4980	.0063	3.18	.4993	.0025
2.29	.4890	.0290	2.59	.4952	.0139	2.89	.4981	.0061	3.19	.4993	.0025
2.30	.4893	.0283	2.60	.4953	.0136	2.90	.4981	.0060	3.20	.4993	.0024
2.31	.4896	.0277	2.61	.4955	.0132	2.91	.4982	.0058	3.21	.4993	.0023
2.32	.4898	.0270	2.62	.4956	.0129	2.92	.4982	.0056	3.22	.4994	.0022
2.33	.4901	.0264	2.63	.4957	.0126	2.93	.4983	.0055	3.23	.4994	.0022
2.34	.4904	.0258	2.64	.4959	.0122	2.94	.4984	.0053	3.24	.4994	.0021
2.35	.4906	.0252	2.65	.4960	.0119	2.95	.4984	.0051	3.25	.4994	.0020
2.36	.4909	.0246	2.66	.4961	.0116	2.96	.4985	.0050	3.26	.4994	.0020
2.37	.4911	.0241	2.67	.4962	.0113	2.97	.4985	.0048	3.27	.4995	.0019
2.38	.4913	.0235	2.68	.4963	.0110	2.98	.4986	.0047	3.28	.4995	.0018
2.39	.4916	.0229	2.69	.4964	.0107	2.99	.4986	.0046	3.29	.4995	.0018
2.40	.4918	.0224	2.70	.4965	.0104	3.00	.4987	.0044	3.30	.4995	.0017
2.41	.4920	.0219	2.71	.4966	.0101	3.01	.4987	.0043	3.40	.4997	.0012
2.42	.4922	.0213	2.72	.4967	.0099	3.02	.4987	.0042	3.50	.4998	.0009
2.43	.4925	.0208	2.73	.4968	.0096	3.03	.4988	.0040	3.60	.4998	.0006
2.44	.4927	.0203	2.74	.4969	.0093	3.04	.4988	.0039	3.70	.4999	.0004
2.45	.4929	.0198	2.75	.4970	.0091	3.05	.4989	.0038	3.80	.49993	.0003
2.46	.4931	.0194	2.76	.4971	.0088	3.06	.4989	.0037	3.90	.49995	.0002
2.47	.4932	.0189	2.77	.4972	.0086	3.07	.4989	.0036	4.00	.49997	.0001
2.48	.4934	.0184	2.78	.4973	.0084	3.08	.4990	.0035			
2.49	.4936	.0180	2.79	.4974	.0081	3.09	.4990	.0034			

Table reproduced by permission of McGraw-Hill Book Company, New York, from G.A Ferguson, *Statistical Analysis in Psychology and Education*, 3rd ed., 1971, and originally from J E. Wert, *Educational Statistics*, both published by McGraw-Hill Book Company.

TABLE – 5

Correlation coefficient, *r*

Degrees of freedom	Probability, P			Degrees of freedom	Probability, P		
	0.05	0.01	0.001		0.05	0.01	0.001
1	0.997	1.000	1.000	16	0.468	0.590	0.708
2	0.950	0.990	0.999	17	0.456	0.575	0.693
3	0.878	0.959	0.991	18	0.444	0.561	0.679
4	0.811	0.917	0.974	19	0.433	0.549	0.665
5	0.755	0.875	0.951	20	0.423	0.537	0.652
6	0.707	0.834	0.925	25	0.381	0.487	0.597
7	0.666	0.798	0.898	30	0.349	0.449	0.554
8	0.632	0.765	0.872	35	0.325	0.418	0.519
9	0.602	0.735	0.847	40	0.304	0.393	0.490
10	0.576	0.708	0.823	45	0.288	0.372	0.465
11	0.553	0.684	0.801	50	0.273	0.354	0.443
12	0.532	0.661	0.780	60	0.250	0.325	0.408
13	0.514	0.641	0.760	70	0.232	0.302	0.380
14	0.497	0.623	0.742	80	0.217	0.283	0.357
15	0.482	0.606	0.725	90	0.205	0.267	0.338
				100	0.195	0.254	0.321

TABLE – 6

Critical values of Mann-Whitney U. α = 0.02 (two-tail) and 0.01 (one-tail).

n	*9*	*10*	*11*	*12*	*13*	*14*	*15*
3	1	1	1	2	2	2	3
4	3	3	4	5	5	6	7
5	5	6	7	8	9	10	11
6	7	8	9	11	12	13	15
7	9	11	12	14	16	17	19
8	11	13	15	17	20	22	24

TABLE – 7

Critical T_o scores for Wieoxon's signed rank test.
n = size of each sample or group. α = level of significance.

n	α for two-tain tests				*n*	α for two-tail tests			
	.10	.05	.02	.01		.10	.05	.02	.01
5	1				28	130	117	102	92
6	2	1			29	141	127	111	100
7	4	2	0		30	152	137	120	109
8	6	4	2	0	31	163	148	130	118
9	8	6	3	2	32	175	159	141	128
10	11	8	5	3	33	188	171	151	138
11	14	11	7	5	34	201	183	162	149
12	17	14	10	7	35	214	195	174	160

13	21	17	13	10	36	228	208	186	171
14	26	21	16	13	37	242	222	198	183
15	30	25	20	16	38	256	235	211	195
16	36	30	24	19	39	271	250	224	208
17	41	35	28	23	40	287	264	238	221
18	47	40	33	28	41	303	279	252	234
19	54	46	38	32	42	319	295	267	248
20	60	52	43	37	43	336	311	281	262
21	68	59	49	63	44	353	327	297	277
22	75	66	56	49	45	371	344	313	292
23	83	73	62	55	46	389	361	329	307
24	92	81	69	61	47	408	379	345	323
25	101	90	77	68	48	427	397	362	339
26	110	98	85	76	49	446	415	380	356
27	120	107	93	84	50	466	434	398	373
	.05	.025	.01	.005		.05	.025	.01	.005
	α for one-tail tests					α for one-tail tests			

Table taken from Frank Wilcoxon and Roberta A. Wilcox, Some *Rapid Approximate Statistical Procedures*, revised ed., 1964, Lederle Laboratories, New York, by the kind permission of American Cyanamid Company.

TABLE – 8

Table. Critical values of Speaman's rho. Levels of significance of two-tail test

n	*.10*	*.02*	*.01*
5	.900	1.000	
6	.829	.943	1.000
7	.714	.893	.929
8	.643	.833	.881
9	.600	.783	.833
10	.564	.746	.794
12	.506	.712	.777
14	.456	.645	.715
16	.425	.601	.665
18	.399	.564	.625
20	.377	.534	.591
22	.359	.508	.562
24	.343	.485	.537
26	.329	.465	.515
28	.317	.448	.496
30	.306	.432	.478
	.05	.01	.005
	Levels of significance for one-tail test		

Reprinted by permission of McGraw-Hill Book Company, New York, from J. P. Guilford and B. Fruchter, *Fundamental Statistics in Psychology and Education*, 6th ed., 1978, and originally from W. J. Dixon and F. J. Massey, Jr., *Introduction to Statistical Analysis*, 1951, both published by McGraw-Hill Book Company.

CHAPTER

NOTATION AND IMPORTANT FORMULAE

NOTATION

α : Usually denotes the level of significance (pre-specified value of the Type-I error probability bound) for a hypothesis testing problem.

$A \bigcup B$: Union of events A and B.

$A \bigcap B$: Intersection of events A and B.

A^c : Complement of event A.

$A \subseteq B$: A is a subset of B.

$B(n, p)$: Binomial experiment where n = number of trials and p = probability of success in a single trial.

X_k^2 : Chi-square distribution with k degrees of freedom (*df*).

$X_{k,\beta}^{2(R)}$: $100(1-\beta)^{th}$ percentage point of the X_k^2 distribution.

$X_{k,\beta}^{2(L)}$: $100\beta^{th}$ percentage point of the X_k^2 distribution.

c_1 : Factor for control limits of $\overline{X}$-chart based on range.

c_2 : Factor for limits of $\overline{X}$-chart based on standard deviation.

c_3 : Factor for lower control limit of R-chart.

c_4 : Factor for upper control limit of R-chart.

c_5 : Factor for lower control limit of s-chart.

c_6 : Factor for upper control limit of s-chart.

ϕ : Empty event (set) or Null event (set).

F_{l_1,l_2} : (Snedecor's) F-distribution (or curve) with l_1 and l_2 degrees of freedom (*df*).

$F_{l_1,l_2,\beta}^{(L)}$: $100\beta^{th}$ percentage point of the F_{l_1,l_2}-distribution (or curve).

$F_{l_1,l_2,\beta}^{(R)}$: $100(1-\beta)^{th}$ percentage point of the F_{l_1,l_2}-distribution (or curve).

H_A : Alternative Hypothesis.

H_0 : Null Hypothesis.

μ : Mean of a probability distribution (or a population).

μ_X : Expected value of a random variable X.

$M_k(n, p_1, \ldots, p_k)$: Multinomial experiment where n = number of trials and p_i probability of getting the ith outcome in a single trial with $p_1 + p_2 + \ldots + p_k = 1$.

n : Sample size.

N : Population size.
R_x : Range of random variable X.
σ : Standard deviation of a probability distribution (or a population).
$\sum$: Stands for summation of certain quantities.
$\sum_{i=1}^{m} a_i$: Sum of the terms $a_1, a_2, ..., a_m$
σ_X : Standard deviation of a random variable X.
S : Sample space of an experiment.
s_x^2 or s^2 : Variance of the observation $X_1, X_2, ... X_n$ (sample variance).
s_x or s : Standard deviation of observation $X_1, X_2, ... X_n$ (sample SD).
$N(\mu, \sigma^2)$: Normal curve with centre (mean) μ and spread (standard deviation) σ
$P(E)$: Probability of an event E.
P-value ; Smallest significance level for which the null hypothesis is rejected.
t_k : t-distribution (or curve) with k degrees of freedom (*df*).
$t_{k,\beta}$: $100(1-\beta)^{th}$ percentage point of the t_k-distribution (or curve).
$X_1, X_2, ... X_n$: Observation in a raw dataset.
$\overline{X}$: Average of observation $X_1, X_2, ..., X_n$ (sample mean or average).
z_β : $100(1-\beta)^{th}$ percentage point of the $N(0, 1)$ distribution.

IMPORTANT FORMULAE

1. Measures of Central Tendency:

(*a*) Arithmetic mean $(\bar{x})$

$$A.M.\ (\bar{x}) = \frac{\sum fx}{\sum f} = \frac{\sum fx}{N}$$

$$\overline{X} = a \times \frac{\sum fd'}{N} \times i$$

(*b*) Median $(M) = L + \dfrac{\frac{N}{2} - C}{fm} \times i$

(*c*) Mode $= L_1 + \dfrac{d_1}{d_1 + d_2} \times i$

$$L_1 + \frac{fm - f_1}{2 \cdot fm - f_2 - f_1}$$

Mode = 3 Median – 2 Mean

(*d*) Relation between A.M. G.M and H.M

A.M. $\geq$ G.M. $\geq$ H.m.

2. Measures of Variation:

(*a*) Range (R) = Largest value (L) – Smallest value (S)

$R = L - S$

(*b*) Coefficient of range $= \dfrac{L-S}{L+S}$

(*c*) Mean deviation (M.D) $= \frac{1}{2}\Sigma f \mid D \mid$

(*d*) Coefficient of mean deviation:

$$\text{C.M.D.} = \frac{MD}{\text{Mean/Median}} \times 100$$

(*e*) Quartile deviation $= \frac{Q_3 - Q_1}{2}$

(*f*) Coefficient of quartile deviation

$$= \frac{\text{Quartile deviation}}{\text{Median}} \times 100$$

(*g*) Standard deviation (S.D.)

(*i*) $S.D. = \sqrt{\frac{\Sigma f(X - \bar{X})^2}{n}}$

(*ii*) $S.D. = \sqrt{\frac{\Sigma fd^2}{n} - \left(\frac{\Sigma fd}{n}\right)^2}$

(*iii*) $S.D. = \sqrt{\frac{\Sigma fd^2}{n} - \left(\frac{\Sigma fd}{n}\right)^2} \times i$

(*h*) Coefficient of variance (*C.V*)

$$C.V = \frac{S.D}{A.M} \times 100$$

(*i*) Standard Error of Mean (*S.E*)

(*i*) $S.E = \frac{S.D}{\sqrt{n}}$

(*ii*) $S.E. = (\bar{X}_1 - \bar{X}_2) = S.D\sqrt{\frac{1}{n_1} + \frac{1}{n_2}}$

3. Theoretical distribution:

I. Normal: $f(x) = \frac{1}{\sigma\sqrt{2x}} e \frac{-(n-\mu)^2}{2\sigma^2} \quad (-a < n < \alpha)$

II. Z score i.e. $Z = \frac{X_m - \bar{X}}{\sigma}$

III. Permutation : ${}^nP_r = \frac{\underline{|n}}{\underline{|n-r}}$

&

Combination ${}^nC_r = \frac{\underline{|n}}{\underline{|n-r}\,\underline{|r}}$

IV. Binomial expansion:

$$(p+q)^n = p^n + np^{n-}1q + \frac{n(n-1)}{1.2}p^{n-2}q^2 + \frac{n(n-1)(n-2)}{1.2.3}p^{n-3}q^3 + ...$$

$$\frac{n(n-1)(n-2)... \times 3.2}{1.2.3......... \times (n-1)} p.q^{n-1} + q^n$$

V. Poison distribution:

$$p(n) = \frac{e^{-m} m^n}{\lfloor n} \quad [n = 0, 1, 2, 3..........n]$$

VI. Coefficient of Skewners

(*a*) Pearson's first measures $= \dfrac{\text{Mean} - \text{Mode}}{S.D}$

(*b*) Pearson's second measure $= \dfrac{3(\text{Mean} - \text{Median})}{S.D}$

(*c*) Bowley's measure $= \dfrac{Q_3 - 2Q_2 + Q_1}{Q_3 - Q_1}$

(*d*) Kelly's coefficient of skewness: $\dfrac{P_{10} + P_{90} - 2\,\text{median}}{P_{90} - P_{10}}$

Coefficient of skewness (based on moments) $= B_1 = \dfrac{{\mu_3}^2}{{\mu_2}^3}$

VII. Measures of kurtosis $= B_2 = \dfrac{\mu_1}{{\mu_2}^2} = \dfrac{\mu^4}{\sigma^4}$

4. Probability:

(*a*) Classical definition: $P(A) = \dfrac{m}{n}$

(*b*) Theories of probabilities:

(*i*) $P(A + B) = P(A) + P(B)$

(*ii*) $P(AB) = P(A).\ P(B)$ when independent.

(*iii*) $P(AB) = P(A).\ P(B/A) = P(B).P(A/B)$

(*c*) Binomial Distribution

$P(r) = {}^{n}c_r p^r q^{n-r}$, $r = 0, 1,2,3, n$

5. Chi-Square test:

(*i*) Test for goodness fit:

$$\text{Chi-square } (x^2) = \frac{\Sigma\left[\,|(O - E)| - \frac{1}{2}\right]^2}{E}$$

(*ii*) Test for contingency chi-square:

$$X^2 = \frac{N\left\{(|\,ad - bc\,|) - \frac{N}{2}\right\}^2}{R_1 \times R_2 \times R_3 \times R_4}$$

6. Student t distribution:

(*i*) $t = \dfrac{|\,X_1 - \overline{X}_2\,|}{SE}$

(*ii*) $t \dfrac{\overline{D}}{SD\sqrt{n}}$

7. F-Test:

$$F = \frac{\text{variance of sample} - 1}{\text{variance of sample} - 2} = \frac{\sigma_1^2}{\sigma_2^2}$$

8. Z-Test:

(*i*) $$Z = \frac{\text{Observation} - \text{Mean}}{S.D} = \frac{X - \overline{X}}{S.D}$$

(*ii*) $$Z = \frac{\overline{X} - \mu}{SE(X)}$$

(*iii*) $$Z = \frac{\overline{X}_1 - \overline{X}_2}{SE(X_1 - X_2)}$$

9. Correlation:

(*i*) $$r(XY) = \frac{\sum dxdy}{\sqrt{\sum dx^2 \times \sum dy^2}}$$

(*ii*) $$r_{xy} = \frac{\sum dxay - \left(\frac{\sum dx \sum dy}{n}\right)}{\sqrt{\sum dx^2 - \frac{(\sum dx)^2}{n}} \times \sqrt{\sum dy^2 - \frac{(\sum dy)^2}{n}}}$$

(*iii*) Rank correlation

$$R = \frac{1 - 6\sum d^2}{n^3 - n}$$

(*v*) Kendall's rank correlation

$$\tan(\tau) = \frac{\text{Total score}}{\text{Maximum Possible score}} = \frac{S}{N_{C_r}}$$

$$= \frac{S}{\frac{N(N-1)}{2}} = \frac{2S}{N(N-1)}$$

(*v*) Partial correlation:

$$r_{1,23} = \frac{r_{12} - r_{13}\, r_{23}}{\sqrt{\left(1 - r_{23}^2\right)\left(1 - r_{23}^2\right)}}$$

(*vi*) Multiple correlation:

$$R_{1,23} = \sqrt{\frac{r_{12}^2 + r_{13}^2 - 2r_{12}\, r_{13}\, r_{23}}{1 - r_{23}^2}}$$

10. Regression:

(*i*) Regression equation of Y on X.

$$Y - \overline{Y} = b_{yx}(X - \overline{X})$$

$$\left[by\,x = \frac{\text{cov}(x, y)}{\sigma^2 y} = r\frac{\sigma x}{\sigma y} \right]$$

$$= r\frac{\sigma y}{\sigma x}\left(X - \overline{X}\right)$$

(*ii*) Regression equation of X on Y

$$X - \overline{X} = b_{xy}\,(Y - \overline{Y})$$

$$\left[b_{xy} = \frac{\text{cov}(xy)}{\sigma x^2} = r\frac{\sigma y}{\sigma n} \right]$$

(*iii*) Multiple linear regression:

$\hat{X}$ is the predicted score of the criterion

I. $B_2 = \dfrac{r_{12} - r_{13}\,r_{23}}{1 - r_{23}^2}$ $\qquad B_3 = \dfrac{r_{13} - r_{12}\,r_{23}}{1 - r_{23}^2}$

II. $b_{1,23} = B_2 \times \dfrac{S_1}{S_2}$ $\qquad b_{1,32} = B_3 \times \dfrac{S_1}{S_3}$

III. $a_1,^{23} = \overline{X}_1 - b_{1,23}\,\overline{X}_2 - b_{1,32}\,\overline{X}_3$

IV. $\hat{X}_1 = a_{1,23} + b_{1,23}\,X_2 + b_{1,32}\,X_3$

11. (*i*) Analysis of variance (Anova)

Source of variation	Sum of squares (*SS*)	Degrees of freedom (*df*)	Mean square (*MS*)	F-Values	
				(*Obs*)	Tab
Between groups	*SSB*	$k - 1$	*MSB*	$F = \dfrac{MSB}{MSW}$	$F = .05$
With in groups	*SSW*	$df_w = N - K$	*MSW*		

(*ii*) Bonferroni Adjustment t-test

$$t = \frac{\overline{X}_1 - \overline{X}_2}{S_{\overline{X}_1 - \overline{X}_2}} = \frac{\overline{X}_1 - \overline{X}_2}{\sqrt{\dfrac{s^2 w}{n_1} + \dfrac{s^2 w}{n_2}}}$$

CHAPTER

LOGARITHMS TABLES

LOGARITHMS

	0	1	2	3	4	5	6	7	8	9	Mean Differences								
											1	2	3	4	5	6	7	8	9
10	0000	0043	0086	0128	0170	0212	0253	0294	0334	0374	4	8	12	17	21	25	29	33	37
11	0414	0453	0492	0531	0569	0607	0645	0682	0719	0755	4	8	11	15	19	23	26	30	34
12	0792	0828	0864	0899	0934	0969	1004	1038	1072	1106	3	7	10	14	17	21	24	28	31
13	1139	1173	1206	1239	1271	1303	1335	1367	1399	1430	3	6	10	13	16	19	23	26	29
14	1461	1492	1523	1553	1584	1614	1644	1673	1703	1732	3	6	9	12	15	18	21	24	27
15	1761	1790	1818	1847	1875	1903	1931	1959	1987	2014	3	6	8	11	14	17	20	22	25
16	2041	2068	2095	2122	2148	2175	2201	2227	2253	2279	3	5	8	11	13	16	18	21	24
17	2304	2330	2355	2380	2405	2430	2455	2480	2504	2529	2	5	7	10	12	15	17	20	22
18	2553	2577	2601	2625	2648	2672	2695	2718	2742	2765	2	5	7	9	12	14	16	19	21
19	2788	2810	2833	2856	2878	2900	2923	2945	2967	2989	2	4	7	9	11	13	16	18	20
20	3010	3032	3054	3075	3096	3118	3139	3160	3181	3201	2	4	6	8	11	13	15	17	19
21	3222	3243	3263	3284	3304	3324	3345	3365	3385	3404	2	4	6	8	10	12	14	16	18
22	3424	3444	3464	3483	3502	3522	3541	3560	3579	3598	2	4	6	8	10	12	14	15	17
23	3617	3636	3655	3674	3692	3711	3729	3747	3766	3784	2	4	6	7	9	11	13	15	17
24	3802	3820	3838	3856	3874	3892	3909	3927	3945	3962	2	4	5	7	9	11	12	14	16
25	3979	3997	4014	4031	4048	4065	4082	4099	4116	4133	2	3	5	7	9	10	12	14	15
26	4150	4166	4183	4200	4216	4232	4249	4265	4281	4298	2	3	5	7	8	10	11	13	15
27	4314	4330	4346	4362	4378	4393	4409	4425	4440	4456	2	3	5	6	8	9	11	13	14
28	4472	4487	4502	4518	4533	4548	4564	4579	4594	4609	2	3	5	6	8	9	11	12	14
29	4624	4639	4654	4669	4683	4698	4713	4728	4742	4757	1	3	4	6	7	9	10	12	13
30	4771	4786	4800	4814	4829	4843	4857	4871	4886	4900	1	3	4	6	7	9	10	11	13
31	4914	4928	4942	4955	4969	4983	4997	5011	5024	5038	1	3	4	6	7	8	10	11	12
32	5051	5065	5079	5092	5105	5119	5132	5145	5159	5172	1	3	4	5	7	8	9	11	12
33	5185	5198	5211	5224	5237	5250	5263	5276	5289	5302	1	3	4	5	6	8	9	10	12
34	5315	5328	5340	5353	5366	5378	5391	5403	5416	5428	1	3	4	5	6	8	9	10	11
35	5441	5453	5465	5478	5490	5502	5514	5527	5539	5551	1	2	4	5	6	7	9	10	11
36	5563	5575	5587	5599	5611	5623	5635	5647	5658	5670	1	2	4	5	6	7	8	10	11
37	5682	5694	5705	5717	5729	5740	5752	5763	5775	5786	1	2	3	5	6	7	8	9	10
38	5798	5809	5821	5832	5843	5855	5866	5877	5888	5899	1	2	3	5	6	7	8	9	10
39	5911	5922	5933	5944	5955	5966	5977	5988	5999	6010	1	2	3	4	5	7	8	9	10
40	6021	6031	6042	6053	6064	6075	6085	6096	6107	6117	1	2	3	4	5	6	8	9	10
41	6128	6138	6149	6160	6170	6180	6191	6201	6212	6222	1	2	3	4	5	6	7	8	9
42	6232	6243	6253	6263	6274	6284	6294	6304	6314	6325	1	2	3	4	5	6	7	8	9
43	6335	6345	6355	6365	6375	6385	6395	6405	6415	6425	1	2	3	4	5	6	7	8	9
44	6435	6444	6454	6464	6474	6484	6493	6503	6513	6522	1	2	3	4	5	6	7	8	9
45	6532	6542	6551	6561	6571	6580	6590	6599	6609	6618	1	2	3	4	5	6	7	8	9
46	6628	6637	6646	6656	6665	6675	6684	6693	6702	6712	1	2	3	4	5	6	7	7	8
47	6721	6730	6739	6749	6758	6767	6776	6785	6794	6803	1	2	3	4	5	5	6	7	8
48	6812	6821	6830	6839	6848	6857	6866	6875	6884	6893	1	2	3	4	4	5	6	7	8
49	6902	6911	6920	6928	6937	6946	6955	6964	6972	6981	1	2	3	4	4	5	6	7	8
50	6990	6998	7007	7016	7024	7033	7042	7050	7059	7067	1	2	3	3	4	5	6	7	8

(Contd.)

LOGARITHMS

	0	1	2	3	4	5	6	7	8	9	Mean Differences								
											1	2	3	4	5	6	7	8	9
51	7076	7084	7093	7101	7110	7118	7126	7135	7143	7152	1	2	3	3	4	5	6	7	8
52	7160	7168	7177	7185	7193	7202	7210	7218	7226	7235	1	2	2	3	4	5	6	7	7
53	7243	7251	7259	7267	7275	7284	7292	7300	7308	7316	1	2	2	3	4	5	6	6	7
54	7324	7332	7340	7348	7356	7364	7372	7380	7388	7396	1	2	2	3	4	5	6	6	7
55	7404	7412	7419	7427	7435	7443	7451	7459	7466	7474	1	2	2	3	4	5	5	6	7
56	7482	7490	7497	7505	7513	7520	7528	7536	7543	7551	1	2	2	3	4	5	5	6	7
57	7559	7566	7574	7582	7589	7597	7604	7612	7619	7627	1	2	2	3	4	5	5	6	7
58	7634	7642	7649	7657	7664	7672	7679	7686	7694	7701	1	1	2	3	4	4	5	6	7
59	7709	7716	7723	7731	7738	7745	7752	7760	7767	7774	1	1	2	3	4	4	5	6	7
60	7782	7789	7796	7803	7810	7818	7825	7832	7839	7846	1	1	2	3	4	4	5	6	6
61	7853	7860	7868	7875	7882	7889	7896	7903	7910	7917	1	1	2	3	4	4	5	6	6
62	7924	7931	7938	7945	7952	7959	7966	7973	7980	7987	1	1	2	3	3	4	5	6	6
63	7993	8000	8007	8014	8021	8028	8035	8041	8048	8055	1	1	2	3	3	4	5	5	6
64	8062	8069	8075	8082	8089	8096	8102	8109	8116	8122	1	1	2	3	3	4	5	5	6
65	8129	8136	8142	8149	8156	8162	8169	8176	8182	8189	1	1	2	3	3	4	5	5	6
66	8195	8202	8209	8215	8222	8228	8235	8241	8248	8254	1	1	2	3	3	4	5	5	6
67	8261	8267	8274	8280	8287	8293	8299	8306	8312	8319	1	1	2	3	3	4	5	5	6
68	8325	8331	8338	8344	8351	8357	8363	8370	8376	8382	1	1	2	3	3	4	4	5	6
69	8388	8395	8401	8407	8414	8420	8426	8432	8439	8445	1	1	2	2	3	4	4	5	6
70	8451	8457	8463	8470	8476	8482	8488	8494	8500	8506	1	1	2	2	3	4	4	5	6
71	8513	8519	8525	8531	8537	8543	8549	8555	8561	8567	1	1	2	2	3	4	4	5	5
72	8573	8579	8585	8591	8597	8603	8609	8615	8621	8627	1	1	2	2	3	4	4	5	5
73	8633	8639	8645	8651	8657	8663	8669	8675	8681	8686	1	1	2	2	3	4	4	5	5
74	8692	8698	8704	8710	8716	8722	8727	8733	8739	8745	1	1	2	2	3	4	4	5	5
75	8751	8756	8762	8768	8774	8779	8785	8791	8797	8802	1	1	2	2	3	3	4	5	5
76	8808	8814	8820	8825	8831	8837	8842	8848	8854	8859	1	1	2	2	3	3	4	5	5
77	8865	8871	8876	8882	8887	8893	8899	8904	8910	8915	1	1	2	2	3	3	4	4	5
78	8921	8927	8932	8938	8943	8949	8954	8960	8965	8971	1	1	2	2	3	3	4	4	5
79	8976	8982	8987	8993	8998	9004	9009	9015	9020	9025	1	1	2	2	3	3	4	4	5
80	9031	9036	9042	9047	9053	9058	9063	9069	9074	9079	1	1	2	2	3	3	4	4	5
81	9085	9090	9096	9101	9106	9112	9117	9122	9128	9133	1	1	2	2	3	3	4	4	5
82	9138	9143	9149	9154	9159	9165	9170	9175	9180	9186	1	1	2	2	3	3	4	4	5
83	9191	9196	9201	9206	9217	9217	9222	9227	9232	9238	1	1	2	2	3	3	4	4	5
84	9243	9248	9253	9258	9263	9269	9274	9279	9284	9289	1	1	2	2	3	3	4	4	5
85	9294	9299	9304	9360	9315	9320	9325	9330	9335	9340	1	1	2	2	3	3	4	4	5
86	9345	9350	9355	9360	9365	9370	9375	9380	9385	9390	1	1	2	2	3	3	4	4	5
87	9395	9400	9405	9410	9415	9420	9425	9430	9435	9440	0	1	1	2	2	3	3	4	4
88	9445	9450	9455	9460	9465	9469	9474	9479	9484	9489	0	1	1	2	2	3	3	4	4
89	9494	9499	9504	9509	9513	9518	9523	9528	9533	9538	0	1	1	2	2	3	3	4	4
90	9542	9547	9552	9557	9562	9566	9571	9576	9581	9586	0	1	1	2	2	3	3	4	4
91	9590	9595	9600	9605	9609	9614	9619	9624	9628	9633	0	1	1	2	2	3	3	4	4
92	9638	9643	9647	9652	9657	9661	9666	9671	9675	9680	0	1	1	2	2	3	3	4	4
93	9685	9689	9694	9699	9703	9708	9713	9717	9722	9727	0	1	1	2	2	3	3	4	4
94	9731	9736	9741	9745	9750	9754	9759	9763	9768	9773	0	1	1	2	2	3	3	4	4
95	9777	9782	9786	9791	9795	9800	9805	9809	9814	9818	0	1	1	2	2	3	3	4	4
96	9823	9827	9832	9836	9841	9845	9850	9854	9859	9863	0	1	1	2	2	3	3	4	4
97	9868	9872	9877	9881	9886	9890	9894	9899	9903	9908	0	1	1	2	2	3	3	4	4
98	9912	9916	9921	9926	9930	9934	9939	9943	9948	9952	0	1	1	2	2	3	3	4	4
99	9956	9961	9965	9969	9974	9978	9983	9987	9991	9996	0	1	1	2	2	3	3	3	4

ANTILOGARITHMS

	0	1	2	3	4	5	6	7	8	9	Mean Differences								
											1	2	3	4	5	6	7	8	9
.00	1000	1002	1005	1007	1009	1012	1014	1016	1019	1021	0	0	1	1	1	1	2	2	2
.01	1023	1026	1028	1030	1033	1035	1038	1040	1042	1045	0	0	1	1	1	1	2	2	2
.02	1047	1050	1052	1054	1057	1059	1062	1064	1067	1069	0	0	1	1	1	1	2	2	2
.03	1072	1074	1076	1079	1081	1084	1086	1089	1091	1094	0	0	1	1	1	1	2	2	2
.04	1096	1099	1102	1104	1107	1109	1112	1114	1117	1119	0	1	1	1	1	2	2	2	2
.05	1122	1125	1127	1130	1132	1135	1138	1140	1143	1146	0	1	1	1	1	2	2	2	2
.06	1148	1151	1153	1156	1159	1161	1164	1167	1169	1172	0	1	1	1	1	2	2	2	2
.07	1175	1178	1180	1183	1186	1189	1191	1194	1197	1199	0	1	1	1	1	2	2	2	2
.08	1202	1205	1208	1211	1213	1216	1219	1222	1225	1227	0	1	1	1	1	2	2	2	3
.09	1230	1233	1236	1239	1242	1245	1247	1250	1253	1256	0	1	1	1	1	2	2	2	3
.10	1259	1262	1265	1268	1271	1274	1276	1279	1282	1285	0	1	1	1	1	2	2	2	3
.11	1288	1291	1294	1297	1300	1303	1306	1309	1312	1315	0	1	1	1	2	2	2	2	3
.12	1318	1321	1324	1327	1330	1334	1337	1340	1343	1346	0	1	1	1	2	2	2	2	3
.13	1349	1352	1355	1358	1361	1365	1368	1371	1374	1377	0	1	1	1	2	2	2	3	3
.14	1380	1384	1387	1390	1393	1396	1400	1403	1406	1409	0	1	1	1	2	2	2	3	3
.15	1413	1416	1419	1422	1426	1429	1432	1435	1439	1442	0	1	1	1	2	2	2	3	3
.16	1445	1449	1452	1455	1459	1462	1466	1469	1472	1476	0	1	1	1	2	2	2	3	3
.17	1479	1483	1486	1489	1493	1496	1500	1503	1507	1510	0	1	1	1	2	2	2	3	3
.18	1514	1517	1521	1524	1528	1531	1535	1538	1542	1545	0	1	1	1	2	2	2	3	3
.19	1549	1552	1556	1560	1563	1567	1570	1574	1578	1581	0	1	1	1	2	2	3	3	3
.20	1585	1589	1592	1596	1600	1603	1607	1611	1614	1618	0	1	1	1	2	2	3	3	3
.21	1622	1626	1629	1633	1637	1641	1644	1648	1652	1656	0	1	1	2	2	2	3	3	3
.22	1660	1663	1667	1671	1675	1679	1683	1687	1690	1694	0	1	1	2	2	2	3	3	3
.23	1698	1702	1706	1710	1714	1718	1722	1726	1730	1734	0	1	1	2	2	2	3	3	4
.24	1738	1742	1746	1750	1754	1758	1762	1766	1770	1774	0	1	1	2	2	2	3	3	4
.25	1778	1782	1786	1791	1795	1798	1803	1807	1811	1816	0	1	1	2	2	2	3	3	4
.26	1820	1824	1828	1832	1837	1841	1845	1849	1854	1858	0	1	1	2	2	3	3	3	4
.27	1862	1866	1871	1875	1879	1884	1888	1892	1897	1901	0	1	1	2	2	3	3	3	4
.28	1905	1910	1914	1919	1923	1928	1932	1936	1941	1945	0	1	1	2	2	3	3	4	4
.29	1950	1954	1959	1963	1968	1972	1977	1982	1986	1991	0	1	1	2	2	3	3	4	4
.30	1995	2000	2004	2009	2014	2018	2023	2028	2032	2037	0	1	1	2	2	3	3	4	4
.31	2042	2046	2051	2056	2061	2065	2070	2075	2080	2084	0	1	1	2	2	3	3	4	4
.32	2089	2094	2099	2104	2109	2113	2118	2123	2128	2133	0	1	1	2	2	3	3	4	4
.33	2138	2143	2148	2153	2158	2163	2168	2173	2178	2183	0	1	1	2	2	3	3	4	4
.34	2188	2193	2198	2203	2208	2213	2218	2223	2228	2234	1	1	2	2	3	3	4	4	5
.35	2239	2244	2249	2254	2259	2265	2270	2275	2280	2286	1	1	2	2	3	3	4	4	5
.36	2291	2296	2301	2307	2312	2317	2323	2328	2333	2339	1	1	2	2	3	3	4	4	5
.37	2344	2350	2355	2360	2366	2371	2377	2382	2388	2393	1	1	2	2	3	3	4	4	5
.38	2399	2404	2410	2415	2421	2427	2432	2438	2443	2449	1	1	2	2	3	3	4	4	5
.39	2455	2460	2466	2472	2477	2483	2489	2495	2500	2506	1	1	2	2	3	3	4	5	5
.40	2512	2518	2523	2529	2535	2541	2547	2553	2559	2564	1	1	2	2	3	4	4	5	5
.41	2570	2576	2582	2588	2594	2600	2606	2612	2618	2624	1	1	2	2	3	4	4	5	5
.42	2630	2636	2642	2649	2655	2661	2667	2673	2679	2685	1	1	2	2	3	4	4	5	6
.43	2692	2698	2704	2710	2716	2723	2729	2735	2742	2748	1	1	2	3	3	4	4	5	6
.44	2754	2761	2767	2773	2780	2786	2793	2799	2805	2812	1	1	2	3	3	4	4	5	6
.45	2818	2825	2831	2838	2844	2851	2858	2864	2871	2877	1	1	2	3	3	4	5	5	6
.46	2884	2891	2897	2904	2911	2917	2924	2931	2938	2944	1	1	2	3	3	4	5	5	6
.47	2951	2958	2965	2972	2979	2985	2992	2999	3006	3013	1	1	2	3	3	4	5	5	6
.48	3020	3027	3034	3041	3048	3055	3062	3069	3076	3083	1	1	2	3	4	4	5	6	6
.49	3090	3097	3105	3112	3119	3126	3133	3141	3148	3155	1	1	2	3	4	4	5	6	6
.50	3162	3170	3177	3184	3192	3199	3206	3214	3221	3228	1	1	2	3	4	4	5	6	7

(Contd.)

ANTILOGARITHMS

	0	1	2	3	4	5	6	7	8	9	Mean Differences								
											1	2	3	4	5	6	7	8	9
.51	3236	3243	3251	3258	3266	3273	3281	3289	3296	3304	1	2	2	3	4	5	5	6	7
.52	3311	3319	3327	3334	3342	3350	3357	3365	3373	3381	1	2	2	3	4	5	5	6	7
.53	3388	3396	3404	3412	3420	3428	3436	3443	3451	3459	1	2	2	3	4	5	6	6	7
.54	3467	3475	3483	3491	3499	3508	3516	3524	3532	3540	1	2	2	3	4	5	6	6	7
.55	3548	3556	3565	3573	3581	3589	3597	3606	3614	3622	1	2	2	3	4	5	6	7	7
.56	3631	3639	3648	3656	3664	3673	3681	3690	3698	3707	1	2	3	3	4	5	6	7	8
.57	3715	3724	3733	3741	3750	3758	3767	3776	3784	3793	1	2	3	3	4	5	6	7	8
.58	3802	3811	3819	3828	3837	3846	3855	3864	3873	3882	1	2	3	4	4	5	6	7	8
.59	3890	3899	3908	3917	3926	3936	3945	3954	3963	3972	1	2	3	4	5	5	6	7	8
.60	3981	3990	3999	4009	4018	4027	4036	4046	4055	4064	1	2	3	4	5	6	6	7	8
.61	4074	4083	4093	4102	4111	4121	4130	4140	4150	4159	1	2	3	4	5	6	7	8	9
.62	4169	4178	4188	4198	4207	4217	4227	4236	4246	4256	1	2	3	4	5	6	7	8	9
.63	4266	4276	4285	4295	4305	4315	4325	4335	4345	4355	1	2	3	4	5	6	7	8	9
.64	4365	4375	4385	4395	4406	4416	4426	4436	4446	4457	1	2	3	4	5	6	7	8	9
.65	4467	4477	4487	4498	4508	4519	4529	4539	4550	4560	1	2	3	4	5	6	7	8	9
.66	4571	4581	4592	4603	4613	4624	4634	4645	4656	4667	1	2	3	4	5	6	7	9	10
.67	4677	4688	4699	4710	4721	4732	4742	4753	4764	4775	1	2	3	4	5	7	8	9	10
.68	4786	4797	4808	4819	4831	4842	4853	4864	4875	4887	1	2	3	4	6	7	8	9	10
.69	4898	4909	4920	4932	4943	4955	4966	4977	4989	5000	1	2	3	5	6	7	8	9	10
.70	5012	5023	5035	5047	5058	5070	5082	5093	5105	5117	1	2	4	5	6	7	8	9	11
.71	5129	5140	5152	5164	5176	5188	5200	5212	5224	5236	1	2	4	5	6	7	8	10	11
.72	5248	5260	5272	5284	5297	5309	5321	5333	5346	5358	1	2	4	5	6	7	9	10	11
.73	5370	5383	5395	5408	5420	5433	5445	5458	5470	5483	1	3	4	5	6	8	9	10	11
.74	5495	5508	5521	5534	5546	5559	5572	5585	5598	5610	1	3	4	5	6	8	9	10	12
.75	5623	5636	5649	5662	5675	5689	5702	5715	5728	5741	1	3	4	5	7	8	9	10	12
.76	5754	5768	5781	5794	5808	5821	5834	5848	5861	5875	1	3	4	5	7	8	9	11	12
.77	5888	5902	5916	5929	5943	5957	5970	5984	5998	6012	1	3	4	5	7	8	10	11	12
.78	6026	6039	6053	6067	6081	6095	6109	6124	6138	6152	1	3	4	6	7	8	10	11	13
.79	6166	6180	6194	6209	6223	6237	6252	6266	6281	6295	1	3	4	6	7	9	10	11	13
.80	6310	6324	6339	6353	6368	6383	6397	6412	6427	6442	1	3	4	6	7	9	10	12	13
.81	6457	6471	6486	6501	6516	6531	6546	6561	6577	6592	2	3	5	6	8	9	11	12	14
.82	6607	6622	6637	6653	6668	6683	6699	6714	6730	6745	2	3	5	6	8	9	11	12	14
.83	6761	6776	6792	6808	6823	6839	6855	6871	6887	6902	2	3	5	6	8	9	11	13	14
.84	6918	6934	6950	6966	6982	6998	7015	7031	7047	7063	2	3	5	6	8	10	11	13	15
.85	7079	7096	7112	7129	7145	7161	7178	7194	7211	7228	2	3	5	7	8	10	12	13	15
.86	7244	7261	7278	7295	7311	7328	7345	7362	7379	7396	2	3	5	7	8	10	12	13	15
.87	7413	7430	7447	7464	7482	7499	7516	7534	7551	7568	2	3	5	7	9	10	12	14	16
.88	7586	7603	7621	7638	7656	7674	7691	7709	7727	7745	2	4	5	7	9	11	12	14	16
.89	7762	7780	7798	7816	7834	7852	7870	7889	7907	7925	2	4	5	7	9	11	13	14	16
.90	7943	7962	7980	7998	8017	8035	8054	8072	8091	8110	2	4	6	7	9	11	13	15	17
.91	8128	8147	8166	8185	8204	8222	8241	8260	8279	8299	2	4	6	8	9	11	13	15	17
.92	8318	8337	8356	8375	8395	8414	8433	8453	8472	8492	2	4	6	8	10	12	14	15	17
.93	8511	8531	8551	8570	8590	8616	8630	8650	8670	8690	2	4	6	8	10	12	14	16	18
.94	8710	8730	8750	8770	8790	8810	8831	8851	8872	8892	2	4	6	8	10	12	14	16	18
.95	8913	8933	8954	8974	8995	9016	9036	9057	9078	9099	2	4	6	8	10	12	15	17	19
.96	9120	9141	9162	9183	9204	9226	9247	9268	9290	9311	2	4	6	8	11	13	15	17	19
.97	9333	9354	9376	9397	9419	9441	9462	9484	9506	9528	2	4	7	9	11	13	15	17	20
.98	9550	9572	9594	9616	9638	9661	9683	9705	9727	9750	2	4	7	9	11	13	16	18	20
.99	9772	9793	9817	9840	9863	9886	9908	9931	9954	9977	2	5	7	9	11	14	16	18	20

BIBLIOGRAPHY

	Name of the Author		Name of the Book
1.	Arora P.N. and Malhan P.K.	:	Biostatistics
2.	Banerjee P.K.	:	Problems on Genetics Molecular Genetics & Evolutionary Genetics
3.	Bhattacharya D and Roy Choudhury	:	Statistics: Theory and Practices
4.	Das D & Das A	:	Statistics in Biology and Psychology
5.	Das NG	:	Statistical Methods
6.	De S.N.	:	Business Statistics
7.	Fisher R.A.	:	Statistical Methods for Research Workers
8.	Gritfith A. Gelbart W.M. Miller S.H. & Lewontin R.C.	:	Modern Genetic Analysis
9.	Kowles Richard	:	Solving Problems in Genetics
10.	Mahajan B.K.	:	Methods in Biostatistics
11.	Pillai R.S.N. & Bagavati V.	:	Statistics
12.	Russel P.J.	:	Genetics (5th edition)
13.	Russel P.J.	:	Genetics. A Molecular Approach (International edition) (2nd edition)
14.	Saxena H.C.	:	Elementary Statistics
15.	Strick-Berger M.W.	:	Genetics (3rd edition)
16.	Standfield W.D.	:	Schaums Outline Theory & Problems is Genetics (2rd edition)
17.	Snustad D. Peter and Simmons M.J.	:	Principal of Genetics (4th edition)
18.	Zar. Jerrold H.	:	Biostatistical Analysis (4th edition)